RODD'S CHEMISTRY OF CARBON COMPOUNDS

ELSEVIER PUBLISHING COMPANY
335 JAN VAN GALENSTRAAT, P.O. BOX 211, AMSTERDAM

AMERICAN ELSEVIER PUBLISHING COMPANY, INC.
52 VANDERBILT AVENUE, NEW YORK, N.Y. 10017

ELSEVIER PUBLISHING COMPANY LIMITED
RIPPLESIDE COMMERCIAL ESTATE
BARKING, ESSEX

LIBRARY OF CONGRESS CATALOG CARD NUMBER 51-14658

WITH 16 TABLES

PRINTED IN THE NETHERLANDS

RODD'S CHEMISTRY OF CARBON COMPOUNDS

ADVISORS

RODD'S CHEMISTRY OF CARBON COMPOUNDS

VOLUME I

GENERAL INTRODUCTION

ALIPHATIC COMPOUNDS

★

VOLUME II

ALICYCLIC COMPOUNDS

★

VOLUME III

AROMATIC COMPOUNDS

★

VOLUME IV

HETEROCYCLIC COMPOUNDS

★

VOLUME V

MISCELLANEOUS

GENERAL INDEX

★

RODD'S CHEMISTRY OF CARBON COMPOUNDS

A modern comprehensive treatise

SECOND EDITION

Edited by

S. COFFEY

M.Sc. (London), D.Sc. (Leyden), F.R.I.C.

formerly of

I.C.I. Dyestuffs Division, Blackley, Manchester

VOLUME I PART B

MONOHYDRIC ALCOHOLS
THEIR ETHERS AND ESTERS

SULPHUR ANALOGUES

NITROGEN DERIVATIVES

ORGANOMETALLIC COMPOUNDS

ELSEVIER PUBLISHING COMPANY

AMSTERDAM LONDON NEW YORK

1965

CONTRIBUTORS TO THIS VOLUME

J. G. Buchanan M.A., PH.D.
Department of Chemistry, The University, Newcastle upon Tyne

R. E. Fairbairn, B.SC., PH.D., F.R.I.C.
Dyestuffs Division, I.C.I. Ltd., Hexagon House, Blackley, Manchester 9
(Index)

H. Goldwhite, M.A., PH.D.
Department of Chemistry, California State College at Los Angeles,
Los Angeles 32, Calif.

N. A. Hughes, B.A., PH.D.
Department of Chemistry, The University, Newcastle upon Tyne

C. B. Milne, B.SC., A.R.I.C.
Thornton Research Centre, Chester

G. A. Swan, D.SC., PH.D., F.R.I.C.
Department of Chemistry, The University, Newcastle upon Tyne

E. S. Waight, PH.D., D.I.C.
Department of Chemistry, Imperial College of Science and Technology,
London, S.W.7

A. N. Wright, B.A., PH.D.
Thornton Research Centre, Chester

PREFACE TO VOLUME IB

THE reasons for this revised edition of *Rodd's* "Chemistry of Carbon Compounds", and the general details with regard to Volume I dealing with aliphatic (acyclic) compounds, are set out in the Preface to Volume IA. The original Volume I was published in two rather bulky parts. For the revised edition this arrangement was obviously impracticable and the new Volume I will appear, for the greater convenience of the reader and those concerned with the publication of the book, in a series of slimmer parts, each of which will deal as far as possible with a clearly defined domain of aliphatic chemistry.

The present book, Volume IB, describes in four chapters, the remaining singly-linked, mono-substituted, aliphatic hydrocarbons. Chapter 4 on monohydric alcohols, their ethers and esters, deals also with halogeno- and nitroalkanols, which are better discussed as alcohols rather than as derivatives of glycols as is required by the original *Richter* classification. The sulphur analogues of the alcohols and their derivatives are discussed in Chapter 5, and nitrogen derivatives of the aliphatic hydrocarbons in Chapter 6. Here again, for obvious chemical reasons, halogeno and nitro are regarded as non-functional groups, and di- and poly-nitroalkanes, as well as halogenonitroalkanes, are included in this chapter. Chapter 7, undoubtedly one of the most important chapters in the new Volume I, deals with aliphatic organometallic and organometalloidal compounds. Organometallic chemistry which now has its own specialized journals, has literally exploded during the last 20 years; it is still expanding rapidly and the present account sets out the position at the beginning of 1964. The *Richter* classification is not satisfactory for dealing with these compounds in that it divorces aliphatic from cyclic and aromatic compounds, which are clearly closely related yet show important differences in this domain of organic chemistry. For this reason, some mention of alicyclic and aromatic metal compounds and particularly of carbonyl complexes, is included in this chapter.

S. COFFEY

September, 1964

CONTENTS

Chapter 4. Monohydric Alcohols, their Ethers and Esters

by J. G. BUCHANAN, N. A. HUGHES AND G. A. SWAN

Chapter 5. Sulphur Analogues of the Alcohols and their Derivatives

by H. Goldwhite

Chapter 6. Nitrogen Derivatives of the Aliphatic Hydrocarbons

by H. Goldwhite

VOLUMES IA, C, D, E, F AND G

Vol. IA contains chapters covering the following subjects:

General Introduction; Hydrocarbons; Halogen Derivatives

Further sub-volumes of Volume I, contrary to what is stated on page XVI of Vol. IA, will contain chapters covering the following subjects:

Vol. IC Acyclic Aldehydes, Ketones and Monocarboxylic Acids; Carbon Monoxide, Isocyanides and Fulminic Acid; Carbonic Acid and its Derivatives.

Vol. ID Dihydric Alcohols, Their Oxidation Products and Derivatives

Vol. IE Tri- and Tetra-hydric Alcohols, Their Oxidation Products and Derivatives

Vol. IF Penta- and Higher-hydric Alcohols, Their Oxidation Products and Derivatives; Saccharides

Vol. IG Enzymes; Macromolecules; Cumulative Index Vols. IA-IG

LIST OF COMMON ABBREVIATIONS AND SYMBOLS USED

A	acid
Å	Ångström units
Ac	acetyl
a	axial
as	asymmetrical
at.	atmosphere
B	base
Bu	butyl
b.p.	boiling point
C, mC and μC	curie, millicurie and microcurie
conc.	concentrated
crit.	critical
D	Debye unit, 1×10^{-18} e.s.u.
D	dissociation energy
D	dextro-rotatory
DL	optically inactive (externally compensated)
d	density
dec. or decomp.	with decomposition
deriv.	derivative
E	energy; extinction; electromeric effect
E1, E2	uni- and bi-molecular elimination mechanisms
E1cB	unimolecular elimination in conjugate base
E.S.R.	electron spin resonance
Et	ethyl
e	nuclear charge; equatorial
f	oscillator strength
f.p.	freezing point
G	free energy
G.L.C.	gas liquid chromatography
g	spectroscopic splitting factor, 2.0023
H	applied magnetic field; heat content
h	Planck's constant
I	spin quantum number; intensity; inductive effect
K	dissociation constant
k	Boltzmann constant; velocity constant
kcal.	kilocalories
L	laevorotatory
M	molecular weight; molar; mesomeric effect
Me	methyl
m	mass; mole; molecule; *meta-*

ml	millilitre
m.p.	melting point
[M]	molecular rotation
N	Avogadro number; normal
N.M.R.	nuclear magnetic resonance
n	normal; refractive index; principal quantum number
o	ortho-
P	polarisation; probability; orbital state
Pr	propyl
Ph	phenyl
p	*para-;* orbital
R	clockwise configuration
S	counterclockwise configuration; entropy; net spin of incompleted electronic shells; orbital state
S_N1, S_N2	uni- and bi-molecular nucleophilic substitution mechanisms
S_Ni	internal nucleophilic substitution mechanisms
s	symmetrical; orbital
sec	secondary
soln.	solution
T	absolute temperature
Tosyl	*p*-toluenesulphonyl
Trityl	triphenylmethyl
t	time
temp.	temperature (in degrees centigrade)
tert	tertiary
U	potential energy
v	velocity
α	optical rotation (in water unless otherwise stated)
$[\alpha]$	specific optical rotation
α_A	atomic susceptibility
α_E	electronic susceptibility
ε	dielectric constant; extinction coefficient
μ	microns (10^{-4} cm); dipole moment; magnetic moment
μ_B	Bohr magneton
λ	wavelength
v	frequency; wave number
χ, χ_d, χ_μ	magnetic, diamagnetic and paramagnetic susceptibilities
~	about
(+)	dextrorotatory
(−)	laevorotatory
⊖	negative charge
⊕	positive charge

OFFICIAL PUBLICATIONS

B.P.	British (United Kingdom) Patent
F.P.	French Patent
G.P.	German Patent
Sw.P.	Swiss Patent
U.S.P.	United States Patent
B.I.O.S.	British Intelligence Objectives Sub-Committee Reports, H.M. Stationery Office, London.
C.I.O.S.	Combined Intelligence Objectives Sub-Committee Reports
F.I.A.T.	Field Information Agency, Technical Reports of U.S. Group Control Council for Germany
B.S.	British Standards Specification
A.S.T.M.	American Society for Testing and Materials

SCIENTIFIC JOURNALS AND PERIODICALS

With few obvious and self-explanatory modifications the abbreviations used in references to journals and periodicals comprising the extensive literature on organic chemistry, are those used in the World List of Scientific Periodicals.

Chapter 4

Monohydric Alcohols, their Ethers and Esters

J. G. BUCHANAN, N. A. HUGHES and G. A. SWAN

The alcohols can be regarded as derived from the acyclic hydrocarbons by the replacement of hydrogen atoms by hydroxyl (—OH) groups. When one hydroxyl group is introduced, a *monohydric* alcohol* is produced:

$$CH_3{\cdot}CH_3 \longrightarrow CH_3{\cdot}CH_2OH$$

Introduction of further hydroxyl groups, each being attached to a different carbon atom, leads to di-, tri- and poly-hydric alcohols: CH_3OH, methyl alcohol or methanol (the simplest monohydric alcohol); $CH_2OH{\cdot}CH_2OH$, ethylene glycol or ethane-1,2-diol (the simplest dihydric alcohol or alkane-diol); $CH_2OH{\cdot}CHOH{\cdot}CH_2OH$, glycerol or propane-1,2,3-triol (the simplest trihydric alcohol or alkanetriol), etc.

Further, more than one hydrogen atom attached to the same carbon atom may be replaced by hydroxyl groups, but the resulting compounds, with very few exceptions, are unstable and split off water. There are the following possibilities: (a) two hydrogen atoms of a terminal CH_3 group, (b) two hydrogen atoms of an intermediate CH_2 group, and (c) three hydrogen atoms of a terminal CH_3 group may be replaced by an equal number of hydroxyl groups. Elimination of water leads to the formation of three new classes of compounds:

(a) 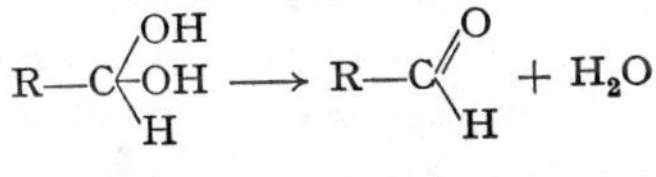

* In systematic IUPAC nomenclature the functional —OH group is denoted by the suffix *-ol* attached to the root name of the parent hydrocarbon.

(b) 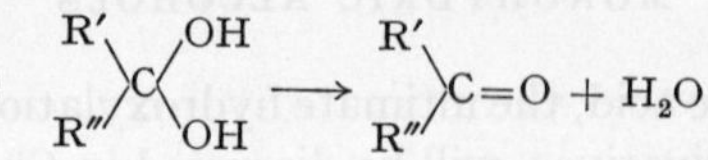

(c) 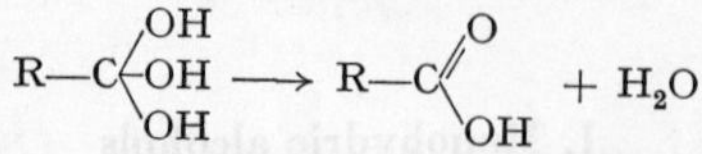

(a) Compounds containing the group —C(=O)H (aldehyde group) are known as *aldehydes*.

(b) Compounds containing the group >C=O in union with two carbon atoms (keto- or oxo-group) are called *ketones*.

(c) Compounds containing the group —C(=O)OH (carboxyl group) are named *carboxylic acids*.

Alcohols, aldehydes and ketones are neutral substances. Carboxylic acids are acidic and form salts in the same way as mineral acids. Aldehydes, ketones and carboxylic acids, as the oxidation products of alcohols, are discussed in subsequent chapters.

Unsaturated hydrocarbons similarly yield unsaturated alcohols, aldehydes, ketones and carboxylic acids. In this and following chapters the unsaturated derivatives will receive attention after the saturated compounds corresponding with them; *i.e.* the unsaturated alcohols will follow the saturated alcohols. Halogenated monohydric alcohols are discussed in the present chapter, since in these compounds the OH is the principal functional group.

Similarly, there are many different oxidation products of di-, tri- and poly-hydric alcohols, which may contain any combination of the different oxygen-containing groups in the same molecule.

Finally, by replacement of the four hydrogen atoms of methane by hydroxyl groups, the hypothetical orthocarbonic acid is formed, which by loss of two molecules of water forms carbon dioxide, the intermediate metacarbonic acid being likewise incapable of free existence. Esters of orthocarbonic acid are known and carbonates are derived from the meta acid.

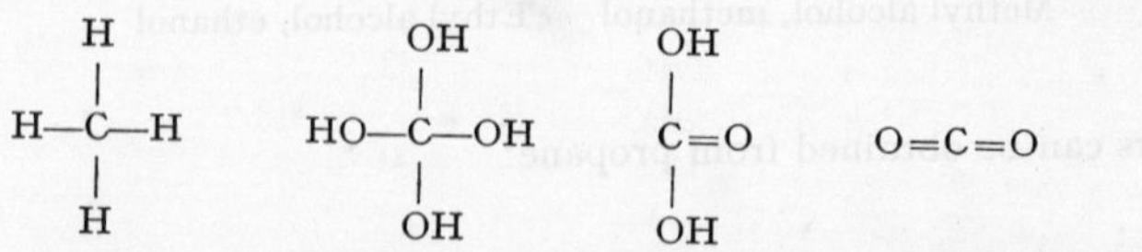

Methane — Orthocarbonic acid — Metacarbonic acid — Carbon dioxide

In this book carbonic acid, the ultimate hydroxylation product of methane, with its numerous derivatives, will be discussed in Chapter 11.

1. Monohydric alcohols

(a) Saturated alcohols, paraffin alcohols or alkanols, $C_nH_{2n+1}OH$

The monohydric alcohols can be regarded as being derived from water by replacement of one hydrogen atom by an alkyl group. If both hydrogen atoms in water are so substituted, the ethers are obtained.

$H\cdot O\cdot H$	$C_2H_5\cdot O\cdot H$	$C_2H_5\cdot O\cdot C_2H_5$
Water	Ethyl alcohol	Diethyl ether

The presence of the hydroxyl group in the monohydric alcohols determines their properties; the hydrogen atom in this group can be replaced by acid residues, forming *esters*:

$$C_2H_5OH + HOSO_3H \longrightarrow C_2H_5OSO_3H + H_2O$$

(the mechanism of esterification will be discussed later in the chapter dealing with carboxylic acids); by alkyl groups and by metals to form *ethers* and *alkoxides*:

$C_2H_5\cdot O\cdot CH_3$	$C_2H_5\cdot ONa$
Ethyl methyl ether	Sodium ethoxide

Structure and isomerism. Methane and ethane can each give rise to only one alcohol;

CH_3OH	$CH_3\cdot CH_2OH$
Methyl alcohol, methanol	Ethyl alcohol, ethanol

Two isomers can be obtained from propane:

$CH_3\cdot CH_2\cdot CH_2OH$	$CH_3\cdot CHOH\cdot CH_3$
n-Propyl alcohol, *n*-propanol, propan-1-ol	isopropyl alcohol, isopropanol propan-2-ol

Two isomeric alcohols may be obtained from each of the isomeric butanes:

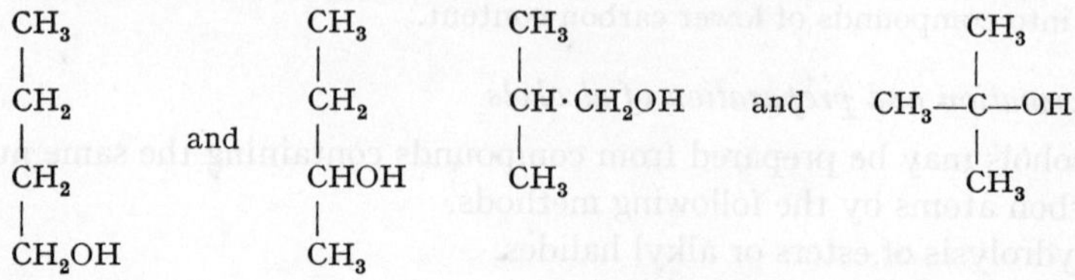

n-Butyl alcohol, butan-1-ol; *sec*-Butyl alcohol, butan-2-ol; Isobutyl alcohol, 2-methylpropan-1-ol; *tert*-Butyl alcohol, 2-methylpropan-2-ol

From higher paraffins correspondingly more isomeric alcohols are derived.

Kolbe regarded the alcohols as derivatives of methyl alcohol, for which he proposed the name *carbinol*. By replacing one hydrogen atom attached to carbon in carbinol by an alkyl group, primary alcohols result:

$$CH_3 \cdot CH_2OH \qquad C_2H_5 \cdot CH_2OH$$

Methylcarbinol or ethanol — Ethylcarbinol, *n*-propyl alcohol or propan-1-ol

If the replacing group has an unbranched carbon chain, the primary alcohols are said to be normal (indicated by the prefix *n*-). On oxidation, primary alcohols yield aldehydes which may be further oxidised to the carboxylic acids:

$$R \cdot CH_2OH \longrightarrow R \cdot CHO \longrightarrow R \cdot CO(OH)$$

Primary alcohol — Aldehyde — Acid

In secondary alcohols (denoted by *sec*-), two hydrogen atoms in carbinol are replaced by alkyl groups:

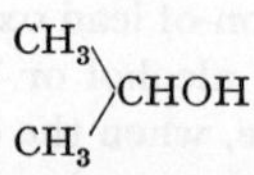

Dimethylcarbinol, isopropanol, propan-2-ol

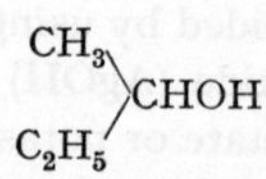

Ethylmethylcarbinol, *sec*-butyl alcohol, butan-2-ol

On oxidation, secondary alcohols give ketones:

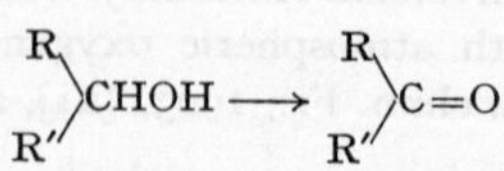

Secondary alcohol — Ketone

Replacement of all three hydrogen atoms of the methyl group of carbinol by alkyl groups results in tertiary alcohols, such as *tert*-butyl alcohol (above), which may be termed trimethylcarbinol. Kolbe's nomenclature is particularly useful for tertiary alcohols. When tertiary alcohols are oxidised, they break down into compounds of lower carbon content.

(i) Formation and preparation of alcohols

Alcohols may be prepared from compounds containing the same number of carbon atoms by the following methods.

(1) Hydrolysis of esters or alkyl halides.
(2) The reduction of polyhydric alcohols.
(3) The action of nitrous acid on primary amines.
(4) The reduction of aldehydes, ketones and carboxylic acids, esters, chlorides and amides.
(5) The hydration of olefins.

They may be synthesised from compounds having a smaller number of carbon atoms as follows.

(6) The use of organometallic compounds.
(7) The action of metal alkoxides on alcohols.
(8) Reppe's synthesis from olefins, carbon monoxide and water.

Formation from compounds containing the same number of carbon atoms

(1a) *From alkyl halides.* The alkyl halides afford a means of passing from the paraffins and olefins to the alcohols. They may be hydrolysed by dilute alkali, *e.g.* sodium, potassium, calcium or barium hydroxides or alkali carbonates. In some cases, water itself is sufficient to effect hydrolysis. The ease of hydrolysis increases from chloride to iodide:

$$C_2H_5Br + NaOH \longrightarrow C_2H_5OH + NaBr$$

Sometimes olefins are also formed, particularly when an organic solvent is used. This can be avoided by using a suspension of lead oxide or freshly precipitated moist silver oxide (AgOH) in aqueous alcohol or by treating the alkyl halide with silver acetate or potassium acetate, when the corresponding acetate is formed (*B. K. Mereshkowsky*, Ann., 1923, **431**, 231) and may be hydrolysed as described in (1b):

$$C_2H_5Br + CH_3\cdot CO_2K \longrightarrow CH_3\cdot CO_2C_2H_5 + KBr$$

Another method is the conversion of the alkyl halide into the Grignard reagent, followed by oxidation with atmospheric oxygen and subsequent hydrolysis (cf. *L. Bouveault*, Bull. Soc. chim. Fr., 1903, [iii], **29**, 1053):

$$C_2H_5Br \xrightarrow{Mg} C_2H_5MgBr \xrightarrow{O_2} C_2H_5OMgBr \longrightarrow C_2H_5OH$$

(1b) *From esters of carboxylic acids.* These can be readily hydrolysed by aqueous sodium or potassium hydroxide:

$$CH_3{\cdot}CO_2C_2H_5 + KOH \longrightarrow CH_3{\cdot}CO_2K + C_2H_5OH$$

The reaction is known as *saponification*, because the soaps, *i.e.* the alkali salts of the higher fatty acids, and glycerol (*q.v.*) are obtained from the glyceryl esters of the fatty acids—the fats—in this way.

(1c) *From esters of sulphuric acid.* Hydrolysis of the ester occurs on boiling with water:

$$C_2H_5OSO_3H + H_2O \longrightarrow C_2H_5OH + H_2SO_4$$

This reaction constitutes a transition from the olefins to the alcohols, since sulphate esters may be easily obtained from the corresponding olefins, cf. (5).

(2) *The reduction of polyhydric alcohols* with hydriodic acid yields first allylic iodides and then the iodides of the secondary alcohols, which are eventually converted by method (1a), into the alcohols themselves:

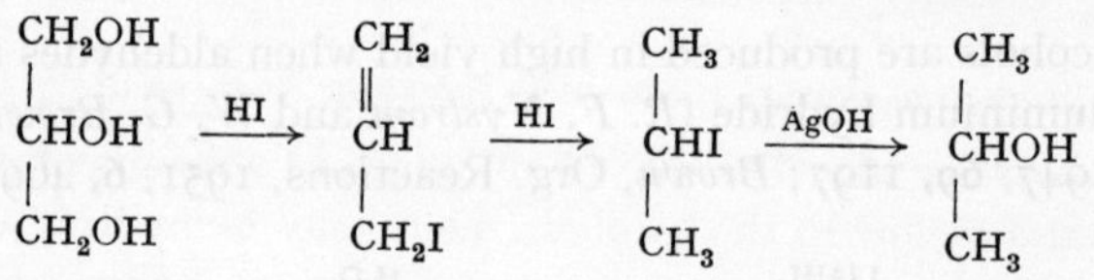

Glycerol Allyl iodide Isopropyl iodide Isopropyl alcohol

Similarly the chlorohydrins of the polyhydric alcohols may be reduced, *e.g.*, monochlorohydrin giving ethanol and 1,3-dichlorohydrin (from glycerol), isopropanol:

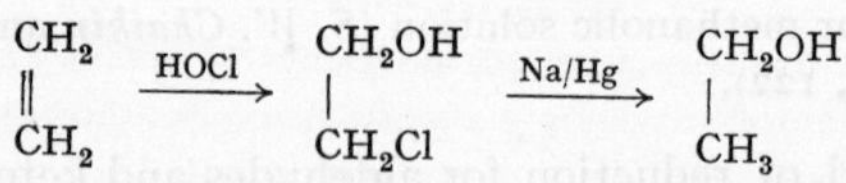

(3) *Action of nitrous acid on the primary amines*:

$$C_2H_5NH_2 + HNO_2 \longrightarrow C_2H_5OH + N_2 + H_2O$$

This is not a good preparative method.

Although ethylamine is said to afford ethanol in yields as high as 60%, *n*-propylamine yields a mixture of propan-1-ol, propan-2-ol and propene and the higher amines behave similarly. The reaction fails with methylamine, apparently because methylamine nitrite is hydrolysed too rapidly (*F. C. Whitmore* and *R. S. Thorpe*, J. Amer. chem. Soc., 1941, **63**, 1118) if the solution is too acid (*A. T. Austin*, Nature, 1960, **188**, 1086).

(4) *The reduction of* (a) *aldehydes*, (b) *ketones*, (c) *carboxylic esters*, (d) *carboxylic acids*, (e) *carboxylic anhydrides and* (f) *carboxylic acid chlorides*.

(a) Aldehydes may be reduced to primary alcohols with sodium amalgam and hydrochloric acid (*A. Wurtz*, Ann., 1862, **123**, 140):

$$CH_3{\cdot}CHO + 2H \longrightarrow CH_3{\cdot}CH_2OH$$

Reduction can also be carried out catalytically. Platinum oxide in the presence of a ferrous salt (*W. H. Carothers* and *R. Adams*, J. Amer. chem. Soc., 1923, **45**, 1071; *ibid.*, 1924, **46**, 1675), nickel (*H. Adkins* and *H. R. Billica*, *ibid.*, 1948, **70**, 695) and especially "copper chromite" (cf. *Adkins* and *R. Connor*, *ibid.*, 1931, **53**, 1091) have been used as catalysts (*R. Schröter*, "Newer Methods of Preparative Organic Chemistry", Interscience, Publishing Co., New York and London, 1948, p. 83; *C. Grundmann*, *ibid.*, p. 112). For the industrial preparation of alcohols, the catalytic hydrogenation of aldehydes prepared by the Oxo Synthesis is important.

Primary alcohols are produced in high yield when aldehydes are reduced by lithium aluminium hydride (*R. F. Nystrom* and *W. G. Brown*, J. Amer. chem. Soc., 1947, **69**, 1197; *Brown*, Org. Reactions, 1951, **6**, 469):

$$RCHO \xrightarrow{LiAlH_4} LiAl(OCH_2R)_4 \xrightarrow{H_2O} RCH_2OH$$

This procedure is of wide application and is also used in the reduction of ketones and carboxylic acids and their derivatives to alcohols (see below). Sodium borohydride can also be used to reduce aldehydes or ketones, although it has little action on most other carbonyl compounds, except acid chlorides. It may be used in aqueous or methanolic solution (*S. W. Chaikin* and *Brown*, J. Amer. chem. Soc., 1949, **71**, 122).

A specific method of reduction for aldehydes and ketones is to heat the compound with aluminium isopropoxide in isopropyl alcohol, when the latter is oxidised to acetone:

$$\underset{\text{Aluminium isopropoxide}}{[(CH_3)_2CHO{\cdot}]_3Al} + \underset{\text{Acetaldehyde}}{3\,CH_3{\cdot}CHO} \rightleftharpoons \underset{\text{Acetone}}{3(CH_3)_2CO} + \underset{\text{Aluminium ethoxide}}{(CH_3{\cdot}CH_2O{\cdot})_3Al}$$

This *Meerwein-Ponndorf-Verley reduction* is widely applicable (*H. Meerwein* and *R. Schmidt*, Ann., 1925, **444**, 221; *A. Verley*, Bull. Soc. chim. Fr., 1925, [iv], **37**, 537; *H. Lund*, Ber., 1937, **70**, 1520; *A. L. Wilds*, Org. Reactions, 1944,

2, 178; *T. Bersin*, "Newer Methods of Preparative Organic Chemistry", Interscience, 1948, p. 125).

The Cannizzaro reaction, treatment of an aldehyde with alkali, is occasionally of value for the preparation of primary alcohols:

$$2\ R{\cdot}CHO \xrightarrow{OH^{\ominus}} R{\cdot}CH_2OH + R{\cdot}CO_2^{\ominus}$$

(b) Reduction of ketones occurs under similar conditions to that of aldehydes. The products are secondary alcohols, but in many of the older methods appreciable quantities of pinacols are also formed (*C. Friedel*, Ann., 1862, **124**, 324):

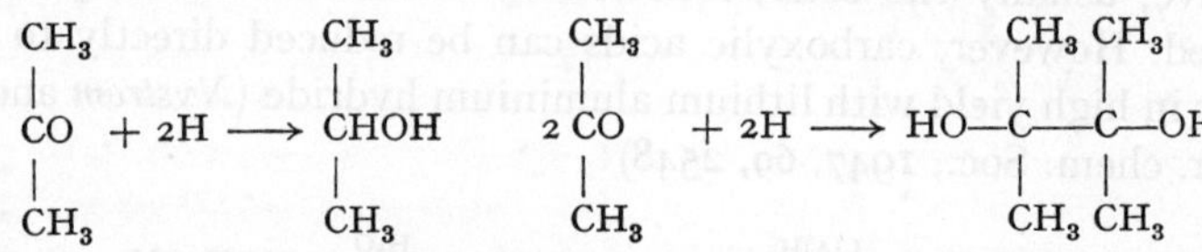

Acetone — Isopropyl alcohol — Pinacol

When ketones are reduced catalytically or by the Meerwein-Ponndorf-Verley method there are few side reactions, and high yields of the secondary alcohols are obtained. The conditions employed are similar to those for aldehydes.

Both ketones and aldehydes can be reduced by fermenting yeast (*C. Neuberg* and *H. Steenbock*, Biochem. Z., 1913, **52**, 494; *C. J. Lintner* and *H. J. von Liebig*, Z. physiol. Chem., 1911, **72**, 449). Although the method is of little preparative value, it can be used for the production of certain optically active secondary alcohols (*Neuberg* and *F. F. Nord*, Ber., 1919, **52**, 2237).

(c) Carboxylic esters are reduced to primary alcohols by sodium and alcohol (*L. Bouveault* and *G. Blanc*, Compt. rend., 1903, **136**, 1676; *ibid.*, 1903, **137**, 60; Bull. Soc. chim. Fr., 1904, [iii], **31**, 748). An important modification is that of *H. J. Prins* (Rec. Trav. chim., 1923, **42**, 1050) who carried out the reaction at low temperatures in the presence of acid. Sodium hydride can replace sodium in the Bouveault-Blanc method (*G. Darzens*, Compt. rend., 1947, **224**, 570).

Reduction of amides by the Bouveault-Blanc method leads to the simultaneous production of amines (*Bouveault* and *Blanc*, *ibid.*, 1904, **138**, 148).

Catalytic hydrogenation of esters to alcohols takes place over "copper chromite" or Raney nickel, but the conditions required are more drastic than for aldehydes or ketones (*H. Adkins*, Org. Reactions, 1954, **8**, 1; *Grundmann*, *loc. cit.*, p. 114).

Lithium aluminium hydride reduces esters to alcohols in excellent yield, under very mild conditions (*Nystrom* and *Brown, loc. cit.*):

$$RCO_2R' \xrightarrow{LiAlH_4} LiAl(OR')_2(OCH_2R)_2 \xrightarrow{H_2O} RCH_2OH + R'OH$$

The dialkylaluminium hydrides are similar to lithium aluminium hydride in their reducing properties (*K. Ziegler et al.*, Angew. Chem., 1955, **67**, 424 and 425). Calcium borohydride can also be used for the reduction of esters and is more selective in its action than is lithium aluminium hydride (*J. Kollonitsch, O. Fuchs* and *V. Gábor*, Nature, 1954, **173**, 125; 1955, **174**, 346).

(d) Carboxylic acids are not easily reduced, and conversion into some derivative, usually the ester, followed by reduction, has frequently been employed. However, carboxylic acids can be reduced directly to primary alcohols in high yield with lithium aluminium hydride (*Nystrom* and *Brown*, J. Amer. chem. Soc., 1947, **69**, 2548):

$$RCO_2H \xrightarrow{LiAlH_4} LiAl(OCH_2R)_4 \xrightarrow{H_2O} RCH_2OH$$

The reaction is widely applicable.

(e) Acid anhydrides are converted into primary alcohols in low yield by reduction with sodium amalgam (*E. Linneman*, Ann., 1872, **161**, 178). Lithium aluminium hydride gives much better yields.

(f) Acid chlorides behave similarly; sodium amalgam gives primary alcohols and again high yields are obtained by using lithium aluminium hydride. Sodium hydrotrimethoxyborate and lithium borohydride allow the reduction of an acid chloride to a primary alcohol, without affecting ester groups present in the same molecule (*H. C. Brown* and *E. J. Mead*, J. Amer. chem. Soc., 1953, **75**, 6263; *Kollonitsch, Fuchs* and *Gábor, loc. cit.*).

(5) *The hydration of olefins* is important commercially for the preparation of alcohols (*B. T. Brooks*, Ind. Eng. Chem., 1935, **27**, 282). It is usually brought about using sulphuric acid, addition occurring according to Markownikoff's rule with olefins containing more than two carbon atoms, leading eventually to secondary alcohols:

$$\underset{\text{Propylene}}{CH_3\cdot CH{=}CH_2} \xrightarrow[(H_2O)]{(H_2SO_4)} \underset{\text{Isopropyl alcohol}}{CH_3\cdot CHOH\cdot CH_3} \text{ (Markownikoff addition)}$$

Anti-Markownikoff addition can be achieved through hydroboronation, which is effected by reaction of the olefin at room temperature with sodium borohydride and aluminium chloride in diethylenediglycol dimethyl ether ("diglyme"):

$$9\,R{\cdot}CH{:}CH_2 + AlCl_3 + 3NaBH_4 \longrightarrow 3\,(R{\cdot}CH_2{\cdot}CH_2{\cdot})_3B + AlH_3 + 3NaCl$$

The resulting organoborane is readily oxidised by alkaline hydrogen peroxide to the corresponding alcohol (*H. C. Brown* and *B. C. S. Rao*, J. Amer. chem. Soc., 1959, 81, 6423):

$$(R{\cdot}CH_2{\cdot}CH_2{\cdot})_3B \xrightarrow{H_2O_2} (R{\cdot}CH_2{\cdot}CH_2{\cdot}O{\cdot})_3B \xrightarrow{H_2O} R{\cdot}CH_2{\cdot}CH_2{\cdot}OH$$

Moreover, internal olefins can be converted into primary alcohols, *e.g.*:

$$\text{2-Hexene} \longrightarrow \text{Organoborane} \xrightarrow[\text{2) Oxidation}]{\text{1) Reflux in "diglyme"}} \text{Hexan-1-ol}$$

the double bond passing along the carbon chain in the process (*idem, ibid.*, p. 6434; *G. Zweifel* and *H. C. Brown*, Org. Reactions, 1963, 13, 1).

Another method of achieving anti-Markownikoff hydration, due to *K. Ziegler et al.* (Ann., 1954, 589, 91; 1960, 629, 241) consists in treating the olefin with aluminium hydride, then allowing the resulting trialkylaluminium to react with air:

$$2Al + 6\,RCH{:}CH_2 + 3H_2 \longrightarrow 2\,(RCH_2{\cdot}CH_2{\cdot})_3Al$$

$$2(RCH_2{\cdot}CH_2{\cdot})_3Al + 3O_2 \longrightarrow 2(RCH_2{\cdot}CH_2{\cdot}O{\cdot})_3Al$$

On decomposition with water, the resulting aluminium alkoxide yields a primary alcohol.

The older method for effecting anti-Markownikoff hydration was through the peroxide-catalysed addition of hydrogen bromide, followed by hydrolysis.

Synthetic methods

(6) *From organometallic compounds.* (a) *Grignard reagents*, which for discussion of the following reactions may be formulated as RMgX, formed from magnesium and alkyl halides (see Chap. 7), react with a large number of compounds to form products which by hydrolytic cleavage yield alcohols. With aldehydes and ketones, the following reactions take place:

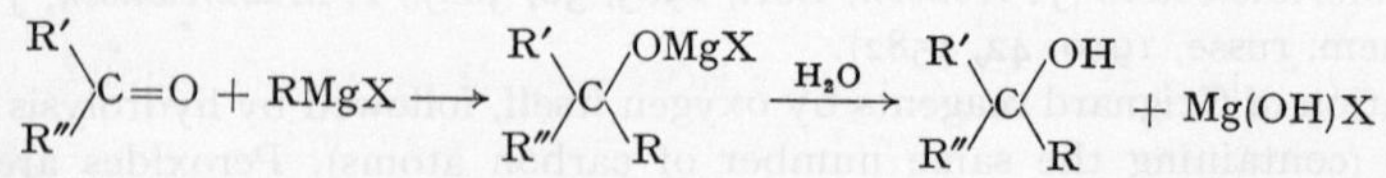

Formaldehyde, used in the form of the polymer trioxymethylene, gives primary alcohols (*V. Grignard* and *L. Tissier*, Compt. rend., 1902, 134, 107); other aldehydes give secondary, and ketones tertiary alcohols (*Grignard, ibid.*, 1900, 130, 1322).

Tertiary alcohols are the final products when an excess of a Grignard reagent reacts with carboxylic esters, acid chlorides or anhydrides, acids, amides and nitriles (*esters, A. D. Petrow*, Bull. Acad. Sci. U.S.S.R. 1938, 347; *acid anhydrides, H. Fournier*, Bull. Soc. chim. Fr., 1910, [iv], **7**, 836; *acid halides, H. Gilman, R. E. Fothergill* and *H. Parker*, Rec. Trav. chim., 1929, **48**, 748; *G. Darzens* and *H. Rost*, Compt. rend., 1911, **153**, 772; *R. C. Fuson* and *J. W. Robinson, Jr.*, J. Amer. chem. Soc., 1940, **62**, 358; *sodium salts, I. Salkind* and *T. Beburischwili*, Ber., 1909, **42**, 4500). Derivatives of formic acid, when available, give secondary alcohols (*R. T. Dillon* and *H. J. Lucas*, J. Amer. chem. Soc., 1928, **50**, 1711):

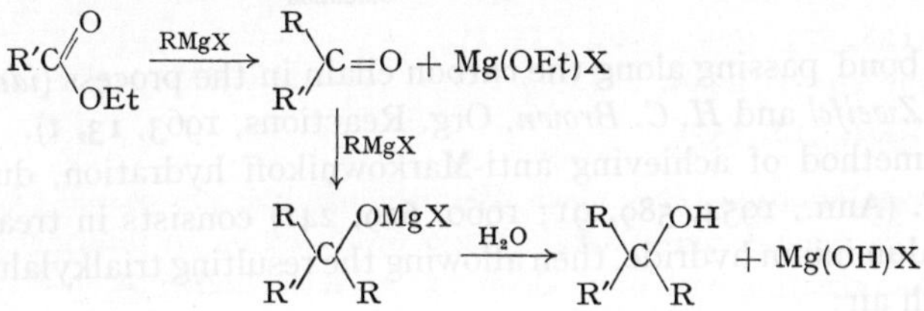

When the higher alkylmagnesium compounds are used, secondary alcohols are formed as well as the tertiary, a reduction process taking place (*P. Sabatier* and *A. Mailhe*, Compt. rend., 1905, **141**, 298; *Letellier, ibid.*, 1908, **146**, 343). In many reactions the tertiary alcohols, which are first formed, lose water and yield unsaturated hydrocarbons, which may even constitute the main product of the reaction (*A. Béhal, ibid.*, 1901, **132**, 480; *V. Grignard*, Chem. Ztbl., 1901, II, 622). Reduction may also occur as a side reaction (*K. Hess* and *W. Wustrow*, Ann., 1924, **437**, 256).

Primary alcohols may also be obtained by hydrolysis of the addition products of ethylene oxide with alkylmagnesium halides (cf. *L. Henry*, Compt. rend., 1907, **144**, 308):

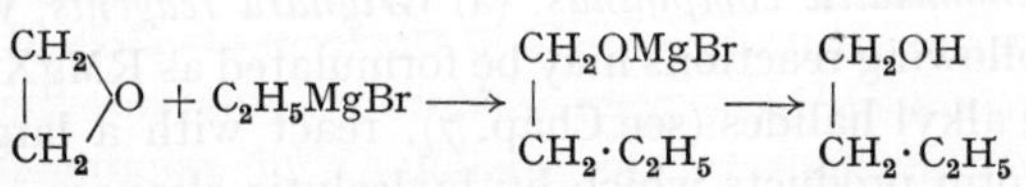

Tertiary alcohols are formed by the action of alkylmagnesium compounds on chloroformic esters (*J. Houben*, Ber., 1903, **36**, 3087; *I. Mazurewitsch*, J. Soc. phys. chem. russe, 1910, **42**, 1582).

Oxidation of Grignard reagents by oxygen itself, followed by hydrolysis gives alcohols (containing the same number of carbon atoms). Peroxides are also effective and the reaction is accompanied by luminescence (*L. Bouveault*, Bull. Soc. chim. Fr., 1903, [iii], **29**, 1051; *H. Gilman* and *C. E. Adams*, J. Amer. chem. Soc., 1925, **47**, 2816; *H. Hock, H. Kropf* and *F. Ernst*, Angew. Chem., 1959, **71**, 541).

(b) *Zinc alkyls*, the use of which in the synthesis of alcohols from acid chlorides was discovered by A. Butlerow in 1864 and developed by *E. E. Blaise* (Bull. Soc. chim. Fr., 1911, [iv], **9**, i–xxvi), have now been replaced by the Grignard reagents (above) and organocadmium compounds which are less inflammable and give better yields. The reactions undergone by zinc alkyls in the synthesis of alcohols from ketones, aldehydes or esters are analogous to those of the Grignard reagents described above. The lithium alkyls are not very frequently used for the synthesis of alcohols.

(7) *Sodium or barium alkoxides* when heated with the corresponding alcohol, especially those of high molecular weight, produce monohydric alcohols with two or three times as many carbon atoms in the molecule (*W. Markownikoff* and *P. Zuboff*, Ber., 1901, **34**, 3246; *M. Guerbet*, Compt. rend., 1899, **128**, 511; *ibid.*, 1902, **135**, 172); *e.g.* isoamyl alcohol gives a decyl alcohol of the constitution $CHMe_2 \cdot CH_2 \cdot CH_2 \cdot CH(CHMe_2) \cdot CH_2OH$ (*J. U. Nef*, Ann., 1901, **318**, 157).

A similar condensation takes place when the alcohols are passed over heated magnesium, the magnesium alkoxide being produced as an intermediate compound and then reacting with more alcohol to form the higher alcohol, *e.g.* propyl alcohol gives 2-methylpentanol in 30% yield (*E. Terentieff*, Bull. Soc. chim. Fr., 1924, [iv], **35**, 1145).

(8) In *Reppe's alcohol synthesis*, a mixture of an olefin and carbon monoxide is treated with a tertiary amine and iron carbonyl in the presence of water:

$$CH_2{:}CH_2 + 3CO + 2H_2O \longrightarrow CH_3 \cdot CH_2 \cdot CH_2OH + 2CO_2$$

(*W. Reppe* and *H. Vetter*, Ann., 1953, **582**, 133; *N. von Kutopow* and *H. Kindler*, Angew. Chem., 1960, **72**, 802).

In addition to the above general methods alcohols are formed by various other reactions. Methanol and certain higher alcohols are produced by the combination of carbon monoxide and hydrogen ("synthesis gas") (see p. 16). Some alcohols occur naturally, chiefly in the form of esters of organic acids. Fermentation processes have been used to manufacture certain alcohols. The formation of alcohols by the breakdown of α-amino-acids by yeast fermentation and the action of other micro-organisms is of biological interest (see formation of fusel oil, p. 22).

(ii) Interconversion of primary, secondary and tertiary alcohols

Conversion of primary into secondary and tertiary alcohols. By the elimination of water primary alcohols become olefins C_nH_{2n}, which yield iodides corresponding to secondary alcohols when treated with concentrated hydriodic acid, the iodine becoming attached to the carbon atom bearing the least number of hydrogen atoms (Vol. IA, p. 419). Secondary alcohols result when these iodides are treated with moist silver oxide:

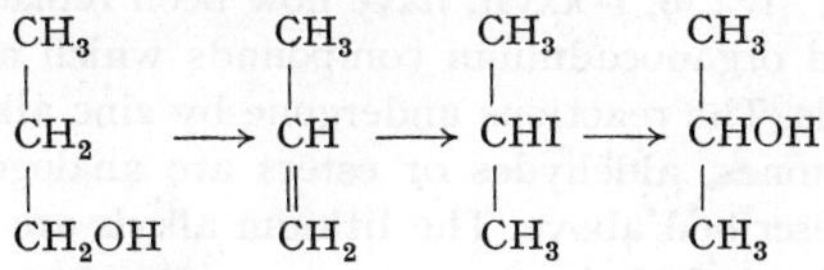

n-Propyl alcohol — Propylene — Isopropyl iodide — Isopropyl alcohol

In a similar manner, primary alcohols in which the group CH_2OH is joined to a secondary radical, pass into tertiary alcohols:

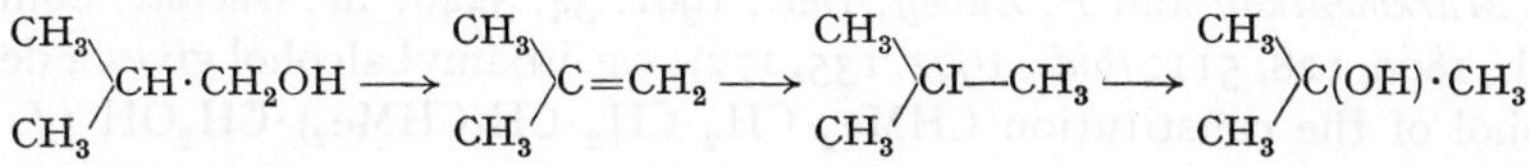

Isobutyl alcohol — Isobutylene — *tert*-Butyl iodide — *tert*-Butyl alcohol

The change, in this case, is better effected by using sulphuric acid in place of hydriodic acid.

(iii) Properties and reactions

Physical properties. In physical properties alcohols exhibit a gradation corresponding with their increase in molecular weight like other compounds belonging to homologous series. The lower alcohols are mobile liquids, dissolving readily in water, and possessing the characteristic alcoholic odour and burning taste. As their carbon content increases, their solubility in water grows rapidly less. The normal alcohols, containing from one to sixteen carbon atoms, are liquid at ordinary temperature, whilst the higher members are crystalline solids, without odour or taste, resembling the fats. The boiling points of alcohols of similar structure increase gradually in proportion to the increase of their molecular weights, the rise being about 20° for each CH_2 increment. The primary alcohols have higher boiling points than the isomeric secondary, and the latter higher than the tertiary alcohols. Generally the boiling point is lowered by an accumulation of methyl groups. The higher members can be distilled only under reduced pressure.

Chemical properties and reactions. The alcohols are neutral compounds. In many respects the first members of the series resemble water and combine with many salts as alcohols of crystallisation.

Some of the more important reactions are:

(1) The hydroxyl hydrogen is easily replaced by sodium, potassium and other metals, yielding thereby alkoxides (see p. 33) from which ethers may be prepared.

(2) Esters are produced by interaction with inorganic acids (see p. 53 *et seq.*) or carboxylic acids, particularly in the latter case, in the presence of an acidic catalyst. With carboxylic acids there is no more than a formal

similarity to the neutralisation of an acid by a base. An equilibrium is established in accordance with the law of mass action:

$$RCO_2H + R'OH \rightleftharpoons RCO_2R' + H_2O$$

Sulphuric acid, zinc chloride or hydrogen chloride are commonly used as catalysts (*E. Fischer* and *A. Speier*, Ber., 1895, **28**, 3252). Primary alcohols are more easily esterified than secondary, and tertiary are not esterified by this treatment (for mechanism see Chap. 9). Alcohols of all three classes are readily esterified by treatment with carboxylic acids in the presence of toluene-*p*-sulphonyl chloride and pyridine (*J. H. Brewster* and *C. J. Ciotti*, J. Amer. chem. Soc., 1955, **77**, 6214). Reactions with keten also results in acetylation (*G. Quadbeck*, Angew. Chem., 1956, **68**, 361).

(3) Alkyl halides (see Vol. IA, Chap. 3, p. 478) are produced when the alcohols are heated with halogen acids. In contrast to esterification by organic acids, tertiary alcohols are the most easily converted into halides, the reaction often occurring in the cold with concentrated aqueous solutions of the appropriate acid. An alternative method for their formation consists in heating the alcohols with phosphorus halides, sometimes in the presence of zinc chloride (*R. H. Clark* and *H. R. L. Streight*, Trans. roy. Soc. Canada, 1929, [3], **23**, 77). Reduction of the alkyl halides with an acid and dissolving metal, catalytically or with lithium aluminium hydride affords a means of converting alcohols into the corresponding hydrocarbons.

(4) The primary saturated alcohols, on being passed as vapours over heated, finely divided metals (Cu, Ni, Zn, Al) or metallic oxides are dehydrogenated to aldehydes (*P. Sabatier* and *J. B. Senderens*, Compt. rend., 1903, **136**, 738, 921, 983; *W. Ipatiew*, Ber., 1903, **36**, 1990; *L. Bouveault*, Bull. Soc. chim. Fr., 1908, [iv], **4**, 119; *H. Adkins et al.*, J. Amer. chem. Soc., 1933, **55**, 2992; *R. E. Dunbar* and *M. R. Arnold*, J. org. Chem., 1945, **10**, 501). Similarly, secondary alcohols give rise to ketones. Dehydrogenation can also be effected by refluxing in the presence of Raney nickel. Tertiary alcohols are dehydrated to olefins.

Oxidation may also be effected by the Oppenauer method (*R. V. Oppenauer*, Rec. Trav. chim., 1937, **56**, 137; *C. Djerassi*, Org. Reactions, 1951, **6**, 207; *T. Bersin*, "Newer Methods of Preparative Organic Chemistry", Interscience, 1948, p. 125), which consists in heating the alcohol with a ketone in the presence of aluminium *tert*-butoxide or isopropoxide. This is particularly useful for the oxidation of secondary alcohols; tertiary alcohols do not react.

tert-Butyl chromate is a fairly specific reagent for the oxidation of primary alcohols to aldehydes (*R. V. Oppenauer* and *H. Oberrauch*, Anales asoc. quím. argentina, 1949, **27**, 246).

Potassium dichromate in dilute sulphuric acid can also be used for the oxidation of alcohols; in the case of compounds containing groups sensitive to acid, chromium trioxide in pyridine may be employed (cf. *G. I. Poos et al.*, J. Amer. chem. Soc., 1953, 75, 422).

Oxidation to carboxylic acids can be achieved by the action of oxygen in the presence of platinum under alkaline conditions (*K. Heyns* and *H. Paulsen*, Angew. Chem., 1957, 69, 600).

(5) Energetic dehydrating agents convert alcohols, especially tertiary alcohols, into olefins.

(6) Alcohols combine with aldehydes at room temperature, with evolution of heat, to form hydroxy-ethers or hemi-acetals; these compounds revert to their components at higher temperatures:

$$CH_3 \cdot CHO + HOEt \rightleftharpoons CH_3 \cdot CH(OH) \cdot OEt$$

(*H. L. de Leeuw*, Z. physik. Chem., 1911, 77, 284).

In the presence of hydrochloric acid, acetals are formed.

(7) Alcohols can be converted into their methyl ethers by diazomethane in the presence of fluoroboric acid (*M. Neeman et al.*, Tetrahedron, 1959, 6, 36).

Reactions distinguishing primary, secondary and tertiary alcohols.

(1) As already mentioned, primary alcohols, on oxidation, yield aldehydes and carboxylic acids; secondary alcohols form ketones with the same number of carbon atoms, and tertiary alcohols break down. Oxidation with chromium trioxide can be used as a test-tube reaction to distinguish the three classes of alcohols (*H. Roth et al.*, "Methoden der Organischen Chemie", *Houben-Weyl*, 4th Ed., Ed. *Müller*, Thieme, Stuttgart, 1953, Vol. II, p. 337).

(2) When the alcohols are converted into their iodides, and the latter are changed by silver nitrite into nitroparaffins, these can be characterised by their colour reactions, according as they contain a primary, secondary, or tertiary alkyl group (*ibid.*, p. 619).

(3) Acetates are formed when primary and secondary alcohols are heated with glacial acetic acid to 155°. Tertiary alcohols, under similar treatment, lose water and form olefins.

(4) When primary alcohols are heated with soda lime they form sodium salts of their corresponding acids:

$$RCH_2OH + NaOH \longrightarrow RCO_2Na + 2H_2$$

(5) When an alcohol is shaken with a solution of zinc chloride in concentrated hydrochloric acid a precipitate of the alkyl chloride appears immediately with tertiary alcohols and after five minutes with secondary alcohols. Primary alcohols lower than hexyl are soluble in the reagent (*H. J. Lucas*, J. Amer. chem. Soc., 1930, 52, 802).

(6) Primary and secondary alcohols yield acetates with acetyl chloride; the tertiary alcohols, on the other hand, give rise to tertiary alkyl halides (*L. Henry*, Bull. Acad. roy. Belg., 1906, 261).

(7) Alcohols react with phthalic anhydride to form alkyl hydrogen phthalates. Tertiary alcohols, which are usually unreactive towards phthalic anhydride in pyridine, readily react in the presence of triethylamine (*A. G. Davies, J. Kenyon* and *L. W. F. Salamé*, J. chem. Soc., 1957, 3148). The times required for half reaction of 3-nitrophthalic anhydride at 4° with primary, secondary and tertiary alcohols are 20–57, 63–450 and 2700–3600 minutes, respectively (*K. Freudenberg* and *G. Achtzehn*, Ber., 1955, **88**, 10).

(iv) *Individual alcohols*

The most important monohydric alcohols are methyl alcohol, CH_3OH and ethyl alcohol, C_2H_5OH.

(1) **Methyl alcohol,** *methanol, carbinol,* CH_3OH

History. R. Boyle discovered wood spirit in 1661 among the products of the dry distillation of wood. In 1812 P. Taylor recognised it as being similar to but not identical with spirits of wine (ethyl alcohol). *J. Dumas* and *E. Peligot* (Ann., 1835, **15**, 1) made the first study of it. It is formed in the dry distillation of wood; the name methyl is derived from μεθυ (wine), and ὕλη (wood). The total synthesis of methyl alcohol from carbon by way of carbon disulphide, methane and methyl chloride is of historical interest (*M. Berthelot*, Ann. Chim., 1858, [iii], **52**, 97).

The aqueous product obtained in the distillation of wood at 500° in iron retorts contains methyl alcohol, acetone, acetic acid, methyl acetate and other compounds. It is distilled over quick lime or soda to remove acetic acid. Purification of the crude wood spirit by distillation is described by *D. W. Huebner, R. J. Young* and *G. H. Payn*, Dict. appl. Chem., 4th Edn., Vol. VIII, p. 12. Methyl alcohol is also produced in the dry distillation of molasses. Esters of methyl alcohol occur in nature, *e.g.* methyl salicylate, $C_6H_4(OH)\cdot CO_2CH_3$, wintergreen oil, derived from *Gaultheria procumbens*.

Methanol is now produced almost entirely by synthesis from carbon monoxide and hydrogen, "synthesis gas":

$$CO + 2H_2 \rightleftharpoons CH_3OH + 24{\cdot}6 \text{ kcal}$$

The exothermic reaction proceeds with diminution of volume and is therefore carried out under pressure (150–300 atmospheres). Zinc oxide catalysts, promoted with copper oxide and chromium oxide, are used at a temperature of 300–400°. By working at more elevated temperatures and using a catalyst containing an alkali metal salt (alkoxide), higher alcohols are produced as well (cf. p. 12). For patent literature, see *Huebner, Young* and *Payn, loc. cit.*

Methyl alcohol is also manufactured by oxidation of methane from natural gas.

To detect ethyl alcohol in methyl alcohol chemically, the liquid is heated with concentrated sulphuric acid; ethylene is formed from the ethyl alcohol, whilst dimethyl ether results from the methyl alcohol. The amount of methyl alcohol in wood spirit can be determined by converting it quantitatively with phosphorus tri-iodide into methyl iodide which is distilled off and measured (*G. Krell*, Ber., 1873, **6**, 1310; *G. Kraemer* and *M. Grodzki*, *ibid.*, 1876, **9**, 1928). Alternatively, it can be converted by nitrous acid into methyl nitrite which is readily volatile, and when distilled into acid potassium iodide solution, liberates iodine, which can be titrated with thiosulphate (*W. M. Fischer* and *A. Schmidt*, *ibid.*, 1924, **57**, 693). In the presence of ethyl alcohol, oxidation by acid dichromate to CO_2 is the best method (*T. E. Thorpe* and *J. Holmes*, J. chem. Soc., 1904, **85**, 1). The proportions of water and methyl alcohol in a mixture can be determined by means of the Karl Fischer reagent (*D. M. Smith*, *W. M. D. Bryant* and *J. Mitchell*, J. Amer. chem. Soc., 1939, **61**, 2407; 1940, **62**, 1).

Physical properties. Methyl alcohol is a mobile liquid with a characteristic smell and a burning taste. It is miscible with water, other alcohols, ether and many other organic liquids. B.p. 64·509°/760 mm. (*M. Wojciechowski*, J. Res. Nat. Bur. Stand., 1936, **17**, 721); f.p.—97·49° (*D. R. Stull*, J. Amer. chem. Soc., 1937, **59**, 2726); d_4^{15} 0·79609 (*J. Timmermans* and *Hennaut-Roland*, J. chim. phys., 1930, **27**, 401); n_D^{15} 1·33057.

Uses. Methyl alcohol is employed mainly as a solvent, *e.g.* in varnishes, as a denaturing agent for ethyl alcohol and chemically, as the primary intermediate for making methyl compounds of many kinds by "methylation" reactions. It also finds some use as a fuel for special purposes.

Chemical properties. (1) Methyl differs from all other primary alcohols in that it contains the CH_2OH group joined to hydrogen. It can therefore be oxidised easily to carbon dioxide. Formaldehyde is made industrially by catalytic oxidation of methyl alcohol by air over a silver or copper catalyst. Controlled oxidation by acid permanganate gives formaldehyde, whilst chromic acid gives carbon dioxide quantitatively.

(2) Methyl alcohol is dehydrated to dimethyl ether by passing its vapour over alumina at 350° (*J. B. Senderens*, Bull. Soc. chim. Fr., 1909, [iv], **5**, 480) or an acid-catalyst (*Idem*, Compt. rend., 1930, **190**, 1169); higher alcohols yield olefins.

(3) It gives good yields of synthesis gas, useful for pilot-plant work, when the vapour is heated over copper-nickel oxide at 260–320° (*H. S. Seelig* and *R. F. Marschner*, Ind. Eng. Chem., 1948, **40**, 583).

(4) When heated with soda lime it is converted into sodium formate and hydrogen is evolved (*E. E. Reid*, *H. Worthington* and *A. W. Larchar*, J. Amer. chem. Soc., 1939, **61**, 99).

(5) Chlorine and bromine do not react so readily with methyl as with ethyl alcohol (*C. A. Lobry de Bruyn*, Ber., 1893, **26**, 268). Chlorine attacks aqueous methyl alcohol, however, quite easily giving bis-chloromethyl ether, $(ClCH_2)_2O$, which is converted by water into formaldehyde and hydrochloric acid (*A. Brochet*, Compt. rend., 1895, **121**, 130).

(6) Mono-, di- and tri-methylamine are formed when methanol and ammonia are passed over silica gel at 480° (*A. B. Brown* and *E. E. Reid*, J. phys. Chem.,

1924, **28**, 1067), or when ammonium chloride, methyl alcohol and zinc chloride are heated at 300° under pressure (*W. D. Turner* and *A. M. Howald*, J. Amer. chem. Soc., 1920, **42**, 2663).

In addition methanol undergoes the reactions common to all alcohols already referred to on p. 13 *et seq.*

(2) **Ethyl alcohol,** *ethanol, spirits of wine*, CH_3CH_2OH

This compound is commonly called simply alcohol. Because of its formation in the fermentation of saccharine plant juices, alcohol, in an impure form, has been known from ancient times. It was, however, only at the end of the eighteenth century that it was discovered how to obtain it in anhydrous condition. Saussure determined its composition in 1808.

Occurrence. Ethyl alcohol occurs mainly as a metabolic product of microorganisms. It seldom occurs in green plants, but is found, together with ethyl butyrate, in the unripe seeds of *Heracleum giganteum* and *H. spondylium*, and as esters in several other plants.

Formation. It may be obtained by the general methods previously described for primary alcohols: (1) From ethyl chloride; (2) from ethyl hydrogen sulphate; (3) from ethylene chlorohydrin; (4) from ethylamine; (5) from acetaldehyde; and (6) from acetyl chloride or acetic esters. These syntheses, which are now mainly of academic and historical interest, and some other inter-relationships are shown in the following diagram.

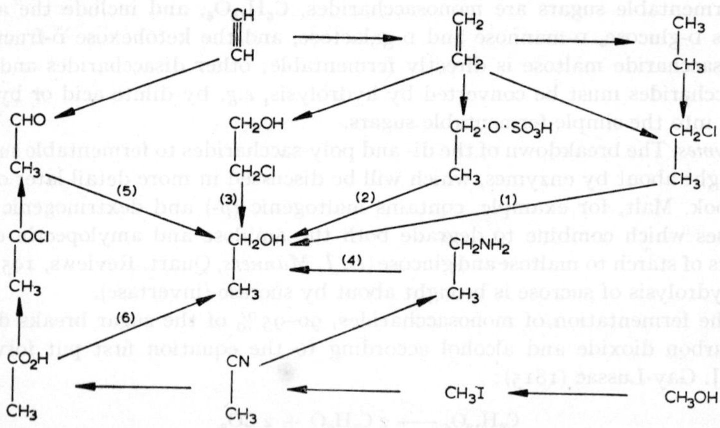

Preparation. Ethyl alcohol is manufactured either (1) by hydration of ethylene, or (2) by fermentation of sugars. An earlier synthetic route (3) by reduction of acetaldehyde obtained from acetylene, is no longer economic.

(1) Ethylene, obtained by the cracking of petroleum, is either dissolved in concentrated sulphuric acid under pressure, the solution diluted, and the ethyl hydrogen sulphate formed hydrolysed by boiling to give ethyl alcohol, or it is hydrated directly with steam at high pressures over a catalyst, *e.g.* celite-

phosphoric acid. (For production and process details: *T. C. Carle* and *D. M. Stewart*, Chem. and Ind., 1962, 830.)

(2) Ethyl alcohol is produced by fermentation for industrial use as well as in the preparation of alcoholic beverages. The discovery that alcoholic fermentation was caused by yeast cells is due to T. Schwann (1837), F. Kützing (1837) and C. Caignard-Latour (1838). This idea, as opposed to J. Liebig's theory of mechanical fermentation (1839), only found acceptance from 1857 onwards, through L. Pasteur's investigations. In 1897, E. Buchner showed that the expressed liquid from mechanically ruptured yeast cells could also bring about alcoholic fermentation. Thus it was shown that fermentation was a purely chemical reaction and not a "vital" process. The active agent in the fermentation is a complex of enzymes and coenzymes and was called zymase.

Conditions of alcoholic fermentation. The yeast grows by budding in dilute, warm (5–30°) sugar solutions; the growth is most rapid at 20–30°. Its development requires salts, especially phosphates, and a source of organic nitrogen which is usually present in sugar-containing liquids suitable for fermentation. When the quantity of alcohol in a fermenting liquid reaches a certain amount, fermentation ceases, since yeast cells cannot grow in liquids containing 14% of alcohol. They are also destroyed by heating to 60° and by small quantities of disinfectants, *e.g.* phenol, mercuric chloride or salicylic acid.

The sugars occurring in ripening fruits—grapes, apples, cherries—and in cane and beet, as well as in many other plants, belong to the class of carbohydrates. The fermentable sugars are monosaccharides, $C_6H_{12}O_6$, and include the aldohexoses D-glucose, D-mannose and D-galactose, and the ketohexose D-fructose. The disaccharide maltose is directly fermentable; other disaccharides and the polysaccharides must be converted by hydrolysis, *e.g.* by dilute acid or by enzymes, into the simple fermentable sugars.

Enzymes. The breakdown of the di- and poly-saccharides to fermentable sugars is brought about by enzymes, which will be discussed in more detail later on in this book. Malt, for example, contains maltogenic (β-) and dextrinogenic (α-) amylases which combine to degrade both the amylose and amylopectin components of starch to maltose and glucose (*D. J. Manners*, Quart. Reviews, 1955, **9**, 73). Hydrolysis of sucrose is brought about by sucrase (invertase).

In the fermentation of monosaccharides, 90–95% of the sugar breaks down into carbon dioxide and alcohol according to the equation first put forward by L. J. Gay-Lussac (1815):

$$C_6H_{12}O_6 \longrightarrow 2\ C_2H_6O + 2\ CO_2$$

At the same time, glycerol (2–5%), fusel oil, and some succinic acid are formed, especially towards the end of the fermentation (*J. Effront*, Compt. rend., 1894, **119**, 92). Fusel oil contains *n*-propyl alcohol, isobutyl alcohol and most important, "fermentation amyl alcohol", which is a mixture of isobutyl carbinol and (—)*sec*-butyl carbinol.

Mechanism of alcoholic fermentation. The elucidation of the mechanism of alcoholic fermentation has attracted the attention of biochemists for many years.

The analysis was made possible by the discovery that cell-free extracts of yeast could be prepared which were capable of fermenting glucose (*E. Buchner*, Ber., 1897, **30**, 117, 1100; 1898, **31**, 568). In 1905, *A. Harden* and *W. J. Young* found that fermentation by yeast juice was accompanied by the esterification of inorganic phosphate (Proc. roy. Soc., 1906, **77B**, 405). They later showed that dialysis of yeast juice resulted in the loss of its activity and that the activity could be restored either by the addition of the dialysate or of boiled yeast juice (*ibid.*, 1906, **78B**, 369).

These fundamental discoveries have led, during the intervening period, to an understanding not only of alcoholic fermentation in yeast but also of the breakdown of glycogen in animal tissues. The earlier work is described by *Harden* in his monograph, "Alcoholic Fermentation", 4th Edn., London, 1932, and more recent work by *O. Meyerhof*, Experientia, 1948, **4**, 169 and *J. S. Fruton* and *S. Simmonds*, "General Biochemistry", 2nd. Edn., Wiley, New York, 1958..

The modern theory of fermentation, known as the Meyerhof-Embden scheme, is shown in Table I. The clarification of the mechanism is due to many workers

TABLE 1

THE MEYERHOF-EMBDEN SCHEME OF ALCOHOLIC FERMENTATION

Reaction	Intermediates	Enzyme system*
	D-Glucose	
1	⇅	Hexokinase/ATP
	Glucose 6-phosphate	
2	⇅	Phosphoglucose isomerase
	Fructose 6-phosphate	
3	⇅	6-Phosphofructokinase/ATP
	Fructose 1,6-diphosphate	
4	⇅	Aldolase
	Glyceraldehyde 3-phosphate	
5	⇅	Phosphoglyceraldehyde dehydrogenase/NAD⊕
	Glyceric acid 1,3-diphosphate	
6	⇅	Phosphoglycerate kinase/ADP
	Glyceric acid 3-phosphate	
7	⇅	Phosphoglyceromutase/glyceric acid 2,3-diphosphate
	Glyceric acid 2-phosphate	
8	⇅	Enolase
	Phosphoenol pyruvate	
9	⇅	Pyruvate phosphokinase/ADP
	Pyruvic acid	
10	↓	α-Carboxylase/thiamine pyrophosphate
	Acetaldehyde + CO_2	
11	⇅	Alcohol dehydrogenase/NADH
	Ethyl alcohol	

* There is some variation in the nomenclature of the enzyme systems. The present nomenclature is that of *M. Dixon* and *E. C. Webb*, "Enzymes", Longmans, London, 1958, pp. 586–589.

notably Neuberg, Robison, Meyerhof, Lohmann, Parnas, Cori and Warburg. An account of the methods used in the study, and the steps by which the present scheme has been reached, is given in the references above.

Several important facts emerge from a study of the various steps. Yeast ferments glucose in order to form adenosine triphosphate (ATP) from adenosine diphosphate (ADP). ATP plays an essential role in metabolism; its phosphate anhydride linkage has a standard free energy of hydrolysis of about —8,000 cal/mole, and it is utilised for many synthetic reactions in the cell which are energetically unfavourable (*F. Lipmann*, Adv. in Enzymology, Interscience, 1941, **1**, 99; 1946, **6**, 231: *M. Dixon*, "Multienzyme Systems", Cambridge Univ. Press, 1949). The overall reaction is

$$C_6H_{12}O_6 + 2\ ADP + 2\ phosphate \longrightarrow 2\ C_2H_5OH + 2\ CO_2 + 2\ ATP$$

The individual reactions are given in Table 1.

ATP is required initially to phosphorylate glucose (reaction 1). In reaction 5, glyceraldehyde 3-phosphate is oxidised by an enzyme containing glutathione as its prosthetic group. Phosphorolysis of the thiolester gives glyceric acid 1,3-diphosphate, whose anhydride phosphate group is transferred to ADP (reaction 6). Phosphoenol pyruvate, formed by dehydration (reaction 8), transfers its phosphate group to ADP (reaction 9). All the reactions except the decarboxylation of pyruvate are reversible.

During the fermentation both oxidation and reduction occur. Glyceraldehyde 3-phosphate is oxidised to glyceric acid 1,3-diphosphate by nicotinamide adenine dinucleotide ($NAD^{\oplus}$, Coenzyme I, Cozymase, DPN) and phosphoric acid in the presence of phosphoglyceraldehyde dehydrogenase (reaction 5). The coenzyme is reduced to the dihydro form (NADH). At a later stage, acetaldehyde is reduced to ethyl alcohol by the reduced coenzyme, the enzyme being alcohol dehydrogenase (reaction 11). The coupling of the oxidation-reduction steps by nicotinamide adenine dinucleotide enables the breakdown of glucose to take place in the absence of oxygen.

Other products of fermentation. When sodium sulphite is added to the fermenting liquid the acetaldehyde is converted into its bisulphite compound as rapidly as it is formed and reaction 11 is blocked. This prevents the regeneration of $NAD^{\oplus}$ by oxidation of the NADH. Under these conditions the NADH is reoxidised by dihydroxyacetone phosphate. The latter is produced, together with glyceraldehyde 3-phosphate in reaction 4 by cleavage of fructose 1,6-diphosphate, and is in equilibrium with glyceraldehyde 3-phosphate through the enzyme triose phosphate isomerase. The product is D-glycerol 1-phosphate, which is hydrolysed to give glycerol (*C. Neuberg*, Ber., 1919, **52**, 1677; *W. Connstein* and *K. Ludecke*, *ibid.*, 1919, **52**, 1385).

By the addition of alkalis, the reaction is changed in another way and glycerol, carbon dioxide, alcohol and acetic acid are formed. The acetaldehyde formed under alkaline conditions is not reduced directly to ethyl alcohol, but undergoes a "dismutation" to ethyl alcohol and acetic acid. As in the case of fermentation in the presence of sulphite, the NADH is oxidised by dihydroxyacetone phosphate, and glycerol is formed (*Neuberg* and *E. Färber*, Biochem. Z., 1917, 78, 238).

The formation of fusel oil in normal fermentations has been explained by the work of *F. Ehrlich* (Ber., 1906, 39, 4072; 1907, 40, 1027; 1911, 44, 139; 1912, 45, 883). These compounds are not produced by the breakdown of the sugar, but by the decarboxylation and deamination of amino-acids which always occur in plant juices and yield the nitrogen necessary for the growth of the yeast cells. Free ammonia is not produced and the following sequence has been suggested (*S. SentheShanmuganathan* and *S. R. Elsden*, Biochem. J., 1958, 69, 210):

$$R{\cdot}CH(NH_2)CO_2H \xrightarrow{\text{Transamination}} R{\cdot}CO{\cdot}CO_2H$$

$$R{\cdot}CO{\cdot}CO_2H \longrightarrow R{\cdot}CHO + CO_2$$

$$R{\cdot}CHO \longrightarrow R{\cdot}CH_2OH$$

Isobutyl carbinol (isoamyl alcohol) for example, is derived from L-leucine.

Alcoholic beverages. The materials used in the preparation of alcoholic liquids by means of fermentation are: (1) saccharine plant juices; (2) starch-containing substances, seeds of grain and potatoes. The fermented liquids are directly consumed (wine, beer) or they are first distilled to produce the various kinds of spirits, the alcohol content of which may exceed 50 %.

(1) By the fermentation of saccharine juices:

(a) without subsequent distillation: from grapes, wine; apples, cider; currants, currant wine, etc.;

(b) with subsequent distillation: from wine, cognac; molasses, rum; cherries, "kirschwasser" (Baden), etc.

(2) By the fermentation of starch-containing substances, after converting the starch into sugar with malt:

(a) without subsequent distillation: from barley, beer; wheat, weissbier (Berlin); rice, saké (Japan).

(b) with subsequent distillation: from barley, rye, wheat, maize, oats: corn whisky of various kinds, gin, vodka; potatoes: potato spirit, etc.

Manufacture of alcohol from blackstrap molasses. Industrially, the most important raw material is molasses, a residual syrup from sugar refining, containing about 50 % of fermentable sugar. Other important raw materials are wood hydrolysates (*E. E. Harris et al.*, Ind. Eng. Chem., 1948, 40, 1216), sulphite waste liquor from paper manufacture, and potatoes [*L. A. Underkofler* and *R. J. Hickey* (Eds.), "Industrial Fermentations", Chemical Publishing Co., New York, 1954, Vol. I].

Water is added to the molasses, giving a sugar concentration of 14–18 %. The mash is brought to a pH 4 to 5 and this renders sterilization unnecessary. A pure culture of a strain of *Saccharomyces cerevisiae* is added. After 36–72 hours at a temperature not exceeding 35° the alcohol content is 6–9 % and the fermented mash or "beer" is ready for distillation. Continuous stills are employed; aldehydes and esters and fusel oil are the important organic contaminants. The product is

95 % aqueous alcohol, "rectified spirit". The fermentation residue has been used as an ingredient of livestock feeds.

Pure absolute alcohol. Anhydrous alcohol is obtained commercially from rectified spirit by azeotropic distillation *e.g.* with benzene (*S. Young* and *E. C. Fortey*, J. chem. Soc., 1902, **81**, 739) or trichloroethylene (*R. Fritzweiler* and *K. R. Dietrich*, Angew. Chem., 1932, **45**, 605). It can also be obtained by drying with chemical dehydrating agents, *e.g.* potassium carbonate, anhydrous copper sulphate, quicklime or barium oxide. Commercial absolute alcohol (about 99 %) can be freed from traces of water and other impurities by several methods [*A. Weissberger et al.*, "Technique of Organic Chemistry", Vol. VII (Organic Solvents), Interscience, New York, 1955, pp. 337–343].

Detection of water in alcohol. Absolute alcohol dissolves barium oxide, assuming a yellow colour at the same time, and does not restore the blue colour to anhydrous copper sulphate. It is soluble without turbidity in a little benzene; when more than 3 % of water is present cloudiness ensues. Aqueous spirit generates acetylene from calcium carbide (*E. R. Weaver*, J. Amer. chem. Soc., 1914, **36**, 2462; *H. Hartley* and *H. R. Raikes*, J. chem. Soc., 1925, **127**, 524). Water may be determined by the Karl Fischer reagent (*D. M. Smith*, *W. M. D. Bryant* and *J. Mitchell*, J. Amer. chem. Soc., 1939, **61**, 2407; 1940, **62**, 1).

Detection of alcohol. Traces of alcohol in solution are detected and determined by oxidation to acetaldehyde or by converting it with dilute potassium hydroxide and iodine into iodoform or by converting it into ethyl benzoate, by shaking with benzoyl chloride and sodium hydroxide.

Physical properties. Pure alcohol is a colourless liquid with an agreeable ethereal odour. It boils at 78·325°/760 mm. (*M. Wojciechowski*, J. Res. Nat. Bur. Stand., 1936, **17**, 721) and solidifies to a resinous mass, which melts at —114·5° (*R. R. Dreisbach* and *R. A. Martin*, Ind. Eng. Chem., 1949, **41**, 2875), n_D^{20} 1·36139 (*C. P. Smyth* and *W. N. Stoops*, J. Amer. chem. Soc., 1929, **51**, 3312, 3330), d_4^{20} 0·78934 (Nat. Bur. Stand., Circular No. 19, Standard Density and Volumetric Tables, 6th Edn., Washington, 1924). It burns with a non-luminous flame and is very hygroscopic. It is miscible with water in all proportions; mixing is accompanied by a decrease in volume and evolution of heat. The amount of alcohol in aqueous solutions can be found by determination of the specific gravity, using an *alcoholometer*, a type of hydrometer (*F. G. H. Tate*, Dict. appl. Chem., 4th Edn., Vol. I, p. 185). The alcohol in alcoholic beverages is first distilled off and then determined in the distillate.

Alcohol is a good solvent for many mineral salts, alkali hydroxides, hydrocarbons, resins, fatty acids and many other carbon compounds. Many gases are more readily soluble in it than in water; 100 volumes of alcohol dissolve 7 volumes of hydrogen, 25 volumes of oxygen and 16 volumes of nitrogen. Ethyl alcohol forms crystalline compounds with some salts, *e.g.* calcium chloride and magnesium chloride.

Chemical properties and reactions. (1) Potassium and sodium react with ethanol, yielding the soluble ethoxides.

(2) With sulphuric acid it yields ethyl hydrogen sulphate, and, on heating,

ethylene or diethyl ether according to the conditions:

$$CH_3{\cdot}CH_2OH + H_2SO_4 \rightleftarrows \underset{\text{Ethyl hydrogen sulphate}}{CH_3{\cdot}CH_2{\cdot}O{\cdot}SO_3H} \xrightarrow{\text{Heat}} \begin{cases} \underset{\text{Ethylene}}{CH_2{=}CH_2} \\ \underset{\text{Diethyl ether}}{CH_3{\cdot}CH_2{\cdot}O{\cdot}CH_2{\cdot}CH_3} \end{cases}$$

With sulphur trioxide carbyl sulphate, $\begin{matrix} CH_2{\cdot}O{\cdot}SO_2 \\ | \\ CH_2{-\!-}SO_2 \end{matrix}\!\!>\!O$, is formed.

(3) Phosphorus trihalides convert it into the corresponding ethyl halide.

(4) Oxidising agents like manganese dioxide and sulphuric acid, chromic acid, platinum black, etc., convert it into acetaldehyde and acetic acid.

(5) Chlorine and bromine oxidise alcohol to acetaldehyde, which forms the acetal with excess of alcohol. Further chlorination leads to the formation of mono- and di-chloroacetal, and finally to chloral alcoholate (see Chap. 8). Bleaching powder converts alcohol into chloroform, and iodine and potassium hydroxide convert it to iodoform (see Vol. IA, pp. 499 *et seq*).

(6) Nitric acid, free from nitrous fumes, gives ethyl nitrate. The oxidation of alcohol with nitric acid involves both carbon atoms and is difficult to control, oxalic acid being an end product.

(7) Mercury fulminate is produced when alcohol acts on mercury and an excess of nitric acid. Boiling with mercuric oxide and sodium hydroxide gives rise to a basic, explosive compound, $C_2Hg_6O_4H_2$ called mercarbide (*K. A. Hofmann*, Ber., 1900, **33**, 1328).

(8) Thermal decomposition of ethyl alcohol at temperatures above 800° gives ethylene and water, and acetaldehyde and hydrogen. The aldehyde in turn forms methane and carbon monoxide (*W. Ipatiew*, *ibid.*, 1901, **34**, 3579). By passing alcohol vapour over a catalyst, *e.g.* finely divided metals or alumina, decomposition occurs at a much lower temperature (*P. Sabatier* and *J. B. Senderens*, Compt. rend., 1903, **136**, 738; *Ipatiew*, Ber., 1903, **36**, 1990). These reactions are sometimes used in the manufacture of ethylene and of acetaldehyde, when required on a relatively small scale.

(3) **Propyl alcohols,** *propanols*, C_3H_7OH

n-*Propyl alcohol, propan-1-ol*, $CH_3{\cdot}CH_2{\cdot}CH_2{\cdot}OH$, b.p. 97·15° (*J. Timmermans* and *Y. Delcourt*, J. Chim. phys., 1934, **31**, 85), n_D^{20} 1·38556 (*A. I. Vogel*, J. chem. Soc., 1948, 1814), d_4^{15} 0·80749 (*Timmermans* and *Delcourt*, *loc. cit.*), occurs in fusel oil from which it is obtained by fractional distillation. It is an agreeable smelling liquid, miscible with water in all proportions, but is insoluble in cold saturated calcium chloride solution, whereby it can be distinguished from ethyl alcohol. It can be prepared from ethyl magnesium chloride and trioxymethylene, and by reduction of propionaldehyde. It is manufactured by catalytic reduction of propylene oxide or propionaldehyde. Oxidation converts it first into propionaldehyde and then into propionic acid. It is dehydrated *e.g.* catalytically or by the action of sulphuric acid giving propylene, from which *n*-propyl alcohol can be prepared by the hydroboronation reaction and subsequent hydrolysis (p. 9), or isopropyl alcohol by hydration with sulphuric acid.

Isopropyl alcohol, secondary propyl alcohol, dimethylcarbinol, isopropanol, propan-2-ol, b.p. 82·40° (*Timmermans* and *Delcourt, loc. cit.*), $n_D^{25°}$ 1·3747 (*R. F. Brunel*, J. Amer. chem. Soc., 1923, **45**, 1334), d_4^{25} 0·78095 (*K. K. Kelley, ibid.*, 1929, **51**, 1145), was prepared in 1855 by M. Berthelot from propylene and sulphuric acid and in 1862 by C. Friedel by the reduction of acetone. On an industrial scale it is made by the hydration of propylene, either by sulphuric acid (*B. T. Brooks*, Ind. Eng. Chem., 1935, **27**, 282), or by water at 220–250°/200 atm. in the presence of a tungsten oxide-zinc oxide-silica catalyst (F.I.A.T. Final Report No. 968). It can also be prepared by hydrolysis of isopropyl iodide; by reduction of propylene oxide with sodium amalgam; from ethyl formate or acetaldehyde by reaction with zinc and methyl iodide or with methyl magnesium iodide. It is formed together with *n*-propyl alcohol and propylene when *n*-propylamine is treated with nitrous acid (*V. Meyer* and *F. Forster*, Ber., 1876, **9**, 535).

Oxidation changes it into acetone; industrially this is carried out by catalytic dehydrogenation, resulting also in the formation of very pure hydrogen.

Chlorine converts it into 1,1,1,3-tetrachloroacetone.

(4) Butyl alcohols, C_4H_9OH

The four possible isomers are known, two primary, a secondary, and a tertiary (see Table 2).

TABLE 2

ISOMERIC BUTYL ALCOHOLS

Name	Formula	B.p.°	n_D^{20}	d_4^{20}
n-Butyl alcohol	$CH_3\cdot(CH_2)_2\cdot CH_2OH$	117·726	1·39922	0·80978
sec-Butyl alcohol*	$(CH_3\cdot CH_2)(CH_3)CH\cdot OH$	99·529	1·39780	0·80674
Isobutyl alcohol	$(CH_3)_2CH\cdot CH_2\cdot OH$	107·89	1·39768 (at 15°)	0·80576
tert-Butyl alcohol	$(CH_3)_3C\cdot OH$	82·41 (m.p. 25·66)	1·38468	0·78581

* These figures refer to the DL-compound.

n-*Butyl alcohol,* n-*propylcarbinol, butan-1-ol*, is formed by the action of sodium amalgam on *n*-butyraldehyde (*A. Lieben* and *A. Rossi*, Ann., 1871, **158**, 137), *n*-butyryl chloride (*A. Saytzeff*, J. pr. Chem., 1871, [ii], **3**, 76) or *n*-butyric anhydride (*E. Linnemann*, Ann., 1872, **161**, 178), or by the reduction of methyl butyrate with sodium and alcohol (*L. Bouveault* and *G. Blanc*, Compt. rend., 1903, **137**, 60). It is also produced by reduction of crotonaldehyde or trichlorobutyraldehyde (*A. Lieben* and *S. Zeisel*, Monatsh., 1880, **1**, 823, 840; *R. Douris*, Bull. Soc. chim. Fr., 1911, [iv], **9**, 922), from ethyl magnesium bromide and ethylene oxide (*V. Grignard*, Compt. rend., 1903, **136**, 1261), or by deamination of *n*-butylamine (*F. C. Whitmore* and *D. P. Langlois*, J. Amer. chem. Soc., 1932, **54**, 3441). On an industrial scale it is made by fermentation and by synthesis. Various strains of *Clostridium acetobutylicum* produce *n*-butyl alcohol, together with acetone and ethyl alcohol, by fermentation of carbohydrates. The process was developed by *C. Weizmann* (B.P., 4,845/1915; for reviews see *C. L. Gabriel* and *F. M. Crawford*, Ind. Eng. Chem., 1930, **22**, 1163; *W. N. McCutchan* and *R. J. Hickey*, in "Industrial Fermentations", Chemical Publishing Co., New York, 1954, Vol. I, p. 347). Synthetically, acetylene, from calcium carbide or petroleum, is converted to acetaldehyde and thence to crotonaldehyde, hydrogenation of which gives *n*-butyl alcohol. Uses include that of solvent for antibiotic recovery; its esters are important as solvents for lacquers and plasticisers.

sec-*Butyl alcohol, ethylmethylcarbinol, butan*-2-*ol*, is the simplest racemic alcohol. The *dextro* form has $[\alpha]_D^{20}$ $+13{\cdot}87^0$ and is best resolved by means of the brucine salt of its acid phthalate (*R. H. Pickard* and *J. Kenyon*, J. chem. Soc., 1911, **99**, 45; 1913, **103**, 1923).

It is obtained from ethyl methyl ketone by reduction, by the action of zinc ethyl on acetaldehyde, or by deamination of *n*-butylamine with nitrous acid (*Whitmore* and *Langlois*, *loc. cit.*). *n*-Butyl alcohol can be dehydrated to 1-butene, and the latter converted by hydriodic acid to *sec*-butyl iodide which on hydrolysis gives *sec*-butyl alcohol. The same iodide is formed on heating erythritol, $CH_2OH{\cdot}[CH(OH)]_2{\cdot}CH_2OH$, with hydriodic acid (*V. de Luynes*, Bull. Soc. chim. Fr., 1864, [ii], **2**, 3). *sec*-Butyl alcohol decomposes at 140–250° into water and 2-butene. The relations existing between *n*-butyl and *sec*-butyl alcohols and the butenes are shown below:

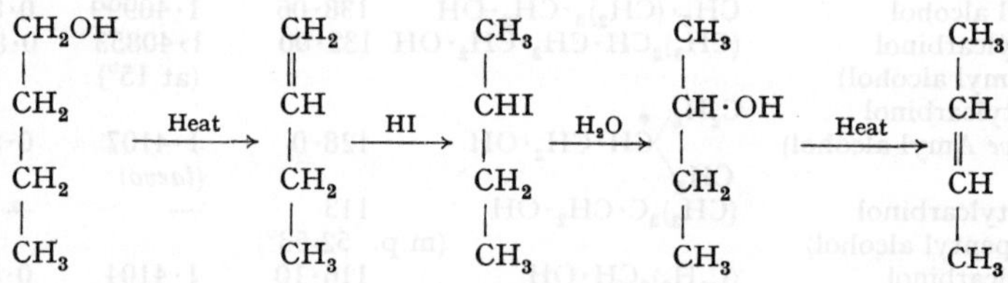

sec-Butyl alcohol is manufactured by the hydration of 1-butene (*B. T. Brooks*, Ind. Eng. Chem., 1935, **27**, 282).

Isobutyl alcohol, isopropylcarbinol, 2-methylpropan-1-ol, occurs in fusel oils and especially in that from potato spirit. It is a liquid with a characteristic odour. Dehydration gives isobutene, $(CH_3)_2C{=}CH_2$ from which by addition of halogen acids derivatives of *tert*-butyl alcohol can be obtained. For the action of chlorine on isobutyl alcohol, see *A. Brochet*, Compt. rend., 1894, **118**, 1280; Ann. Chim., 1897, [vii], **10**, 363.

tert-*Butyl alcohol, trimethylcarbinol, 2-methylpropan-2-ol* was prepared by *A. Butlerow* (Ann., 1867, **144**, 1) from acetyl chloride and zinc methyl, and was the first representative of the tertiary alcohols predicted by Kolbe. *L. Tissier* and *V. Grignard* obtained it from methylmagnesium iodide and acetyl chloride or acetic anhydride (Compt. rend., 1901, **132**, 684). It can be made by several methods involving molecular rearrangements, *e.g.* from isobutyl iodide by the action of acetic acid and silver oxide or by treatment of isobutylamine with nitrous acid (*E. Linnemann*, Ann., 1872, **162**, 12). In both these reactions the carbonium ion $(CH_3)_2CH \cdot CH_2^{\oplus}$ is first formed; this rearranges to the ion $(CH_3)_3C^{\oplus}$ which subsequently gives *tert*-butyl alcohol. *tert*-Butyl alcohol is made commercially by hydration of isobutene (butene-2) (*Brooks, loc. cit.*).

Oxidation of *tert*-butyl alcohol by chromic acid produces acetone, acetic acid and a little isobutyric acid, $(CH_3)_2CH \cdot CO_2H$, corresponding with isobutyl alcohol (*A. Butlerow*, Ann., 1877, **189**, 44). The alcohol decomposes thermally to give isobutene and water, over graphite at 480–500° (*W. Ipatiew*, Ber., 1902, **35**, 1057) or with reduced copper between 280° and 400° (*P. Sabatier* and *J. B. Senderens*, Compt. rend., 1903, **136**, 986).

(5) Amyl alcohols, $C_5H_{11}OH$

Theoretically, eight isomers are possible, four primary alcohols, three secondary and one tertiary, all of which are known. Table 3 contains the formulae and physical properties of the eight amyl alcohols. The name amyl is derived from the Greek word for starch, because the amyl alcohol first discovered was found in the fusel oil obtained from potato spirit.

TABLE 3

ISOMERIC AMYL ALCOHOLS

Name	Formula	b.p.°	n_D^{20}	$d_{4°}^{20}$
n-Amyl alcohol	$CH_3 \cdot (CH_2)_3 \cdot CH_2 \cdot OH$	138·06	1·40999	0·8136
Isobutylcarbinol (isoamyl alcohol)	$(CH_3)_2CH \cdot CH_2 \cdot CH_2 \cdot OH$	132·00	1·40853 (at 15°)	0·8139
sec-Butylcarbinol (*active* Amyl alcohol)	$C_2H_5(CH_3)\overset{*}{C}H \cdot CH_2 \cdot OH$	128·0	1·4107 (*laevo*)	0·8193
tert-Butylcarbinol (neopentyl alcohol)	$(CH_3)_3C \cdot CH_2 \cdot OH$	113 (m.p. 52-53°)	—	—
Diethylcarbinol	$(C_2H_5)_2CH \cdot OH$	116·10	1·4104	0·8218
Methyl-*n*-propylcarbinol	$CH_3(CH_3 \cdot CH_2 \cdot CH_2)\overset{*}{C}H \cdot OH$	119·89	1·4127	0·8132
Isopropylmethylcarbinol	$CH_3((CH_3)_2CH)\overset{*}{C}H \cdot OH$	111·5	1·4095	0·819
Ethyldimethylcarbinol (*tert*-amyl alcohol)	$(CH_3)_2C(OH) \cdot C_2H_5$	102·34	1·4058	0·81346 (at 15°)

* Three of the eight alcohols contain an asymmetric carbon atom, indicated in the formulae by a star. The figures given for these refer to the DL-compound unless stated otherwise.

(i) n-*Amyl alcohol, pentan-1-ol*, can be made by the action of nitrous acid on *n*-amylamine, itself obtained from caproic acid (*R. Gartenmeister*, Ann., 1886,

233, 249); by the reduction of *n*-valeramide, ethyl *n*-valerate or ethyl *n*-propylacetoacetate with sodium and alcohol (*R. Scheuble* and *E. Löbl*, Monatsh., 1904, **25**, 1081; *L. Bouveault* and *G. Blanc*, G.P., 164,294/1905); by hydrogenation of ethyl *n*-valerate at 250° and 220 at. over copper chromite (*H. Adkins* and *K. Folkers*, J. Amer. chem. Soc., 1931, **53**, 1095); from ethylene oxide and *n*-propylmagnesium bromide (*R. C. Huston* and *A. H. Agett*, J. org. Chem., 1941, **6**, 123). Industrially it is made by hydrolysis of *n*-amyl chloride, itself obtained by chlorination of *n*-pentane (*R. L. Kenyon et al.*, Ind. Eng. Chem., 1950, **42**, 2388), or by hydrogenation of *n*-valeraldehyde, obtainable from 1-butene by the Oxo process (Vol. IA, p. 426). It is almost insoluble in water, and has an odour resembling fusel oil.

(ii) *Isobutylcarbinol, isoamyl alcohol, 3-methylbutan-1-ol*, constitutes the chief ingredient of the fermentation amyl alcohol obtained from fusel oil and occurs as esters of angelic and crotonic acids in camomile oil (*J. Köbig*, Ann., 1879, **195**, 92). It may be synthesised by the action of isobutylmagnesium bromide on trioxymethylene (*R. Locquin*, Bull. Soc. chim. Fr., 1904, [iii], **31**, 599); by reduction of isovaleric acid or isovaleraldehyde, which can be obtained in good yield from isobutylene in the Oxo synthesis; by heating *tert*-butyl alcohol with synthesis gas under pressure in the presence of cobaltous acetate:

$$(CH_3)_3C{\cdot}OH + 2H_2 + CO \longrightarrow (CH_3)_2CH{\cdot}CH_2{\cdot}CH_2OH + H_2O$$

(*I. Wender*, *R. Levine* and *M. Orchin*, J. Amer. chem. Soc., 1949, **71**, 4160); by Grignard syntheses from formaldehyde and isobutylmagnesium bromide or from ethylene oxide and isopropylmagnesium bromide (*Huston* and *Agett*, *loc. cit.*). The so-called *fermentation amyl alcohol* of fusel oil, b.p. 129–132°, consists mainly of isobutylcarbinol, with some *sec*-butylcarbinol which causes the mixture to be laevorotatory. Isoamyl alcohol arises in the fermentation by deamination of L-leucine. It has a disagreeable odour.

The different solubilities and crystalline forms of the barium salts of the hydrogen sulphates of the two alcohols distinguish them, and assist in their separation. Isobutylcarbinol may be obtained from the more sparingly soluble salt. The alcohols can also be separated by passing hydrogen chloride into the mixture; isobutylcarbinol is converted to isoamyl chloride, while the other alcohol is unchanged (*F. Just*, Ann., 1883, **220**, 146), or preferably by esterification with nitrophthalic acid (*W. Marckwald*, Ber., 1901, **34**, 479; 1904, **37**, 1038). When crude fermentation amyl alcohol is distilled with zinc chloride, the amylene formed consists mainly of $(CH_3)_2C{:}CH{\cdot}CH_3$, resulting from a rearrangement of isobutylcarbinol, together with 2-methyl-1-butene and 1-pentene.

(iii) sec-*Butylcarbinol, active amyl alcohol, 2-methylbutan-1-ol*. The laevorotatory form, $[\alpha]_D^{20}$ $-5{\cdot}90°$, is the optically active constituent of fermentation amyl alcohol, the proportion present varying from 13 to 58 %, according to its origin (*Marckwald*, *ibid.*, 1902, **35**, 1595). It has the same configuration as the alkyl moiety of L-isoleucine, from which it is derived biochemically (p. 22), and is related to (+)-ethylmethylacetic acid. The sodium alkoxide is racemised by heating at 200° (*P. Frankland* and *T. S. Price*, J. chem. Soc., 1897, **71**, 253). The racemic alcohol can be synthesised from *sec*-butylmagnesium bromide and formaldehyde (*C. E. Wood* and *F. Scarf*, J. Soc. chem. Ind., 1923, **42**, 13T); by hydrogenation of tiglic aldehyde over nickel-kieselguhr at 200° and 100 at. (*G. T. Morgan* and *D. V. N. Hardy*, Chem. and Ind., 1933, 518), or hydrogenation of diethyl ethylmalonate or diethyl ethylmethylmalonate at 250° under pressure over copper-chromium oxide (*H. Adkins et al.*, J. Amer. chem. Soc., 1932, **54**, 1145, 4678). *sec*-Butylcarbinol is produced together with methanol and other alcohols from carbon monoxide and hydrogen in the presence of alkaline catalysts (*G. T. Morgan, D. V. N. Hardy* and *R. A. Procter*, J. Soc. chem. Ind., 1932, **51**, 1T). Complete resolution of the racemic alcohol is difficult (*F. von Falkenhausen*, Biochem. Z., 1931, **242**, 472) but the (—)-form is preferentially metabolised by certain moulds. The absolute configuration of (—)-*sec*-butylcarbinol is discussed by *L. Crombie* and *S. H. Harper*, J. chem. Soc., 1950, 2685.

(iv) tert-*Butylcarbinol, neopentyl alcohol, 2,2-dimethylpropan-1-ol,* is formed by reduction of trimethylacetyl chloride (pivalyl chloride) with sodium amalgam (*L. Tissier*, Compt. rend., 1891, **112**, 1065); by the action of *tert*-butylmagnesium chloride on methyl formate (*L. H. Sommer, H. D. Blankman* and *P. C. Miller*, J. Amer. chem. Soc., 1954, **76**, 803); by lithium aluminium hydride reduction of trimethylacetic acid (*R. F. Nystrom* and *W. G. Brown*, *ibid.*, 1947, **69**, 2548) or catalytic hydrogenation of its ethyl ester (*Adkins* and *Folkers*, *loc. cit.*); by treatment of di-isobutylene, $(CH_3)_3C{\cdot}CH_2{\cdot}C(CH_3){:}CH_2$, with hydrogen peroxide and sulphuric acid (*J. Hoffman*, Org. Synth., 1960, **40**, 76). Nitrous acid converts the corresponding amine into ethyldimethylcarbinol (*Tissier*, *loc. cit.*; *M. Freund* and *F. Lenze*, Ber., 1891, **24**, 2150). The corresponding alkyl halides, which are not made from the alcohol, have been studied as examples of sterically hindered halides (*I. Dostrovsky, E. D. Hughes* and *C. K. Ingold*, J. chem. Soc., 1946, 157, *et seq.*) (see Vol. IA, p. 277).

(v) *Diethylcarbinol, pentan-3-ol,* is formed by the action of zinc and ethyl iodide on ethyl formate or of the latter on ethylmagnesium bromide (*V. Grignard*, Compt. rend., 1901, **132**, 336). It may be made from diethyl ketone by reduction with aluminium isopropoxide (*H. Lund*, Ber., 1937, **70**, 1520). Diethylcarbinol can be converted into methyl-*n*-propylcarbinol via 2-pentene, and 2-iodopentane:

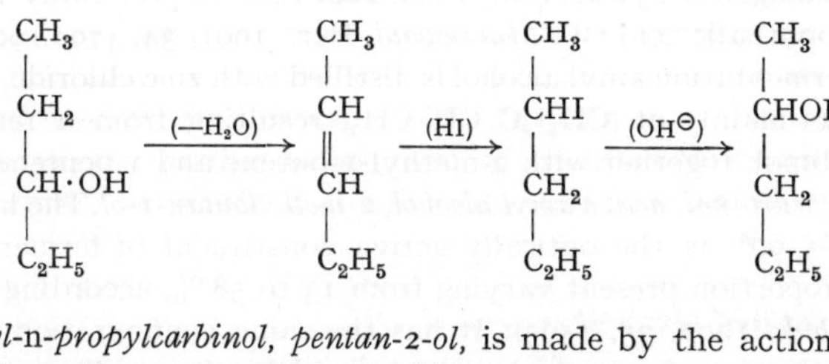

(vi) *Methyl-*n*-propylcarbinol, pentan-2-ol,* is made by the action of *n*-propylmagnesium bromide on acetaldehyde (*Wood* and *Scarf*, *loc. cit.*); by reduction of methyl *n*-propyl ketone catalytically (*C. L. Wilson*, J. Amer. chem. Soc.,

1948, **70**, 1313); by reduction of 2-methyltetrahydrofuran with zinc dust and acetic acid (*W. Hückel* and *W. Gelmroth*, Ann., 1934, **514**, 233); by treatment of 2-pentene with sulphuric acid. The *laevo*-form can be obtained by the action of *Penicillium glaucum* on the racemic mixture (*J. A. Le Bel*, Compt. rend., 1879, **89**, 312), and the *dextro*-form by crystallisation of the brucine or strychnine salts of the acid phthalates. The *dextro*-form has b.p. 118·5–119·5°, n_D^{20} 1·4053, d_4^{20} 0·8103, $[\alpha]_D^{20}$ +13·70° (*R. H. Pickard* and *J. Kenyon*, J. chem. Soc., 1911, **99**, 45).

(vii) *Methylisopropylcarbinol, 3-methylbutan-2-ol*, is formed from isopropylmagnesium chloride and acetaldehyde (*Pickard* and *Kenyon*, J. chem. Soc., 1912, **101**, 620); by reduction of methyl isopropyl ketone or methyl isopropenyl ketone (*E. M. McMahon et al.*, J. Amer. chem. Soc., 1948, **70**, 2971); by the action of zinc methyl on bromoacetyl bromide (*W. Winogradow*, Ann., 1878, **191**, 125) or from chloroacetone, chloroacetyl chloride or isobutylene oxide by the action of methylmagnesium bromide (*L. Henry*, Compt. rend., 1907, **145**, 21; *R. C. Huston, R. I. Jackson* and *G. B. Spero*, J. Amer. chem. Soc, 1941, **63**, 1459). The alcohol has been resolved using the brucine salt of the hydrogen phthalate (*Pickard* and *Kenyon*, *loc. cit.*). The dextro-form has b.p. 110–112°, n_D^{20} 1·3973, d_4^{20} 0·8180, $[\alpha]_D^{20}$ +4·85°. It is converted into *tert*-amyl halides on treatment with hydrogen halides or phosphorus pentachloride, 2-methylbutene-2 being formed as an intermediate:

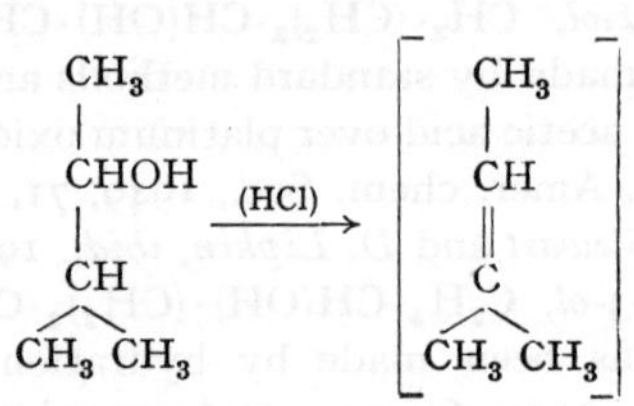

(viii) tert-*Amyl alcohol, ethyldimethylcarbinol, 2-methylbutan-2-ol*, is a liquid with an odour like that of camphor and is hypnotic. It is produced by treatment of trimethylethylene, (2-methylbutene-2), present in commercial amylene, with sulphuric acid (*R. Adams, O. Kamm* and *C. S. Marvel*, J. Amer. chem. Soc., 1918, **40**, 1950). It is also formed by the action of nitrous acid on the amine corresponding to *tert*-butylcarbinol (*Freund* and *Lenze*, *loc. cit.*), from propionyl chloride and zinc methyl (*A. Wischnegradsky*, Ann., 1878, **190**, 328) and from 2,3-epoxybutane and methylmagnesium bromide (*D. L. Cottle* and *L. S. Powell*, J. Amer. chem. Soc., 1936, **58**, 2267). When heated it decomposes into water and trimethylethylene, especially in the presence of reduced copper or nickel (*W. Ipatiew*, Ber., 1901, **34**, 600; *P. Sabatier* and *J. B. Senderens*, Compt. rend., 1903, **136**, 986).

(6) Higher saturated alcohols, $C_nH_{2n+1}\cdot OH$

Many representatives of the higher members of this series are known. As the series is ascended, the number of theoretically possible compounds increases, but the known representatives become fewer in number. Some of the more important members are described below.

Soluble derivatives of the higher alcohols, especially the alkyl sodium sulphates, have been introduced into industry as wetting agents (C_{10}—C_{12}) and detergents (C_{16}—C_{18}) (see p. 68). The higher primary alcohols are manufactured by reduction of esters or glycerides of the fatty acids or by the hydrolysis of waxes, *e.g.* spermaceti wax, gives cetyl alcohol. The petroleum industry manufactures synthetic detergents which are *sec*-alkyl sodium sulphates obtained from olefin fractions from cracked petroleum. Soluble surface-active substances are also manufactured by condensing higher alcohols with ethylene oxide whereby polyglycol ethers soluble in water, of the general formula $RO(CH_2{\cdot}CH_2O)_nCH_2{\cdot}CH_2OH$ (R = higher alkyl group), are formed.

A large number of secondary alcohols of the general formula R·CH(OH)·R′ have been resolved by means of the brucine, strychnine or cinchonidine salts of their acid phthalates (*Pickard* and *Kenyon*, J. chem. Soc., 1911, **99**, 45; 1913, **103**, 1923). Their absolute configurations are discussed by *J. A. Mills* and *W. Klyne* in "Progress in Stereochemistry", Butterworth, London, 1954, Vol. I, p. 205.

n-*Hexyl alcohol, hexan-1-ol*, b.p. 157·47°, n_D^{20} 1·41816, d_4^{20} 0·8186, occurs as acetate and butyrate in the oil of the seed of *Heracleum giganteum*, and is itself present in certain other essential oils, *e.g.* Java citronella oil (*H. Bohnsack*, Ber., 1943, **76**, 564).

n-*Butylmethylcarbinol, hexan-2-ol*, $CH_3{\cdot}(CH_2)_3{\cdot}CH(OH){\cdot}CH_3$, b.p. 139·8°, n_D^{20} 1·4144, d_4^{25} 0·8108, can be made by standard methods and also by hydrogenation of 2,5-dimethylfuran in acetic acid over platinum oxide under pressure (*H. A. Smith* and *J. F. Fuzek*, J. Amer. chem. Soc., 1949, **71**, 415). The *dextro*-form has $[\alpha]_D^{18}$ +11·68° (*T. D. Stewart* and *D. Lipkin, ibid.*, 1939, **61**, 3297).

Ethyl-n-*propylcarbinol, hexan-3-ol*, $C_2H_5{\cdot}CH(OH){\cdot}(CH_2)_2{\cdot}CH_3$, b.p. 134·5–135°, n_D^{20} 1·4168, d_4^{20} 0·8193, has been made by hydration of 3-hexene (*L. Spiegler* and *J. M. Tinker, ibid.*, 1939, **61**, 940) or from *n*-butyraldehyde and ethylmagnesium chloride (*F. C. Whitmore* and *H. M. Woodburn, ibid.*, 1933, **55**, 361). The *dextro*-form has $[\alpha]_D^{20}$ +6·81° (*R. S. Airs, M. P. Balfe* and *J. Kenyon*, J. chem. Soc., 1942, 18).

3-*Methyl-pentan-1-ol*, $CH_3{\cdot}CH_2{\cdot}CH(CH_3){\cdot}CH_2{\cdot}CH_2OH$. The *dextro*-form has been isolated from Reunion Geranium oil and the *laevo*-form from Citronella oil (*Bohnsack, loc. cit.*).

tert-**Butylmethylcarbinol**, *pinacolyl alcohol*, 3,3-*dimethylbutan-2-ol*, $(CH_3)_3C{\cdot}CH(OH){\cdot}CH_3$, m.p. 5·3°, b.p. 120·4°, n_D^{20} 1·4148, d_4^{20} 0·8185, results from the reduction of pinacolone (*tert*-butyl methyl ketone) or from the action of *tert*-butylmagnesium bromide on acetaldehyde. It has a camphor-like odour. The *dextro* form has b.p. 120°, $[\alpha]_D^{20}$ +7·71°, n_D^{20} 1·4146, d_4^{25} 0·810 (*Pickard* and *Kenyon*, J. chem. Soc., 1914, **105**, 1115).

Dimethylisopropylcarbinol, 2,3-*dimethylbutan-2-ol*, $(CH_3)_2C(OH){\cdot}CH(CH_3)_2$, m.p. —10·6°, b.p. 118·70°, n_D^{20} 1·4176, d_4^{20} 0·8236, is prepared from acetone and isopropylmagnesium bromide (*M. Delacre*, Bull. Soc. chim. Fr., 1906, [iii], **35**, 811).

n-*Heptyl alcohol, heptan-1-ol,* $CH_3\cdot(CH_2)_6\cdot OH$, b.p. 176·81°, n_D^{20} 1·4241, d_4^{20} 0·8219, is present in fusel oil from potato spirit (*P. Schorigin et al.*, Ber., 1933, **66**, 1087). It has been made from *n*-heptaldehyde (oenanthal) by reduction (*F. Jourdan*, Ann., 1879, **200**, 101) and from *n*-pentylmagnesium bromide and ethylene oxide (*Huston* and *Agett, loc. cit.*).

*Dimethyl-*tert*-butylcarbinol, pentamethylethyl alcohol, 2,3,3-trimethylbutan-2-ol,* $(CH_3)_3C\cdot C(OH)(CH_3)_2$, m.p. 83–84° (hemihydrate), b.p. 130·3–130·8°, d_4^{25} 0·8380, has been obtained by various syntheses from organomagnesium compounds (*Delacre, loc. cit.*).

n-*Octyl alcohol, octan-1-ol,* $CH_3\cdot(CH_2)_7\cdot OH$, m.p. —15·0°, b.p. 195·28°, n_D^{20} 1·4295, d_4^{20} 0·8256, occurs as the acetate in the volatile oil of *Heracleum spondylium*, as the butyrate in the oil of *Pastinaca sativa*, in the oil of *Heracleum giganteum* (*W. Moslinger*, Ann., 1877, **185**, 26) and in potato spirit fusel oil (*Schorigin et al., loc. cit.*). It has been made by reduction of ethyl caprylate, with sodium and alcohol (*L. Bouveault* and *G. Blanc*, Compt. rend., 1903, **136**, 1677) or with hydrogen over copper-chromium oxide (*K. Folkers* and *H. Adkins*, J. Amer. chem. Soc., 1932, **54**, 1145).

sec-*Octanol* and sec-*nonyl alcohols* are made commercially, from iso-octenes, *e.g.* 2,4,4-trimethylpentene-1 and -2, by hydration and oxo-processes and used as intermediates for making wetting agents, detergents and plasticisers.

n-*Nonyl alcohol, nonan-1-ol,* m.p. —5·5°, b.p. 213·5°, n_D^{20} 1·4323, d_4^{20} 0·8274, occurs as the caprylate in orange-flower oil (*K. Stephan*, J. pr. Chem., 1900, [ii], **62**, 532) and in potato spirit fusel oil (*Schorigin et al., loc. cit.*).

n-*Decyl alcohol, decan-1-ol,* m.p. 6·88°, b.p. 229°, n_D^{20} 1·43682, d_4^{20} 0·8292, is present in the blubber of sperm whales (*S. Ueno* and *R. Koyama*, J. chem. Soc. Japan, 1936, **57**, 1).

n-*Dodecyl alcohol, lauryl alcohol, dodecan-1-ol,* $CH_3\cdot(CH_2)_{11}\cdot OH$, m.p. 23·4°, b.p. 153·5°/25 mm., is present in the alcohols derived by hydrolysing the glycerides occurring in sperm blubber oil (*Ueno* and *Koyama, loc. cit.*) and coconut oil. The sulphate is used as a wetting agent.

n-*Tetradecyl alcohol, myristyl alcohol, tetradecan-1-ol,* $CH_3\cdot(CH_2)_{13}\cdot OH$, m.p. 35·0°, b.p. 111–113°/0·1 mm., is present in sperm head oil (*T. P. Hilditch* and *J. A. Lovern*, J. Soc. chem. Ind., 1929, **48**, 365T) and in the head oil of the porpoise (*Lovern*, Biochem. J., 1934, **28**, 394). It can be made by hydrogenation of esters of myristic acid.

n-*Hexadecyl alcohol, cetyl alcohol, hexadecan-1-ol,* $CH_3\cdot(CH_2)_{15}\cdot OH$, m.p. 49·5°, b.p. 179–181°/10 mm., was prepared in 1817 by M. E. Chevreul from cetyl palmitate, the chief ingredient of spermaceti wax from the head oil of the sperm whale, by saponification with potassium hydroxide. The alcohol is also present in other fish oils. When fused with potassium hydroxide it yields palmitic acid:

$$C_{15}H_{31}\cdot CH_2OH + KOH \longrightarrow C_{15}H_{31}\cdot CO_2K + 2H_2$$

The sulphate is used as a detergent.

n-*Eicosyl alcohol, eicosan-1-ol,* $CH_3\cdot(CH_2)_{19}\cdot OH$, m.p. 66·5–67·5°, b.p. 220–225°/3 mm., is present in sperm blubber oil (*Y. Toyama*, J. chem. Soc. Japan,

1938, **59,** 810) and in the unsaponifiable fraction of an acetone extract of plantation rubber (*I. M. Heilbron et al.*, J. chem. Soc., 1941, 344).

n-*Hexacosyl alcohol, ceryl alcohol*, $CH_3\cdot(CH_2)_{25}\cdot OH$, m.p. $79\cdot5^0$, is found in the wax from blades of cocksfoot grass, n-*octacosyl alcohol*, $CH_3\cdot(CH_2)_{27}\cdot OH$, m.p. $83\cdot4^0$, in that from wheat blades and n-*triacontyl alcohol*, $CH_3\cdot(CH_2)_{29}\cdot OH$, m.p. $86\cdot5^0$, in that from lucerne leaf (*A. C. Chibnall et al.*, Biochem. J., 1934, **28,** 2189).

Other plant cuticle waxes contain mixtures of C_{26}, C_{28} and C_{30} alcohols (*e.g.* apple cuticle wax). Carnauba wax contains all the "even-number" alcohols from C_{26} to C_{34}, and the waxes from candelilla and cotton contain those from C_{28} to C_{34}. The waxes of plant cuticles and insects are discussed by *Chibnall* and *S. H. Piper, ibid.*, p. 2209).

(b) Derivatives of saturated alcohols

Compounds discussed in this section are (1) the alkoxides, in which the hydrogen atom in the functional OH group is replaced by a metal, and (2) those substituted alkanols containing "non-functional" groups, *i.e.* halogens and nitro (nitroso and azoxy). The latter are strictly bi-functional compounds and in the first edition of this book they were regarded as derivatives of ethylene glycol. They are, however, essentially monohydric alcohols and so are more conveniently discussed here.

(i) Alkoxides

The alkoxides, or alcoholates as they were called earlier, are usually obtained by allowing a metal *e.g.* sodium (or alternatively, sodium hydride, sodamide or a sodio-aryl) to react with an alcohol:

$$2\,CH_3OH + 2Na \longrightarrow 2\,CH_3ONa + H_2$$

The higher members can also be prepared by dissolving sodium or potassium hydroxides in the alcohol and removing the resultant water by azeotropic distillation (see below).

Sodium ethoxide, $NaOC_2H_5$, is the most important alkoxide and is employed in many synthetic reactions of the "aldol" type, *i.e.* those involving carbanion intermediates. It may be prepared by treating sodium in xylene with the calculated quantity of alcohol (*J. W. Brühl*, Ber., 1902, **35,** 3516; 1904, **37,** 2067). For many purposes, an alcoholic solution is prepared by dissolving sodium in alcohol just prior to use; it is difficult to recover the resultant alkoxide from such solutions due to the stability of crystallisate containing additional molecules of ethanol. An anhydrous methyl alcoholic solution of a catalytic quantity of *sodium methoxide* is often used for de-*O*-acylation of carbohydrate esters by trans-esterification (*G. Zemplén, A. Gerecs* and *I. Hadacsy, ibid.*, 1936, **69,** 1827).

An excess of water converts the ethoxides to alcohol and sodium hydroxide. The ethoxides also result on dissolving potassium or sodium hydroxides in alcohol. In such solutions the equilibrium of the reaction favours the formation of ethoxide ion (*E. F. Caldin* and *G. Long*, J. chem. Soc., 1954, 3737). Water can be removed completely by azeotropic distillation with benzene:

$$OH^{\ominus} + EtOH \rightleftharpoons H_2O + EtO^{\ominus}$$

Calcium ethoxide, $Ca(OC_2H_5)_2$, is formed by the solution of metallic calcium in alcohol, or by the decomposition of calcium carbide by absolute alcohol with the aid of heat (*R. de Forcrand*, Compt. rend., 1894, **119**, 1266; *L. W. Winkler*, Ber., 1905, **38**, 3612).

Aluminium ethoxide, $Al(OC_2H_5)_3$, m.p. 140°, b.p. 200°/6–8 mm. (*H. Adkins*, J. Amer. chem. Soc., 1922, **44**, 2178), *aluminium* n-*propoxide*, m.p. 106°, b.p. 240°/5 mm. and *aluminium isopropoxide*, m.p. 118°, b.p. 125–130°/4 mm. are remarkable in that they can be distilled under reduced pressure without decomposition. *Aluminium methoxide* is decomposed by heat under these conditions. These compounds are prepared by the action of the respective alcohols on aluminium amalgam. The aluminium alkoxides, particularly the isopropoxide, have been used as reagents for the reduction of carbonyl groups (see p. 7.)

The alkoxides, including those of other metals, have been reviewed by *D. C. Bradley*, Prog. in inorg. Chem., 1960, **2**, 303.

*(ii) Halogenohydrins (halohydrins); halogenoalkanols**

Saturated alcohols in which one of the hydrogen atoms attached to carbon is replaced by halogen, halogenoalkanols, are known also as halo- or halogenohydrins.

Those carrying the halogen atom on the carbon atom carrying the hydroxyl group also, are formed transiently in certain reactions but are usually so unstable that, with the exception of 1,2-dibromoethanol (see below), they are known only in the form of derivatives, *e.g.* as ethers or esters. They are usually referred to as aldehyde (ketone) halogenohydrins; they are discussed later in Chapter 8 as derivatives of aldehydes (ketones) with which they are closely related chemically.

Thus the term halogenohydrin as usually applied refers to halogenoalkanols in which the halogen and hydroxyl groups are attached to different carbon atoms.

They are obtained as follows:

(1) By the action of hydrochloric or hydrobromic acids on glycols, giving chloro- and bromo-hydrins, respectively (Org. Synth., Coll. Vol. III, 446).

* This and section *(iii)* on nitroalkanols was contributed by *E. S. Waight*.

The reaction proceeds readily at —15° if acetic acid is present (*R. Boschan* and *S. Winstein*, J. Amer. chem. Soc., 1956, **78**, 4921):

$$HOCH_2 \cdot CH_2OH + HCl \longrightarrow HOCH_2 \cdot CH_2Cl + H_2O$$

Iodohydrins and fluorohydrins may be obtained from the corresponding chloro- or bromo-compounds by the action of inorganic iodides and fluorides.

(2) By the addition of HOX (where X=Cl, Br, I) to olefins; the addition follows Markownikoff's rule (see Vol. IA, p. 301):

$$CH_2{:}CH \cdot CH_3 + HOCl \longrightarrow CH_2Cl \cdot CHOH \cdot CH_3$$

For a discussion of the mechanism of addition of hypochlorous acid to an ethylenic double bond, see *P. Ballinger* and *P. B. D. de la Mare* (J. chem. Soc., 1957, 1481). Addition of hypobromous acid may be accomplished using *N*-bromosuccinimide in dilute acetic acid (*R. A. Raphael, ibid.*, 1949, S 44). Addition of hypochlorous acid in the anti-Markownikoff sense, *i.e.* to give $RCHCl \cdot CH_2OH$, is achieved by the use of chromyl chloride (*S. J. Cristol* and *K. R. Eilar*, J. Amer. chem. Soc., 1950, **72**, 4353).

(3) By the action of hydrogen or metallic halides on epoxides (Org. Synth., Coll. Vol. I, p. 117; II, p. 571).

(4) By selective hydrolysis of dihalides (*P. Höring*, Ber., 1905, **38**, 3464).

(5) By hydration of unsaturated halides (*L. Henry*, Rec. Trav. chim., 1903, **22**, 209).

(6) From halogenated aldehydes, ketones, and esters, by reaction with Grignard reagents or lithium alkyls (*J. W. Cornforth, R. H. Cornforth,* and *K. K. Mathew*, J. chem. Soc., 1959, 112).

Properties. Halogenohydrins undergo reactions expected for compounds containing hydroxyl groups and halogens, thus the hydroxyl may be esterified, oxidized, and alkylated. Nucleophilic displacements of halogen are, however, complicated by the tendency to form epoxides. Thus 2-chloropropanol fails to give the expected product with alcoholic potassium cyanide, but forms instead the isomer, $CH_3CH(OH)CH_2CN$, by way of propylene oxide.

(1) Reduction in general gives the alcohol. Halogenohydrins, other than 1,2-halogenohydrins, which are rapidly hydrogenolysed, are relatively stable to lithium aluminium hydride and can thus be obtained from the corresponding halogeno-acids by the use of this reagent (*C. E. Sroog et al.*, J. Amer. chem. Soc., 1949, **71**, 1710).

(2) Oxidation gives halogenated aldehydes, ketones, and acids.

(3) With water glycols are formed (for a kinetic study of the hydrolysis

of ethylene halohydrins, see *H. D. Cowan, C. L. McCabe,* and *J. C. Warner, ibid.*, 1950, **72**, 1194. Dilute alkalis give epoxides, which may react further, *e.g.* to give the glycol.

Ethylene chlorohydrin, *2-chloroethanol*, $CH_2Cl{\cdot}CH_2OH$, b.p. 128·6°/760 mm., d_4^{25} 1·1947, n_D^{20} 1·4422, is manufactured by method (2) above. *Ethylene bromohydrin*, b.p. 51·0–51·4°/4 mm., d_4^{20} 1·177, *iodohydrin*, b.p. 75·0–75·2°/5 mm., d_4^{20} 2·2197, and *fluorohydrin*, b.p. 103·5°, d_4^{20} 1·104 (*B. C. Saunders et al.*, J. chem. Soc., 1949, 773) may be prepared from the chlorohydrin.

3-*Chloropropan-1-ol, trimethylene chlorohydrin*, $CH_2Cl{\cdot}CH_2{\cdot}CH_2OH$, b.p. 65°/10 mm., d_4^{20} 1·1309, may be obtained from propane-1,3-diol and hydrogen chloride; with alkali it gives trimethylene oxide. Higher homologues are prepared similarly. *Tetramethylene chlorohydrin*, b.p. 81–82°/14 mm., is conveniently prepared from tetrahydrofuran, either by the direct action of hydrogen chloride (Org. Synth., 1943, Coll. Vol. II, p. 571), or through the acetate, $Cl(CH_2)_4O_2C{\cdot}CH_3$ (*C. Crisan*, Ann. Chim., 1956, [13], **1**, 436, see p. 5).

DL-1-*Chloropropan-2-ol*, $CH_3{\cdot}CHOH{\cdot}CH_2Cl$, b.p. 127°, is prepared by the acid-catalysed hydration of allyl chloride, or by the action of hypochlorous acid or *tert*-butyl hypochlorite (*S. L. Langedijk*, U.S.P., 2,106,353/1938) on propylene. DL-2-*Chloropropan-1-ol*, $CH_3{\cdot}CHCl{\cdot}CH_2OH$, b.p. 133–134°, is formed along with 1-chloropropan-2-ol from which it may be obtained as follows:

$$CH_3{\cdot}CHOH{\cdot}CH_2Cl \xrightarrow{OAc^{\ominus}} CH_3{\cdot}CHOH{\cdot}CH_2OAc \xrightarrow[ZnCl_2]{HCl}$$

$$CH_3{\cdot}CHCl{\cdot}CH_2OAc \xrightarrow[CH_3OH]{HCl} CH_3{\cdot}CHCl{\cdot}CH_2OH$$

(+)-2-*Chloropropan-1-ol*, b.p. 40°/15 mm., $[\alpha]_D^{18}$ 9·3° (*E. Abderhalden* and *E. Eichwald*, Ber., 1918, **51**, 1317). For other physical properties of the propylene chlorohydrins, see *G. Forsberg* and *L. Smith*, Acta Chem. Scand., 1947, **1**, 577.

2,2-*Difluoroethyl alcohol*, b.p. 95·5–96°, $n_\alpha^{11.8}$ 1·3345, d^{17} 1·30839, is formed by the action of water and yellow mercuric oxide on difluorobromoethane at 160° (*F. Swarts*, Bull. Acad. roy. Belg., 1902, **11**, 731); 2,2-*dichloroethyl alcohol*, b.p. 146°, d^{15} 1·145, by the action of zinc ethyl on dichloroacetaldehyde (*M. de Lacre*, Compt. rend., 1887, **104**, 1184). 1,2-*Dibromoethyl alcohol*, $CH_2Br{\cdot}CHBr(OH)$, b.p. 31°/17 mm., n_D^{20} 1·5492, d_4^{20} 2·1782, is one of the bromination products of paraldehyde (*A. Stepanow, N. Preobraschensky* and *M. Schtschukina*, Ber., 1926, **59**, 2533); 2,2-*dibromoethyl alcohol*, $CHBr_2{\cdot}CH_2OH$, b.p. 70–72°/10 mm., d^0 2·35 (*E. Demole*, Ber., 1876, **9**, 45) is obtained by phytochemical reduction of bromal (*R. Willstätter* and *W. Duisberg*, Ber., 1923, **56**, 2283).

2,2,2-*Trichloroethyl alcohol*, $CCl_3{\cdot}CH_2OH$, m.p. 19°, b.p. 151°, is formed on reduction of chloral by aluminium ethoxide (*W. Chalmers*, Org. Synth. Coll. Vol. II, p. 598), by Grignard reagents (*P. M. Dean* and *W. I. Wolf*, J. Amer. chem. Soc., 1936, **58**, 332) or by yeast (*Willstätter* and *Duisberg, loc. cit.*). Its carbamate, *Voluntal*, m.p. 64–65°, has hypnotic properties (*R. Willstätter et al.*, Chem. Ztbl., 1923, I, 1196); 2,2,2-*tribromoethyl alcohol*, m.p. 80°, b.p. 92–93°/10 mm.

1,2-**Dichlorohydrin,** 2,3-*dichloropropanol*-1, $CH_2Cl·CHCl·CH_2OH$, b.p. 182°, 81°/13·5 mm., d_0^{18} 1.345, n_D^{18} 1·4875 and 1,2-*dibromhydrin*, 2,3-*dibromopropanol*-1, b.p. 219°, 118°/17 mm., are obtained by halogenating allyl alcohol with chlorine and bromine respectively, either in water (*J. Read* and *E. Hurst*, J. chem. Soc., 1922, 989) or carbon bisulphide and 1,2-*di-iodohydrin*, m.p. 43°, is made similarly (*A. Fairbourne* and *D. W. Stephens*, *ibid.*, 1932, 1973). 1,3-**Dichlorohydrin,** 1,3-*dichloropropanol*-2, b.p. 175–175°, d_4^{17} 1·3506, $n_D^{16.9}$ 1·48025, is prepared (1) from glycerol and hydrogen chloride in the presence of acetic acid (Org. Synth., Coll. Vol. III, 446) at 100–110°; (2) by the action of hydrogen chloride on epichlorhydrin (*A. J. Hill* and *E. J. Fischer*, J. Amer. chem. Soc., 1922, **44**, 2586; Org. Synth., Coll. Vol. I, 117; II, 571); (3) together with other products when glycerol is treated with thionyl chloride in pyridine. 1,3-*Dibromohydrin*, b.p. 110–112°/20 mm., d^{20} 2·14 is obtained from the action of bromine and phosphorus on glycerol and 1,3-*di-iodohydrin*, m.p. —16° to —20° from dichlorohydrin and sodium iodide.

1,1,1-*Trichloroisopropyl alcohol*, 1,1,1-*trichloropropanol*-2, $CCl_3·CHOH·CH_3$, m.p. 50–51°, b.p. 161·8°/773 mm., and 2,2,3-*trichlorobutanol*, m.p. 61–62°, b.p. 120°/45 mm., are prepared from diethylzinc and chloral and butylchloral, respectively (*K. Garzarolli-Thurnlakh*, Ann., 1881, **210**, 77).

Trichloro-tert-*butyl alcohol*, 2-*trichloromethylpropanol*-2, (chloretone) m.p. 97°, b.p. 167°; *hydrate*, $\frac{1}{2}H_2O$, m.p. 80–81°, obtained by condensing chloroform with acetone in presence of alkali, is a hypnotic and germicide (*C. Willgerodt* and *A. Genieser*, J. pr. Chem., 1888, **37**, 364).

(iii) Nitroalkanols

Nitroalkanols may be obtained (1) by reaction of primary or secondary nitroparaffins with aldehydes or ketones in the presence of a base (*B. M. Vanderbilt* and *H. B. Hass*, Ind. Eng. Chem., 1940, **32**, 34):

$$RCHO + R'CH_2NO_2 \longrightarrow RCHOH·CHR'NO_2$$

Formaldehyde gives nitro-derivatives of primary alcohols together with the products of further condensation at the α-carbon atom; thus with nitromethane all the hydrogen atoms can be replaced to produce tri(hydroxymethyl)nitromethane, $(HOCH_2)_3CNO_2$.

(2) By the interaction of nitrogen dioxide (or fuming nitric acid) and olefins; dinitro-compounds or nitro-nitrites may also be formed depending on the conditions employed (*N. Levy et al.*, J. chem. Soc., 1946, 1093, 1096), the latter being readily hydrolysed to nitro-alcohols. In unsymmetrical olefins nitrogen dioxide attacks first the least alkylated carbon:

$$CH_3CH{:}CH_2 \xrightarrow{NO_2} CH_3CH(ONO)CH_2NO_2 \xrightarrow{H_2O} CH_3\cdot CHOH\cdot CH_2NO_2$$

(for a discussion of the mechanism of addition, see *P. Gray* and *A. D. Yoffe*, Quart. Reviews, 1955, 9, 362).

2-Nitroethanol, $O_2NCH_2CH_2OH$, b.p. 195°, n_D^{20} 1·443 (*nitrate*, b.p. 74°/1 mm.); *2-nitropropan-1-ol*, b.p. 100°/12 mm., n_D^{20} 1·4379, *3-nitropropan-1-ol*, b.p. 123–128°/22 mm., *1-nitropropan-2-ol*, b.p. 112°/30 mm.

(c) Unsaturated alcohols

(i) Olefinic alcohols or alkenols, $C_nH_{2n-1}OH$

These bear the same relation to the olefins, C_nH_{2n}, as the normal alcohols do to the paraffins. They combine with the general character of alcohols the property of the olefins of forming addition compounds. The most important member of the class is allyl alcohol, $CH_2{:}CH\cdot CH_2OH$. The reactions of allylic compounds have been reviewed by *R. H. DeWolfe* and *W. G. Young* (Chem. Reviews, 1956, 56, 753).

General methods of synthesis include: (1) Reaction of Grignard compounds with unsaturated carbonyl compounds, *e.g.*

$$CH_3\cdot CH{:}CH\cdot CHO + C_2H_5MgBr \xrightarrow{\text{hydrolysis}} CH_3\cdot CH{:}CH\cdot CH(OH)\cdot C_2H_5$$

or unsaturated Grignard compounds with aldehydes or ketones, *e.g.*

$$(CH_3)_2CO + CH_2{:}CH\cdot CH_2MgBr \longrightarrow (CH_3)_2C(OH)\cdot CH_2\cdot CH{:}CH_2$$

(2) Addition of lithium alkenyls or alkenylmagnesium halides to aldehydes, ketones, or ethylene oxides, *e.g.*

$$CH_2{:}CHLi + CH_3\cdot CH_2\cdot CHO \longrightarrow CH_2{:}CH\cdot CHOH\cdot CH_2\cdot CH_3$$

(*H. Normant*, Compt. rend., 1955, 240, 314; *idem*, "Advances in Organic Chemistry", Interscience, 1960, Vol. II, p. 20).

(3) Reduction of olefinic carbonyl compounds by the Ponndorf method with aluminium isopropoxide (*A. L. Wilds*, Org. Reactions, 1944, 2, 178) or by lithium aluminium hydride (*R. F. Nystrom* and *W. G. Brown*, J. Amer. chem. Soc., 1947, 69, 1197) or sodium borohydride.

(4) Reduction of esters of olefinic acids (not $\alpha\beta$-unsaturated) either by the Bouveault-Blanc method (*V. L. Hansley*, Ind. Eng. Chem., 1947, 39, 55)

or by lithium aluminium hydride (*Nystrom* and *Brown, loc. cit.* and p. 2548). The latter reagent may also be used to reduce acids and acid chlorides and sometimes leaves an $\alpha\beta$-double bond unattacked.

(5) Certain substituted allyl alcohols may be obtained by anionotropic rearrangements, *e.g.*:

$$R{\cdot}CH(OH){\cdot}CH{:}CH_2 \xrightarrow[\text{and hydrolysis}]{PBr_3\text{: then } CH_3{\cdot}CO_2K} R{\cdot}CH{:}CH{\cdot}CH_2OH$$

(*C. Prévost*, Ann. Chim., 1928, **10**, 152; *M. Bouis*, *ibid.*, 1928, **9**, 402).

(6) Olefinic alcohols are often conveniently obtained by semi-hydrogenation of their acetylenic analogues (*A. W. Johnson*, "The Acetylenic Alcohols", E. Arnold and Co., London, 1946, p. 89; *L. F. Hatch* and *S. S. Nesbitt*, J. Amer. chem. Soc., 1950, **72**, 727). The reduction is usually carried out catalytically, when *cis* alcohols result; but lithium aluminium hydride reduction may yield *trans* products (cf. *W. Oroshnik, G. Karmas* and *A. D. Mebane*, *ibid.*, 1952, **74**, 3807).

(7) Certain dienes can be converted into unsaturated alcohols by monohydroboronation with bis-3-methyl-2-butylborane, followed by oxidation with alkaline hydrogen peroxide, *e.g.* 1,5-hexadiene yields 5-hexene-1-ol (*G. Zweifel, K. Nagase* and *H. C. Brown*, *ibid.*, 1962, **84**, 190).

Vinyl alcohol, $CH_2{:}CHOH$. Reactions which might be expected to yield this compound invariably give acetaldehyde of which vinyl alcohol is the tautomeric enolic form, the equilibrium almost entirely favouring the keto form:

$$CH_2{:}CHOH \longrightarrow CH_3{\cdot}CHO$$

The mercury oxychloride compound of vinyl alcohol has been described (*T. Poleck* and *T. Thümmel*, Ber., 1889, **22**, 2863). However, polyvinyl alcohol, a white powder prepared by hydrolysis of a methanol solution of polyvinyl acetate with sodium methoxide, has industrial uses as a plastic material and as an emulsifying agent.

Vinyl acetate is prepared in 90 % yield by the catalytic vapour phase addition of acetic acid to acetylene, a typical catalyst being charcoal impregnated with 70 % zinc acetate. Other vinyl esters, *e.g.* the *benzoate*, may also be prepared by this method. Vinyl acetate may be polymerised by a variety of methods to give *polyvinyl acetates* of varying properties which are used for adhesives, paints, lacquers, etc., as well as for hydrolysis to polyvinyl alcohol. The reaction of polyvinyl alcohol with aldehydes yields the polyvinyl acetals (Mowitals, Formvar, Alvar, etc.) which are used commercially for safety glass and for solvent resisting tubing.

Allyl alcohol, *2-propenol*, $CH_2{:}CH{\cdot}CH_2OH$, f.p. —50°; b.p. 96–97°; d^{20} 0·854; n_D^{20} 1·4134, is manufactured by hydrolysis of allyl chloride (Vol. IA, p. 520) or from propene, either by epoxidation and hydrolysis, or by oxidation to acraldehyde by oxygen in the presence of steam and a cuprous oxide catalyst at 300–400°, followed by Ponndorf-type reduction, using isopropanol with a catalyst consisting of a mixture of magnesium and zinc oxides at 400°. It can be prepared also from glycerol by heating it with oxalic acid or preferably, formic acid (Org. Synth. Coll., Vol. I, p. 42); monoformin, formed in the reaction, decomposes into allyl alcohol, carbon dioxide and water. Allyl sulphide and diallyl trisulphide occur in oil of garlic and allyl isothiocyanate in oil of mustard.

Properties. Allyl alcohol, a mobile liquid with a pungent odour, is miscible with water and burns with a bright flame. In the presence of alumina or zinc oxide, it is largely rearranged to propionaldehyde (*P. E. Weston* and *H. Adkins*, J. Amer. chem. Soc., 1928, **50**, 1930 and 1929, **51**, 2589). Oxidation with air over a platinum catalyst or with silver oxide yields acraldehyde and acrylic acid, with chromic acid formic acid, with permanganate glycerol, and with perbenzoic acid, the corresponding epoxide, $\overset{\frown O \frown}{CH_2\cdot CH}\cdot CH_2OH$. On catalytic hydrogenation allyl alcohol gives *n*-propanol. Allylic alcohols are selectively dehydrogenated to $\alpha\beta$-unsaturated aldehydes by manganese dioxide in organic solvents (*R. M. Evans*, Quart. Reviews, 1959, **13**, 61; *R. J. Gritter* and *T. J. Wallace*, J. org. Chem., 1959, **24**, 1051) and by tetrachloro-1,2-benzoquinone (*E. A. Braude*, *R. P. Linstead* and *K. R. Wooldridge*, J. chem. Soc., 1956, 3070), and on treatment with alkali metals in liquid ammonia or alkylamines they yield olefins (*A. J. Birch*, Quart. Reviews, 1950, **4**, 69; *A. S. Hallsworth*, *H. B. Henbest* and *T. I. Wrigley*, J. chem. Soc., 1957, 1969). Hydrogen halides and allyl alcohol yield mainly the allyl halides, although addition to the double bond occurs slowly:

$$CH_2{:}CH\cdot CH_2OH \xrightarrow{HX} CH_2{:}CH\cdot CH_2X \xrightarrow{HX} CH_2X\cdot CH_2\cdot CH_2X$$

Allyl halides also result from the action of phosphorus trihalides or thionyl chloride, preferably in the presence of pyridine. When allyl alcohol is heated at 150° with potassium hydroxide, formic acid, *n*-propanol and other products are formed. Some allylic alcohols, on treatment with alkali at 180–240°, undergo fission at the $\alpha\beta$-carbon-carbon bond (*G. H. Hargreaves* and *L. N. Owen*, J. chem. Soc., 1947, 756). The addition of hypochlorous acid gives a mixture of the two possible, isomeric chlorohydrins, from which the epoxide can be obtained; hydrolysis yields glycerol. This reaction is applied commercially.

Halogen-substituted allyl alcohols are obtained by hydrolysis of the corresponding dihalogenopropenes. 2,3-Dichloro-1-propene is a by-product from the manufacture of allyl chloride and yields 2-*chloroallyl alcohol*, $CH_2{:}CCl\cdot CH_2OH$, b.p. 47°/10 mm. on hydrolysis (*G. Kremer*, Bull. Soc. chim. Fr., 1948, 165; *W. E. Noland* and *B. N. Bastion*, J. Amer. chem. Soc., 1955, **77**, 3395). The latter is also obtained by the addition of hydrogen chloride to propargyl alcohol in the presence of mercuric chloride (*W. Reppe et al.*, Ann., 1955, **596**, 1). The action of alcoholic potassium hydroxide on 2-chloroallyl alcohol gives propargyl alcohol (p. 43). 3-*Chloroallyl alcohol*, 3-*chloro*-2-*propenol* $CHCl{:}CH\cdot CH_2OH$ occurs in two geometrically isomeric forms, α-form, b.p. 101°/150 mm.; β-form, b.p. 108°/150 mm. (*L. F. Hatch* and *A. C. Moore*, J. Amer. chem. Soc., 1944, **66**, 285; *Hatch* and *G. B. Roberts*, *ibid.*, 1946, **68**, 1197; *L. J. Andrews* and *R. E. Kepner*, *ibid.*, 1947, **69**, 2230).

2-*Butenol, crotyl alcohol*, $CH_3 \cdot CH{:}CH \cdot CH_2OH$, occurs in two geometrically isomeric forms. *cis*-, f.p. —90·15°; b.p. 123·6°/760 mm.; d_4^{20} 0·8662; n_D^{20} 1·4342, is obtained by hydrogenation of 2-butynol in the presence of palladium (*S. F. Birch* and *D. T. McAllan*, J. chem. Soc., 1951, 2556), while *trans*-crotyl alcohol, b.p. 121·2°/760 mm.; d_4^{20} 0·8521; n_D^{20} 1·4288, is obtained by reduction of crotonaldehyde by lithium aluminium hydride or by the Ponndorf method (*idem, ibid.*; *L. F. Hatch* and *S. S. Nesbitt*, J. Amer. chem. Soc., 1950, **72**, 727). It has also been prepared by the anionotropic rearrangement of 3-buten-2-ol (see below) either as the bromide (action of phosphorus tribromide) or as the trichloroacetate.

3-*Buten-2-ol, methylvinylcarbinol*, $CH_2{:}CH \cdot CHOH \cdot CH_3$. The (±)-form, b.p. 96–96·5°/745 mm.; d_4^{26} 0·831; n_D^{20} 1·4137, is a constituent of wood spirit oil. It is formed by the action of methylmagnesium bromide on acraldehyde and also by the partial hydrogenation of 3-butyn-2-ol (p. 43). It may be dehydrated to buta-1,3-diene and rearranged anionotropically to crotyl alcohol. The carbinol has been resolved as the acid phthalate (*J. Kenyon* and *D. R. Snellgrove*, J. chem. Soc., 1925, **127**, 1174).

3-*Butenol, allylcarbinol*, $CH_2{:}CH \cdot CH_2 \cdot CH_2OH$, b.p. 112·5–113·5°/748 mm.; d_4^{20} 0·8475; n_D^{20} 1·4224, is obtained by the action of allylmagnesium bromide on trioxymethylene (*A. Juvala*, Ber., 1930, **63**, 1989) and by the action of vinylmagnesium bromide on ethylene oxide (*H. Normant*, Compt. rend., 1955, **240**, 440).

(±)-3-*Penten-2-ol, methylpropenylcarbinol*, $CH_3 \cdot CH{:}CH \cdot CHOH \cdot CH_3$, b.p. 120°/740 mm.; d_4^{21} 0·836; n_D^{20} 1·4280, is obtained from crotonaldehyde and methylmagnesium chloride (Org. Synth., **27**, 65). Higher homologues are obtained by use of the appropriate Grignard reagent (*K. v. Auwers* and *H. Westermann*, Ber., 1921, **54**, 2996).

4-*Pentenol*, $CH_2{:}CH(CH_2)_2 \cdot CH_2OH$, b.p. 138·5–139·5°/766 mm.; d_4^{20} 0·8457; n_D^{15} 1·4305, is prepared by reduction of allylacetic ester (*R. Paul*, Ann. Chim., 1932, **18**, 303) and by the action of sodium on tetrahydrofurfuryl chloride (Org. Synth., **25**, 84).

(±)-4-*Penten-2-ol, allylmethylcarbinol*, $CH_2{:}CH \cdot CH_2 \cdot CHOH \cdot CH_3$, b.p. 114°/740 mm.; d_4^{20} 0·8367; n_D^{20} 1·4225, is prepared from allylmagnesium bromide and acetaldehyde (*L. Stöhr*, Ber., 1939, **72**, 1138; *W. H. Yanko, H. S. Mosher* and *F. C. Whitmore*, J. Amer. chem. Soc., 1945, **67**, 664).

2-*Methyl-4-penten-2-ol, allyldimethylcarbinol*, $(CH_3)_2C(OH) \cdot CH_2 \cdot CH{:}CH_2$, b.p. 117–119°/775 mm., d_4^{17} 0·8326; n_D^{17} 1·4277, is obtained from allylmagnesium bromide and acetone (*R. G. R. Bacon* and *E. H. Farmer*, J. chem. Soc., 1937, 1071; *F. G. Fisher*, Ber., 1943, **76**, 734).

cis-3-*Hexen-1-ol, "leaf alcohol"*, $CH_3 \cdot CH_2 \cdot CH{:}CH \cdot CH_2 \cdot CH_2OH$, b.p. 58–58·5°/12 mm.; d_4^{20} 0·8495; n_D^{20} 1·4380, is used in perfumery, having the characteristic smell of green leaves and is obtained by catalytic hydrogenation of 3-hexynol (*F. Sondheimer*, J. chem. Soc., 1950, 877; *L. Crombie* and *S. H. Harper*, *ibid.*, p. 873).

10-*Undecenol*, $CH_2:CH\cdot(CH_2)_8\cdot CH_2OH$, f.p. —2°; b.p. 131–132°/14 mm.; d^{15} 0·8495; n_D^{25} 1·4481 (*F. Brody* and *M. T. Bogert*, J. Amer. chem. Soc., 1943, 65, 1075) and 9-*octadecenol* (cis-*oleyl alcohol*, f.p. 5·5–7·5°; b.p. 145–146°/0·5 mm.; d^{26} 0·842; d_D^{20} 1·4606) (Org. Synth., **15**, 51) are obtained by Bouveault-Blanc reduction of the corresponding esters.

Certain mono-terpenoid alcohols *e.g.* citronellol, and rhodinol and the diterpenoid alcohol phytol, which fall into this class, will be discussed in a subsequent volume dealing with cyclic terpenoids to which they are related.

(ii) *Alcohols* $C_nH_{2n-3}OH$

The alcohols of the above general formula comprise the acetylenic alcohols and the diolefinic alcohols.

(1) *Acetylenic alcohols*

Acetylenic alcohols of the type $R\cdot C\vdots C\cdot OH$, *i.e.* 1-alkynols, are known only as the isomeric ketens (cf. vinyl alcohol, p. 39). The corresponding acetylenic ethers are known, however (see p. 50).

A review of the preparation and properties of the higher homologues has been given by *A. W. Johnson*, (*Op. cit.*; see also *R. A. Raphael*, "Acetylenic Compounds in Organic Synthesis", Butterworths, London, 1955). The most important general method for the preparation of $\alpha\beta$-acetylenic carbinols is the action of organometallic derivatives of ethynyl hydrocarbons on carbonyl compounds, including the Grignard method, and those made *in situ* by the action of sodamide in ethanol or liquid ammonia solution (*I. M. Heilbron, E. R. H. Jones* and *B. C. L. Weedon*, J. chem. Soc., 1945, 81), the action of potassium hydroxide in ethereal solution or better using acetals or glycol dialkyl ethers as solvents (*Compagnie Alais, Froges et Camargue*, B.P., 544,221/1942; *C. Weizmann*, B.P., 573,527/1945, 580,921/1946, 580,922/1946) and reactions in the presence of the heavy metal acetylides (*I.G. Farben*, B.P., 508,062/1939) as illustrated below under individual compounds.

Properties. The acetylenic alcohols are very reactive and have been widely used in synthesis. Their more important reactions are as follows:

(1) Dehydration with phosphoric acid, sulphuric acid, etc., or catalytically in the vapour phase to yield substituted vinylacetylenes (Vol. IA, p. 470).

(2) Conversion into acetylenic halides by the action of phosphorus trihalides, thionyl chloride, etc., in the presence of pyridine.

(3) Oxidation of the primary and secondary acetylenic carbinols with chromium trioxide in acetone to acetylenic acids and ketones, respectively

(*K. Bowden et al.*, J. chem. Soc., 1946, 39; *Heilbron, Jones* and *F. Sondheimer*, *ibid.*, 1949, 604).

(4) Hydrogenation to olefinic and saturated alcohols.

(5) Hydration to keto-alcohols, *e.g.*

$$(CH_3)_2C(OH)\cdot C⋮CH \longrightarrow (CH_3)_2C(OH)\cdot CO\cdot CH_3$$

(6) Oxidation of the copper derivatives of the ethynyl alcohols with alkaline potassium ferricyanide or atmospheric oxygen to diacetylenic diols:

$$CH_2OH\cdot C⋮CCu \xrightarrow{O} (CH_2OH\cdot C⋮C)_2$$

(7) Rearrangement by acids to unsaturated ketones (cf. *K. H. Meyer* and *K. Schuster*, Ber., 1922, 55, 819), *e.g.*

$$(CH_3)_2C(OH)\cdot C⋮C\cdot(CH_2)_4CH_3 \longrightarrow (CH_3)_2C⋮CH\cdot CO(CH_2)_4CH_3$$

(8) *tert*-Acetylenic alcohols can be readily esterified by treatment with carboxylic acids in the presence of toluene-*p*-sulphonyl chloride and pyridine (*G. F. Hennion* and *S. O. Barrett*, J. Amer. chem. Soc., 1957, 79, 2146; *K. Klosa*, Angew. Chem., 1957, 69, 135).

Propargyl alcohol, 2-*propynol*, $HC⋮C\cdot CH_2OH$, is obtained (1) as a by-product in the manufacture of 2-butyne-1,4-diol by the reaction of formaldehyde with acetylene in the presence of copper acetylide (B.I.O.S. report 367; F.I.A.T. report 926); (2) by reaction of formaldehyde and acetylene in liquid ammonia solution in the presence of sodamide (*A. L. Henne* and *K. W. Greenlee*, J. Amer. chem. Soc., 1945, 67, 484); (3) by the action of sodium hydroxide on β-chloroallyl alcohol (*L. F. Hatch* and *A. C. Moore*, *ibid.*, 1944, 66, 285).

It is an agreeably smelling, colourless liquid, f.p. —51·8°; b.p. 113·6°; d_4^{20} 0·9478; n_D^{20} 1·4320. Hydrogenation yields allyl alcohol and *n*-propanol, successively. Oxidation of the insoluble copper derivative with atmospheric oxygen yields 2,4-hexadiyne-1,6-diol. Propargylic alcohols, like allylic alcohols, are selectively dehydrogenated by manganese dioxide or tetrachloro-1,2-benzoquinone.

3-*Butynol*, $CH⋮C\cdot CH_2\cdot CH_2OH$, f.p. —63·6°; b.p. 128·9°; d_4^{20} 0·9257; n_D^{20} 1·4409, is obtained by the action of sodium acetylide on ethylene oxide (*Henne* and *Greenlee*, *loc. cit.*).

3-*Butyn-2-ol*, $CH_3\cdot CH(OH)\cdot C⋮CH$, b.p. 46°/50 mm.; d^{20} 0·8949; n_D^{20} 1·4256, is prepared from acetaldehyde and acetylene in liquid ammonia solution in the presence of sodamide (*Hennion* and *W. S. Murray*, J. Amer. chem. Soc., 1942, 64, 1220).

2-*Methyl-3-butyn-2-ol*, $(CH_3)_2C(OH)\cdot C⋮CH$, f.p. 3·01°; b.p. 104°; d_4^{20} 0·8614; n_D^{20} 1·4212, from acetone and acetylene or alternatively in cellosolve solution in the presence of potassium hydroxide.

(±)-3-*Methyl-1-pentyn-3-ol*, $C_2H_5 \cdot C(CH_3)(OH) \cdot C{:}CH$, b.p. 121–122°; m.p. 30–31°; d_{20}^{20} 0·8721; n_D^{20} 1·4318, from ethyl methyl ketone and acetylene (*K. N. Campbell, B. K. Campbell* and *L. T. Eby, ibid.*, 1938, **60**, 2882) or ethynylmagnesium bromide (*Jones, L. Skatteböl* and *M. C. Whiting*, J. chem. Soc., 1956, 4765). It is used as a hypnotic (Oblivon) and has been resolved into optically active forms (*J. R. Hickman* and *J. Kenyon*, J. chem. Soc., 1955, 2051).

4-*Pentyn-1-ol*, $CH{:}C \cdot (CH_2)_2 \cdot CH_2OH$, b.p. 77°/37 mm.; $d^{22.5}$ 0·9174; n_D^{15} 1·4464, is prepared by the action of sodamide in liquid ammonia on tetrahydrofurfuryl chloride (*G. Eglinton, Jones* and *Whiting, ibid.*, 1952, 2873).

1-*Heptyn-4-ol*, $CH{:}C \cdot CH_2 \cdot CHOH \cdot (CH_2)_2 \cdot CH_3$, b.p. 58–59°/11 mm.; n_D^{20} 1·4443 is obtained by the Reformatsky reaction, using propargyl bromide and butaldehyde, an example of a general method for the synthesis of $\beta\gamma$-acetylenic carbinols (*H. B. Henbest, Jones* and *I. M. S. Walls, ibid.*, 1949, 2696; 1950, 3646).

(2) *Diolefinic alcohols*

The low members of this series are prepared either by the Ponndorf reduction of the corresponding aldehydes or by the partial hydrogenation of the corresponding vinylacetylenic carbinols (*I. M. Heilbron et al.*, J. chem. Soc., 1945, 84). Non-conjugated diene alcohols of the type $RCH{:}CH \cdot CHOH \cdot CH{:}CHR'$ are rearranged on treatment with dilute acids to conjugated diene alcohols.

1,4-Pentadien-3-ol, *divinylcarbinol*, $(CH_2{:}CH)_2CHOH$, b.p. 64–66°/100 mm.; n_D^{17} 1·4400, is obtained by hydrogenation of ethynylvinylcarbinol.

2,4-*Pentadien-1-ol*, $CH_2{:}CH \cdot CH{:}CH \cdot CH_2OH$, b.p. 95–97°/100 mm.; n_D^{18} 1·4902, by anionotropic rearrangement of the previous carbinol.

1,4-*Hexadien-3-ol*, (*propenylvinylcarbinol*), $CH_3 \cdot CH{:}CH \cdot CHOH \cdot CH{:}CH_2$, b.p. 86–87°/100 mm.; n_D^{19} 1·4501, is obtained by hydrogenation of ethynylpropenylcarbinol.

3,5-*Hexadien-2-ol*, $CH_3 \cdot CHOH \cdot CH{:}CH \cdot CH{:}CH_2$, b.p. 67–68°/26 mm.; n_D^{19} 1·4810, by anionotropic rearrangement of propenylvinylcarbinol.

2,4-*Hexadienol*, *sorbyl alcohol*, m.p. 30·5–31·5°; b.p. 76–77°/12 mm., is prepared from the corresponding aldehyde by Ponndorf reduction (*T. Reichstein et al.*, Helv., 1932, **15**, 264) or from the acid or its chloride by reduction with lithium aluminium hydride (*R. F. Nystrom* and *W. G. Brown*, J. Amer. chem. Soc., 1947, **69**, 1197 and 2548).

Certain terpene alcohols which fall into this class, *e.g.* geraniol, nerol and linalool, will be discussed in the subsequent volume dealing with terpenoids.

(iii) *Alcohols* $C_nH_{2n-5}OH$

The alcohols of the above general formula comprise the vinylacetylenic alcohols and the triethylenic alcohols.

(1) Vinylacetylenic alcohols

There are three main classes of these alcohols (see *Johnson, loc. cit.*) of the general formulae:

(i) RR′C(OH)·C⋮C·CH:CH_2

(ii) HC⋮C·CR(OH)·CH:CH_2

(iii) RR′C(OH)·CH:CH·C⋮CH.

Typical members of these classes are given.

(1) 2-**Methyl**-5-**hexen**-3-**yn**-2-**ol**, $(CH_3)_2C(OH)$·C⋮C·CH:CH_2, b.p. 67°/24 mm.; d_4^{20} 0·8872; n_D^{20} 1·4778 (*I. N. Nazarov*, C.A., 1939, **33**, 5682) and its homologues are prepared by reaction of vinylacetylenes with carbonyl compounds either by the Grignard method or with sodamide in liquid ammonia solution.

(ii) (±)-4-*Hexen*-1-*yn*-3-*ol*, CH⋮C·CH(OH)·CH:CH·CH_3, b.p. 154–156°; d^{23} 0·9148; n_D^{20} 1·4651 (*Heilbron, Jones* and *Weedon, loc. cit.*) and its analogues are obtained from αβ-unsaturated carbonyl compounds by reaction with acetylene in the presence of sodamide in liquid ammonia solution or by reaction with cuprous acetylide (*W. Reppe et al.*, Ann., 1955, **596**, 37). On treatment with acids, they undergo an anionotropic rearrangement to the conjugated vinylacetylenic alcohols of the next group.

(iii) (±)-3-*Hexen*-5-*yn*-2-*ol*, CH_3·CHOH·CH:CH·C⋮CH, occurs in *cis*-, b.p. 65·5°; n_D^{21} 1·4758 and *trans*-forms, b.p. 76°; n_D^{21} 1·4848 (*I. Bell, Jones* and *Whiting*, J. chem. Soc., 1957, 2597). These alcohols are obtained as described in (ii) above. 2-*Penten*-4-*yn*-1-*ol*, CH⋮C·CH:CH·CH_2OH, b.p. 71–73°/19 mm.; n_D^{19} 1·4972, is obtained by the action of sodium acetylide in liquid ammonia on epichlorhydrin (*L. J. Haynes et al.*, *ibid.*, 1947, 1583).

(2) Triolefinic alcohols

2,4,6-**Octatrienol**, $CH_3(CH{:}CH)_3CH_2OH$, m.p. 99·5–100·5°; b.p. 95°/12 mm., is prepared by the Ponndorf reduction of octatrienal (*T. Reichstein et al.*, Helv., 1932, **15**, 261).

4-*Propenylhepta*-2,5-*dien*-4-*ol*, *tripropenylcarbinol*, $(CH_3 \cdot CH : CH \cdot)_3C \cdot OH$, b.p. 78–80°/12 mm.; $d_4^{21.5}$ 0·8660; $n_D^{21.5}$ 1·4645, is obtained by the action of propenylmagnesium bromide on ethyl carbonate (*H. Normant* and *P. Maitte*, Bull. Soc. chim. Fr., 1956, 951).

The triethylenic terpene alcohols, farnesol and nerolidol, will be discussed later.

(iv) Alcohols $C_nH_{2n-7}OH$

This class comprises the diacetylenic and tetraolefinic alcohols.

(1) Diacetylenic alcohols

The most important members of this class are prepared by the reaction of ethynyl hydrocarbons with esters, acid chlorides, anhydrides or acetylenic carbonyl compounds (*Johnson, loc. cit.*, p. 214).

1,4-*Pentadiyn-3-ol*, $(HC:C)_2CHOH$, m.p. 51·5–52°, is formed by the action of ethynylmagnesium bromide on ethyl formate (*Jones, Skatteböl* and *Whiting, loc. cit.*).

7-*Methyl-5,8-tridecadiyn-7-ol*, $[CH_3(CH_2)_3C:C]_2C(OH)CH_3$ b.p. 100°/3 mm.; d_4^{24} 0·8840; n_D^{24} 1·4660, by the action of hexynylmagnesium bromide on acetic anhydride (*J. W. Kroeger* and *J. A. Nieuwland*, J. Amer. chem. Soc., 1936, **58**, 1861).

(2) *Tetraolefinic alcohols*

2,4,6,8-**Decatetraenol,** $CH_3(CH:CH)_4CH_2OH$, is a colourless solid, m.p. 171–172·5°, formed by Ponndorf reduction of the corresponding aldehyde (*T. Reichstein et al.*, Helv., 1932, **15**, 1074).

(v) *Alcohols* $C_nH_{2n-9}OH$

2,4,6,8,10-**Dodecapentaenol,** $CH_3(CH:CH)_5CH_2OH$, a faintly yellowish solid, m.p. 204°, is formed by reduction of the aldehyde (*F. G. Fisher, K. Hultzsch* and *W. Flaig*, Ber., 1937, **70**, 370).

Matricarianol, $CH_3·CH:CH·C:C·C:C·CH:CH·CH_2OH$, occurs (mainly as the acetate of the *trans-trans*-isomer) in some *Compositae* (*D. Holme* and *N. A. Sörensen*, Acta Chem. Scand., 1954, **8** 34); it also occurs in certain fungi (*J. D. Bu'Lock*, Quart. Reviews, 1956, **10** 371).

(vi) *Alcohols* $C_nH_{2n-11}OH$

Triacetylenic alcohols

Reaction of acetylenic hydrocarbons with ethyl carbonate, ethyl chloroformate, phosgene or acetylenic acid chlorides yields triacetylenic alcohols (*Johnson, loc. cit.*, p. 238).

Tripropynylcarbinol, $(CH_3·C:C)_3COH$, m.p. 134–135° (*D. D. Coffman et al.*, J. Amer. chem. Soc., 1933, **55** 3792).

Tri(tert-*butylethynyl*)*carbinol*, $[(CH_3)_2C·C:C]_3$ C·OH, m.p. 100–102° (*P. L. Salzberg* and *C. S. Marvel*, *ibid.*, 1928, **50** 1737).

(vii) *More highly unsaturated alcohols*

Several highly unsaturated alcohols, *e.g.* $CH_2:CH·CH:CH·C:C·C:C·C:C·CH:CH·CH_2OH$ (as the acetate in *Carlina vulgaris*) and a number of others occur naturally in fungi (*E. R. H. Jones*, Proc. chem. Soc., 1960, 199).

2. Ethers

(a) *Saturated ethers; dialkyl ethers* $(C_nH_{2n+1})_2O$

Ethers may be regarded as anhydrides of the alcohols, formed by the elimination of water from two molecules of the alcohol:

$$2\ ROH \longrightarrow R\cdot O\cdot R + H_2O$$

Ethers containing two identical alkyl radicals are termed *simple ethers*; those with different radicals *mixed ethers*.

Preparation. (1) Simple ethers may be obtained by dehydration of the appropriate alcohol, either in the liquid or gaseous phase. In the former method a wide variety of dehydrating agents have been used, the most common is sulphuric acid but aromatic sulphonic acids, arsenic and boric acids, sodium and potassium hydrogen sulphates have been used as well. Gaseous phase etherification involves passage of the alcohol vapour over a heated catalyst such as alum, alumina, titanium dioxide, thorium dioxide, or tungsten pentoxide. While these methods work well with primary alcohols, secondary alcohols and, more so, tertiary alcohols tend to give increasing amounts of olefins arising from simple dehydration of the alcohol.

(2) Sodium alkoxides react with alkyl halides, sulphonates and sulphates to give ethers:

$$RONa + R'X \longrightarrow R\cdot O\cdot R' + NaX$$

The reaction may be performed in a solution of the alcohol in excess or in an inert solvent. The method has been used to produce mixed ethers (*C. B. Carter* and *A. E. Coxe*, U.S.P., 1,459,177/1923).

(3) The action of metallic oxides, especially silver oxide, on alkyl halides yields ethers (*L. Palfray* and *S. Sabetay*, Bull. Soc. chim. Fr., 1928, 43, 895):

$$2\ RX + Ag_2O \longrightarrow R\cdot O\cdot R + 2AgX$$

(4) Halogen-substituted ethers yield homologous ethers on reaction with organo-zinc or magnesium compounds (*J. Hamonet*, Compt. rend., 1904, 138, 813):

$$C_5H_{11}\cdot O\cdot CH_2Br + BrMgC_2H_5 \longrightarrow C_5H_{11}\cdot O\cdot C_3H_7 + MgBr_2$$

(5) The addition of alcohols to olefins under the influence of such catalysts as sulphuric acid is used as an industrial synthesis of ethers (*Socony-Vacuum Oil Co.*, U.S.P., 2,077,042/1937):

$$CH_3CH{:}CH_2 + C_2H_5OH \longrightarrow (CH_3)_2CH\cdot O\cdot C_2H_5$$

(6) Simple ethers can arise from the catalytic hydration of olefins; the corresponding alcohol frequently accompanies the ether.

(7) Addition of monohydric alcohols to epoxides, under the influence of pressure and various catalysts, is utilised for the manufacture of glycol monoethers (*C. O. Young*, U.S.P., 1,696,874/1928; B.P., 322,017).

(8) Recently methyl ethers have been obtained by the reaction of diazomethane with alcohols catalysed by fluoroboric acid (*M. C. Caserio, J. D. Roberts, M. Neeman*, and *W. S. Johnson*, J. Amer. chem. Soc., 1958, **80**, 2584).

See *J. Nelles* (Angew. Chem., 1941, **54**, 77) for a review of preparative methods.

Properties. Ethers are neutral compounds, almost insoluble in water; the lower members are volatile and inflammable.

Reactions. Saturated ethers are chemically inert; drastic conditions are usually required for reaction.

(1) When oxidised they yield the same products as their corresponding alcohols.

(2) They are unattacked by boiling alkalis; substitution occurs with chlorine and bromine; they are oxidised by nitric acid, and converted to alkyl hydrogen sulphates by sulphuric acid.

(3) Alkyl halides are formed by the action of halogen acids, especially by hydrogen iodide at 100°:

$$C_2H_5\cdot O\cdot CH_3 + 2HI \longrightarrow C_2H_5I + CH_3I + H_2O$$

This reaction forms the basis of the Zeisel method for the estimation of methoxyl or ethoxyl groups. Hydrogen iodide in the cold causes cleavage into an alcohol and an alkyl iodide; in a mixed ether the larger proportion of iodine goes to the smaller radical (*A. Michael* and *F. D. Wilson*, Ber., 1906, **39**, 2569). A similar fission occurs, often in the cold, with acyl halides (*E. L. Guetus* and *P. G. Stevens*, J. Amer. chem. Soc., 1933, **55**, 378):

$$C_2H_5\cdot O\cdot C_2H_5 + CH_3\cdot COCl \longrightarrow CH_3\cdot CO_2C_2H_5 + C_2H_5Cl$$

(4) Some ethers, especially those containing branched chains or unsaturated groups, may be hydrolysed by heating with dilute acid.

(5) Addition compounds are formed with halogens and many halogen compounds (see *P. Pfeiffer*, "Organische Molekulverbindungen", 1922, p. 23). These compounds vary in stability; that of ether and bromine dissociates into its components above —40° while "borontrifluoride-etherate" distils unchanged at 125°. They are regarded as being oxonium compounds.

(6) On exposure to air and light, ethers decompose with the formation, amongst other products, of highly explosive peroxides.

(7) Ethers on pyrolysis yield a mixture of water, alcohols and unsaturated hydrocarbons; free radicals are involved.

Methyl ether, *dimethyl ether*, $(CH_3)_2O$, b.p. —24·8°, m.p. —141·5°, is prepared industrially by the catalytic dehydration of methanol (*J. C. Woodhouse*, U.S.P., 1,949,344/1934). It found some commercial use as a refrigerant, but has been replaced by non-inflammable products.

Ethyl ether or **ether,** *diethyl ether,* $(C_2H_5)_2O$, b.p. 34·5°, m.p. —116·3° (stable form), —123·3° (metastable form) is the most important simple ether. Its preparation was first recorded in 1552 by Conrad Gessner. Its constitution was established by Williamson in 1850 when he prepared it from sodium ethoxide and ethyl iodide. Industrially it is prepared by dehydration of ethanol with sulphuric acid at 140°, or metal oxide catalysts, or from ethylene by reaction with ethanol or water under the influence of an acid catalyst, *e.g.* sulphuric acid. It is widely used as a solvent and is a safe surgical anaesthetic. The main hazards involved in its use are the formation of explosive mixtures with air and the production of explosive peroxides by *atmospheric oxidation,* especially in the presence of light. This latter reaction has been the subject of extensive research (*A. M. Clover,* J. Amer. chem. Soc., 1922, **44,** 1107; *F. Reimers,* Quart. J. pharm. Pharmacol., 1946, **19,** 27) and the following compounds have been claimed as products of the reaction: 1,1′-dihydroxyethyl peroxide, $CH_3·CH(OH)·O·O·CH(OH)·CH_3$, ethylidene peroxide polymer $[O·CH(CH_3)·O]_n$, acetic and formic acids, hydrogen peroxide, diethyl peroxide (*q.v.*), acetaldehyde, 1-hydroxyethyl hydroperoxide $CH_3CH(OH)·O·OH$, ethyl vinyl ether and vinyl ether. The initial reaction is apparently attack at the α-position of the ether:

$$O_2 + C_2H_5·O·CH_2·CH_3 \longrightarrow C_2H_5·O·CH(OOH)·CH_3 \longrightarrow [O·CH(CH_3)·O]_n$$

With the exception of diethyl peroxide the peroxides formed are derivatives of acetaldehyde.

The autoxidation of ethers can be inhibited by storing them in the presence of antioxidants such as sodium amalgam, diphenylamine or hydroquinone (see *W. Hunter* and *J. Downing,* J. Soc. chem. Ind., 1949, **68,** 362 for review).

Di-isopropyl ether, b.p. 68°, undergoes autoxidation more readily than ethyl ether, the principal product is the dihydroperoxide, $(CH_3)_2C(O·OH)·O·C(CH_3)_2·O·OH$.

Ethyl methyl ether, b.p. 7·6°; *methyl* n-*propyl ether,* b.p. 39°; *isopropyl methyl ether,* b.p. 32°; tert-*butyl methyl ether,* b.p. 55·1°; n-*propyl ether,* b.p. 90·1°; tert-*butyl ethyl ether,* b.p. 72·7°; tert-*butyl isopropyl ether,* b.p. 87·6°; *isoamyl ether,* b.p. 172·5–173°, 60°/10 mm.; *cetyl ether,* b.p. 300°.

*(b) Substituted dialkyl ethers**

(i) Halogenoalkyl ethers

α- or 1-*Halogenodialkyl ethers* in which the halogen is situated on the carbon atom attached to the oxygen atom are obtained (a) by the action of alcohols and halogen hydracids on aldehydes; (b) by addition of hydrogen halide to unsaturated ethers and (c) by the controlled halogenation of dialkyl ethers, *e.g.* diethyl ether on chlorination gives the following successive products:

* This section is contributed by Dr. *E. S. Waight.*

1-Chloroethyl ethyl ether,	$CH_3 \cdot CHCl \cdot O \cdot C_2H_5$,	b.p. 98°
1,2-Dichloroethyl ethyl ether,	$CH_2Cl \cdot CHCl \cdot O \cdot C_2H_5$,	b.p. 145°
Ethyl 1,2,2-trichloroethyl ether,	$CHCl_2 \cdot CHCl \cdot O \cdot C_2H_5$,	b.p. 170–175°
Perchloroethyl ether,	$(C_2Cl_5)_2O$,	m.p. 68°

As already mentioned on p. 34, with regard to the corresponding halogeno-alkanols, they are closely related chemically to the aldehydes and are therefore discussed in more detail in Chapter 8.

β- or 2-Halogenodialkyl ethers resemble the true ethers and are therefore listed below, although in the 1st edition of this book they were formally regarded as derivatives of glycol.

2-**Chloroethyl ethyl ether,** *2-ethoxyethyl chloride,* $CH_2Cl \cdot CH_2 \cdot O \cdot C_2H_5$ b.p. 108–109°, d_4^{20} 0.9945, n_D^{20} 1·412; *2-bromoethyl ethyl ether, 2-ethoxyethyl bromide,* b.p. 128°/768 mms., d_4^{20} 1·3573, n_D^{20} 1·448, from cellosolve and phosphorus tribromide (Org. Synth., Coll. Vol. III, p. 370).

2,2′-**Dichlorodiethyl ether,** *"Chlorex",* $(ClCH_2 \cdot CH_2)_2O$, b.p. 176°/744 mm., 70°/15 mm., m.p. —24·5°, d^{20} 1·213, n_D^{20} 1·4568, is best prepared in the laboratory by the action of thionyl chloride on diethylene glycol. A colourless oil with a pleasant ethereal odour, it is a useful solvent, particularly for fats, waxes, and greases. It is a by-product in the manufacture of ethylene chlorhydrin. With strong alkalis it gives, variously, dioxan (see *Rodd*, C.C.C., 1st Edn., 1960, Vol. IVC, p. 1529), divinyl ether (*W. L. Ruigh* and *R. T. Major*, J. Amer. chem. Soc., 1931, **53**, 2662), or 2-chloroethyl vinyl ether (*W. Chalmers*, Canad. J. Res., 1932, **7**, 464), while with alkoxides it forms dialkoxyalkyl ethers. With amino-compounds it gives morpholine derivatives (see *Rodd. loc. cit.*, p. 1475). *2,2′-Dibromodiethyl ether,* $(BrCH_2CH_2)_2O$, b.p. 110°/26 mm., is prepared from diethylene glycol and phosphorous tribromide in pyridine; with sodium iodide in acetone it gives *2,2′-di-iododiethyl ether,* b.p. 123·5–124°/10 mm. *2-Chloro-2′-fluorodiethyl ether* (see above) b.p. 142–147°, d_{20}^{20} 1·137, n_D^{20} 1·4170.

(ii) Nitrodialkyl ethers

Nitro-ethers can be prepared by the addition of alcohols to nitro-olefins (*A. Lambert et al.*, J. chem. Soc., 1947, 1474):

$$ROH + CH_2{:}CR'NO_2 \longrightarrow ROCH_2 \cdot CHR'NO_2$$

Methyl 2-nitroethyl ether, $MeOCH_2 \cdot CH_2NO_2$, b.p. 67°/12 mm., n_D^{20} 1·417; *ethyl 2-nitroethyl ether,* b.p. 72°/12 mm.; *methyl 2-nitropropyl ether,* $CH_3 \cdot CHNO_2 \cdot CH_2OCH_3$, b.p. 61–64°/14 mm.

(c) Unsaturated ethers

Vinyl ethers may be considered as mixed anhydrides of a saturated alcohol and the enolic form of an aldehyde or ketone, or dehydration products of hemiacetals. They are generally prepared by the addition of alcohols to acetylenes under pressure in the presence of a catalyst such as an alkali metal hydroxide (see *J. A. Nieuwland* and *R. R. Vogt*, "The Chemistry of Acetylene", New York, 1945, p. 126; and also *C. E. Schildknecht, A. O. Voss* and *C. McKinley*, Ind. Eng. Chem., 1947, 39, 180). Alternatively they may be prepared by removal of the elements of hydrogen halide from a suitable halogeno-ether:

$$CH_3{\cdot}O{\cdot}CH_2{\cdot}CH_2Br \longrightarrow CH_3{\cdot}O{\cdot}CH{:}CH_2 + HBr$$

They are of commercial value in that they give rise to a series of *polyvinyl ethers* useful for such diverse purposes as the manufacture of adhesives, insulators, lacquers and distempers. They are readily hydrolysed to the component alcohol and aldehyde or ketone and lead to a wide range of products by addition of alcohols, carboxylic acids, etc., to the double bond.

Methyl vinyl ether, b.p. 5·5°, prepared from acetylene and methanol in the presence of sodium methoxide, can be distilled at pressures higher than atmospheric without loss by polymerisation.

Ethyl vinyl ether, b.p. 35·7°; *isopropyl vinyl ether*, b.p. 55·6°; *butyl vinyl ether*, b.p. 93·8°; *isobutyl vinyl ether*, b.p. 83°.

Divinyl ether, b.p. 28·5°, prepared from 2,2′-dichlorodiethyl ether yields only acetaldehyde on hydrolysis (*W. L. Ruigh* and *R. T. Major*, J. Amer. chem. Soc., 1931, 53, 2662).

Ethinyl ethers, *alkoxyacetylenes*, first prepared in 1940, are becoming increasingly useful synthetic intermediates. They have been obtained from dibromoaldehyde acetals (*T. L. Jacobs et al., ibid.*, 1940, **62**, 1849; 1942, **64**, 223, 2635):

$$CHBr_2{\cdot}CH(OR)_2 \xrightarrow[EtOH]{Zn} CHBr{:}CH{\cdot}OR \xrightarrow{KOH} CH{\vdots}C{\cdot}OR$$

and monochloroaldehyde acetals (*G. Eglinton, E. R. H. Jones, B. L. Shaw,* and *M. C. Whiting*, J. chem. Soc., 1954, 1860):

$$CH_2Cl{\cdot}CH(OEt)_2 \xrightarrow{NaNH_2} CH{\vdots}C{\cdot}OEt$$

Hydration with dilute acids leads to esters:

$$CH{\vdots}C{\cdot}OR \xrightarrow{H_2O} CH_3{\cdot}CO_2R$$

Their main synthetic applications are two in number, the first is summarised in the following scheme (see *R. A. Raphael*, "Acetylenic Compounds in Organic Synthesis", London, 1955 for examples):

$$>C{=}O + HC{:}C{\cdot}OR \longrightarrow >C(OH){\cdot}C{:}C{\cdot}OR \longrightarrow >C(OH){\cdot}CH_2{\cdot}CO_2R$$

$$\downarrow$$

$$>C(OH){\cdot}CH{:}CH{\cdot}OR \longrightarrow >C{:}CH{\cdot}CHO$$

Their other useful property is that of converting carboxylic acids, sulphonic acids and phosphoric acids to their anhydrides (*G. Eglinton et al., loc. cit.*; *J. F. Arens* and *P. Moddermans*, Proc. Koninkl. Ned. Akad. Wetenschap., 1950, 53, 1163). It appears that the initial acid-alkoxyacetylene adduct then reacts with a further molecule of the acid:

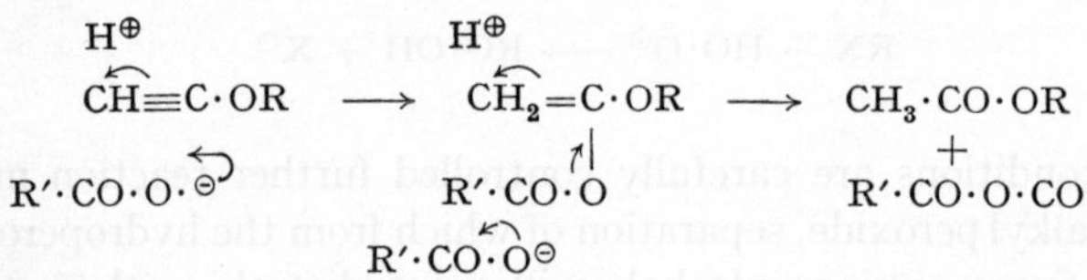

The reaction has been extended to the synthesis of peptides since the acid-acetylene adducts will react with free amino groups (*J. C. Sheehan* and *J. J. Hlavka*, J. org. Chem., 1958, **23**, 635; *H. J. Panneman, A. F. Marx*, and *J. F. Arens*, Rec. Trav. chim., 1959, 78, 487):

$$C_6H_4(CO)_2N{\cdot}CH_2{\cdot}CO_2H \xrightarrow{CH{:}C{\cdot}OEt} C_6H_4(CO)_2N{\cdot}CH_2{\cdot}CO_2{\cdot}C(OEt){:}CH_2$$

$$\xrightarrow{NH_2{\cdot}CH_2{\cdot}CO_2Et} C_6H_4(CO)_2N{\cdot}CH_2{\cdot}CO{\cdot}NH{\cdot}CH_2{\cdot}CO_2Et + CH_3{\cdot}CO_2Et$$

Methoxyacetylene, b.p. 23°; *ethoxyacetylene,* b.p. 50°; *propoxyacetylene,* b.p. 75°; *butoxyacetylene,* b.p. 103°.

3. Alkyl peroxides

These peroxides are derivatives of hydrogen peroxide in which one or both of the hydrogen atoms is substituted by an alkyl group giving rise to alkyl hydroperoxides and dialkyl peroxides respectively. They may be formed in many reactions involving oxidation, such as autoxidation (*C. E. Frank*, Chem. Reviews, 1950, 46, 155), the drying of oils, photochemical processes, etc. With few exceptions the lower members of the series are unstable, explosive substances, but stability increases with molecular weight. Iodometric methods are generally used for their determination, though some dialkyl peroxides may require previous cleavage with hydriodic acid [see *R. Criegee*, "Methoden der Organischen Chemie" (*Houben-Weyl*), Verlag, Stuttgart, 1953, **2**, 568, for this and other methods of estimation].

Vinyl and ethynyl peroxides are not known.

For more complete accounts of organic peroxides see *A. G. Davies*, "Organic Peroxides", Butterworths, London, 1961; and *E. G. E. Hawkins*, Quart. Reviews, 1950, 4, 251.

(a) Alkyl hydroperoxides, RO·OH

These are generally prepared from hydrogen peroxide by reaction with alkyl sulphates or sulphonates in basic solution:

$$RX + HO{\cdot}O^{\ominus} \longrightarrow RO{\cdot}OH + X^{\ominus}$$

Unless the conditions are carefully controlled further reaction may occur to give the dialkyl peroxide, separation of which from the hydroperoxide may be difficult. Some tertiary alcohols will react directly with concentrated hydrogen peroxide in acidic conditions (*R. Criegee* and *H. Dietrich*, Ann., 1948, 560, 135):

$$HO{\cdot}OH + tert\text{-}BuOH \xrightarrow{H^{\oplus}} tert\text{-}BuO{\cdot}OH + H_2O$$

Hydroperoxides have been obtained in low yield and impure state by oxidation of hydrocarbons but the method has little preparative value.

The hydroperoxides give metal salts with alkali though contact with excess alkali causes decomposition to aldehydes, ketones, and acids (*S. S. Medwedew* and *E. Alexejewa*, Ber., 1932, 65, 133). Reaction of the barium salts or the free hydroperoxides in dilute alkali or pyridine with acid halides gives esters (*Medwedew* and *Alexejewa*, *loc. cit.*; *N. A. Milas* and *D. M. Surgenor*, J. Amer. chem. Soc., 1946, 68, 642). With aldehydes and ketones they give monohydroxydialkyl peroxides, R′CH(OH)·O·OR, and with ketones, diperoxy compounds, $R'R''C(O{\cdot}OR)_2$. Reduction of alkyl hydroperoxides yields the corresponding alcohols. Thermal decomposition causes fission of the molecule and the final products depend upon the radicals formed.

Methyl hydroperoxide, MeO·OH, b.p. 38–40°/65 mm., decomposes thermally to methanol, formaldehyde, formic acid, dimethyl ether, oxides of carbon, and water. The *barium* salt is explosive when dry.

Ethyl hydroperoxide, EtO·OH, b.p. 93–97° (41–44°/50 mm.) decomposes at 140–200° into formaldehyde, acetaldehyde and ethanol.

tert-*Butyl hydroperoxide,* $(Me)_3CO{\cdot}OH$, m.p. 3·8–4·8°, b.p. 33–34°/17 mm., is one of the most stable hydroperoxides known, but even so decomposes at temperatures above 100°. Acetone, methanol, *tert*-butanol, and water are formed in proportions varying with temperature.

(b) *Dialkyl peroxides*, RO·OR′

These are obtained by extension of the methods used for preparing alkyl hydroperoxides. Alkylation of hydrogen peroxide with an excess of alkyl sulphate or sulphonate leads to the symmetrical peroxide or an alkyl hydroperoxide may be further alkylated to give a symmetrical or mixed peroxide (*A. Rieche*, Ber., 1929, **62**, 218). Those alcohols which give alkyl hydroperoxides with hydrogen peroxide in the presence of sulphuric acid will also react further to give dialkyl peroxides; both symmetrical and mixed peroxides can be obtained by this method (*N. A. Milas* and *D. M. Surgenor*, J. Amer. chem. Soc., 1946, **68**, 205).

The dialkyl peroxides are explosive liquids and compared with the hydroperoxides are only weak oxidising agents. Again stability increases with molecular weight.

Dimethyl peroxide, MeO·OMe, b.p. 13·5°/740 mm., explodes readily giving formaldehyde.

Ethyl methyl peroxide, MeO·OEt, b.p. 40°/740 mm.; *diethyl peroxide*, EtO·OEt, b.p. 62–63°/740 mm.; *di*-tert-*butyl peroxide*, $Me_3CO·OCMe_3$, b.p. 110° (12–13°/20 mm.) is the most stable simple dialkyl peroxide. It is resistant to catalytic hydrogenation and fails to respond to the usual tests for peroxides. At high temperatures it breaks down into $Me_3CO·$ radicals and is a very active polymerisation catalyst. Thermal decomposition (200–300°) leads to acetone and ethane (see *D. H. Hey*, Ann. Reports, 1948, **45**, 145):

$$(Me_3CO)_2 \longrightarrow 2Me_3CO\cdot \longrightarrow Me_2CO + Me\cdot \longrightarrow C_2H_6$$

4. Esters of mineral or inorganic acids

Esters are produced when an alcohol reacts with an acid with the elimination of water (see p. 13). They may be regarded either as derivatives of alcohols in which the hydroxylic hydrogen atom is replaced by an acid radical (esters of inorganic acids are conveniently regarded in this way and are therefore discussed in this chapter), or as acids in which the acidic hydrogen is replaced by alkyl. Esters of organic, *i.e.* carboxylic, acids are discussed in subsequent chapters as derivatives of the latter. Polybasic acids give *neutral esters* by complete replacement of the acidic hydrogen atoms or *acid esters* by incomplete substitution, *e.g.*, nitric acid gives only mono-esters, while sulphuric acid gives mono- and di-esters and phosphoric acid gives mono-, di-, and tri-esters.

$MeO·PO·(OH)_2$	$(MeO)_2PO·OH$	$(MeO)_3PO$
Methyl dihydrogen phosphate	Dimethyl hydrogen phosphate	Trimethyl phosphate

Alkyl halides could be considered in this section as esters of alcohols and hydrogen halides but they are discussed more appropriately in Chapter 3 (see Vol. IA, p. 478).

This section deals therefore in particular with esters of inorganic oxy-acids. Although details are given later some of the more general methods of preparation and properties of mineral acid esters are given here.

Preparation. (1) Direct reaction between the acid and the alcohol may lead to the ester by elimination of water. The reaction is reversible and to be used as a preparative method the water must be removed to force the reaction to completion. By this method polybasic acids mostly yield alkyl hydrogen esters:

$$EtOH + H_2SO_4 \rightleftharpoons EtO{\cdot}SO_2{\cdot}OH + H_2O$$

(2) Reaction between alkyl halides and silver or alkali salts of the acids gives esters:

$$EtI + Ag_3AsO_4 \longrightarrow (EtO)_3AsO_4 + 3AgI$$

(3) Esters are formed by the action of acid chlorides on alcohols or metal alkoxides:

$$3EtOH + BCl_3 \longrightarrow (EtO)_3B + 3HCl$$

(4) Esters of 2-substituted alkanols are readily obtained by treating the olefin or the derived epoxide with the acid or acid halide.

Properties. The neutral esters are generally insoluble in water and almost all can be distilled. The acid esters are strong acids, soluble in water, giving salts with bases and are involatile. Both types of esters are hydrolysed with varying ease to their component acids and alcohols. The free acid esters may catalyse their own hydrolysis though the alkali salts of these esters are frequently resistant to hydrolysis by bases due to their anionic character.

In the following sections esters of the individual acids are discussed under the various non-metallic elements arranged in order of ascending groups of the Periodic Table.

(a) Esters of boric acid, trialkyl orthoborates

The trialkyl orthoborates are formed by reaction of alcohols with boric acid, boric oxide or boron trichloride. They can also be obtained by ester exchange reactions:

$$3CH_3OH + (C_3H_7O)_3B \rightleftharpoons (CH_3O)_3B + 3C_3H_7OH$$

They are readily hydrolysed, even by damp air, and some form azeotropes with their related alcohol. The azeotrope from methanol and trimethyl borate contains 30% alcohol and 70% ester and forms the basis of a method for the removal of boric acid by distillation with methanol. With sodium alkoxides they give complexes of the type $Na[B(OR)_4]$, these may be formed directly from lithium or sodium borohydride and alcohol (*H. J. Schlesinger, H. C. Brown, H. R. Hoekstra*, and *L. R. Rapp*, J. Amer. chem. Soc., 1953, 75, 199):

$$MBH_4 + 4CH_3OH \longrightarrow M[B(OCH_3)_4] + 4H_2$$

The alkyl borates react with Grignard reagents to give organo-boron compounds.

Trimethyl borate, $B(OCH_3)_3$, b.p. 68·7°; *triethyl borate,* b.p. 117·3°; *tri*-n-*propyl borate,* b.p. 177°; *tri-isopropyl borate,* b.p. 140°; *tri*-n-*butyl borate,* b.p. 115°/15 mm.; *tri-isobutyl borate,* b.p. 212°; *tri*-sec-*butyl borate,* b.p. 194°; *tri*-tert-*butyl borate,* b.p. 55°/10 mm. *Tris(2-fluoroethyl)borate,* b.p. 173°; *tris(2-chloroethyl)borate,* b.p. 70°/0·1 mm., d_4^{20} 1·278, n_D^{20} 1·4556.

See *M. F. Lappert*, Chem. Reviews, 1956, **56**, 959, for a review on organic compounds of boron including derivatives of boric acid and other boron oxyacids.

(b) Esters of silicic acid, tetra-alkoxysilanes

Tetra-alkyl orthosilicates, tetra-alkoxysilanes. These are obtained by the interaction of silicon tetrahalides with alcohols or sodium alkoxides. The reactivity of the intermediates decreases as the halogens are replaced and the intermediates can be isolated by manipulation of the reaction conditions (*J. R. Wright, R. O. Bolt, A. Goldschmidt* and *A. D. Abbot*, J. Amer. chem. Soc., 1958, 80, 1733):

$$ROH + SiCl_4 \longrightarrow ROSiCl_3 \longrightarrow (RO)_2SiCl_2 \longrightarrow (RO)_3SiCl \longrightarrow (RO)_4Si$$

Forcing conditions, involving sodium *tert*-butoxide with silicon tetrachloride or tetrafluoride, are required for the preparation of tetra-*tert*-butoxysilane (*H. Breederveld* and *H. I. Watermann*, Rec. Trav. chim., 1954, 73, 871; *J. F. Hyde* and *J. W. Curry*, J. Amer. chem. Soc., 1955, 77, 3140). Ester exchange can be used to obtain higher members from tetramethoxysilane (*B. Helferich* and *W. Reimann*, Ber., 1947, 80, 163):

$$(MeO)_4Si + 4(CH_3)_2CHOH \xrightarrow{(CH_3)_2CHONa} [(CH_3)_2CHO]_4Si + 4CH_3OH$$

On hydrolysis, the tetra-alkoxysilanes give mixtures of *polymeric siloxanes* of the form, $(RO)_3Si[O{\cdot}Si(OR)_2]_nO{\cdot}Si(OR)_3$. In the case of the ethyl esters specific members of this series, where n = 0, 1, 2 or 4, have been prepared from ethanol and the appropriate polysiliconoxychloride (*W. C. Schumb* and *D. F. Holloway*, J. Amer. chem. Soc., 1941, 63, 2753).

The tetra-alkoxysilanes behave as alkylating agents and the methyl and ethyl esters are useful alternatives to the silicon tetrahalides in reactions with Grignard reagents when compounds containing carbon-silicon bonds are formed:

$$(EtO)_4Si + MeMgCl \longrightarrow (EtO)_3SiMe + (EtO)_2SiMe_2$$

Tetramethoxysilane, $Si(OCH_3)_4$, b.p. 122°; *tetraethoxysilane,* b.p. 168·8°; *tetra-*n-*propoxysilane,* b.p. 224°/750 mm.; *tetraisopropoxysilane,* b.p. 183–186°; *tetra-*n-*butoxysilane,* b.p. 273–277°; *tetra-isobutoxysilane,* b.p. 255–258°; *tetra-*sec-*butoxysilane,* b.p. 248–249°; *tetra-*tert-*butoxysilane,* b.p. 92–95°/5 mm., m.p. 48–49°.

Polyethoxysilanes, $(EtO)_3Si[OSi(OEt)_2]_nO{\cdot}Si(OEt)_3$. n = 0, b.p. 235° (96°/3 mm.); n = 1, b.p. 268–273° (126°/0·5 mm.); n = 2, b.p. 290–295°; n = 4, b.p. 300–310°.

See *C. Eaborn,* "Organosilicon Compounds", Butterworths, London, 1960, p. 292, for further information.

(c) Esters of nitrous acid, alkyl nitrites

Alkyl nitrites (R·O·NO) are isomeric with nitroparaffins, ($R{\cdot}NO_2$), the latter having a direct carbon-nitrogen bond (see p. 93). They are readily distinguished by their reactions; under alkaline conditions the nitrites are hydrolysed to alcohols and nitrite ion, the nitro-paraffins give salts of their *aci*-forms; reduction of nitrite esters also gives the parent alcohols while the nitro-compounds give amines.

Alkyl nitrites result from the esterification of alcohols with nitrous fumes, nitrous acid or sodium nitrite in dilute sulphuric acid (*W. A. Noyes,* J. Amer. chem. Soc., 1933, 55, 3888). The reaction of alkyl halides with silver nitrite leads to both nitrite esters and the isomeric nitroparaffins (*R. B. Reynolds* and *H. Adkins, ibid.,* 1929, 51, 279); other products have also been claimed in this reaction (*N. Kornblum, J. T. Patton* and *J. B. Nordmann, ibid.,* 1948, 70, 746). Nitrites have been prepared also by the action of nitrosyl chloride in pyridine on alcohols (see *Kornblum et al., loc. cit.*).

The alkyl nitrites are very readily hydrolysed and therefore are useful sources of nitrous acid where sodium nitrite cannot be used, *e.g.,* in the preparation of solid diazonium salts; amyl nitrite is the ester most commonly used for this purpose. Ester exchange occurs readily in alcohols so that methyl nitrite is evolved when ethyl nitrite is dissolved in methanol. Under the influence of sodium alkoxide catalysts they react with compounds containing active methylene groups giving oximes; the mechanism is probably similar to that of the reaction of alkyl nitrates with such compounds (following section). An interesting reaction takes place with nitrite esters containing a methyl group suitably juxtaposed to the nitrite group. Under the influence of ultraviolet light an intramolecular rearangement occurs with the formation of an oxime on the original methyl group (*D. H. R. Barton et al.*, J. Amer. chem. Soc., 1960, **82**, 2640):

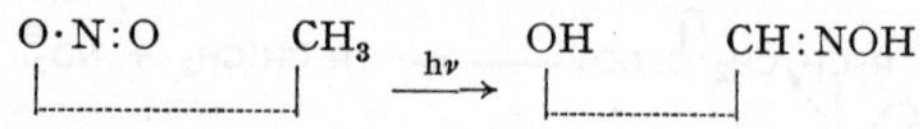

Methyl nitrite, $CH_3·O·NO$, b.p. —16° (*W. H. Hartung* and *F. Crossley*, Org. Synth., Coll. Vol. II, p. 363); *ethyl nitrite*, b.p. 18° (*W. L. Semon* and *V. R. Damerell*, Org. Synth., Coll. Vol. II, p. 204); n-*propyl nitrite*, b.p. 49°; *isopropyl nitrite*, b.p. 40°/745 mm. (*N. Levin* and *W. H. Hartung*, Org. Synth., Coll. Vol. III, p. 192); n-*butyl nitrite*, b.p. 27°/88 mm. (*W. A. Noyes*, Org. Synth., Coll. Vol.II p. 108); *isobutyl nitrite*, b.p. 67°; sec-*butyl nitrite*, b.p. 68°; tert-*butyl nitrite*, b.p. 34°/250 mm.; n-*amyl nitrite*, b.p. 104°; *isoamyl nitrite*, b.p. 99°.

2-*Fluoroethyl nitrite*, b.p. 65–66°/760 mm., d_{20}^{20} 1·1409, n_D^{20} 1·3572; 2-*chloroethyl nitrite*, b.p. 90°, d_{20}^{20} 1·212, n_D^{20} 1·4125; 2-*chloropropyl nitrite*, b.p. 105–106°; 2-*chloroisopropyl nitrite*, b.p. 103–104°.

(d) *Esters of nitric acid*

(i) *Alkyl nitrates*

The most common method of preparation is esterification of the alcohol with concentrated nitric acid, either alone or mixed with concentrated sulphuric acid. A little urea is added to destroy any nitrous acid produced by oxidative side-reactions, as the presence of considerable amounts can lead to decomposition with explosive violence. The reaction is performed at low temperatures (0–10°) to avoid oxidation of the alcohol (*R. L. Shriner* and *E. A. Parker*, J. Amer. chem. Soc., 1933, **55**, 766). Tertiary nitrates are prepared by addition of nitric acid to olefins (*A. Michael* and *G. H. Carlson*, *ibid.*, 1935, **57**, 1268):

$$(CH_3)_2C{:}CH_2 + HNO_3 \longrightarrow (CH_3)_3C·O·NO_2$$

The reaction of finely powdered silver nitrate with alkyl halides in an inert solvent is another useful method of synthesis (*J. W. Baker* and *D. M. Easty*, J. chem. Soc., 1952, 1193). The reaction mixture can be made homogeneous by using methyl cyanide, in which silver nitrate is appreciably soluble, as solvent (*A. F. Ferris et al.*, J. Amer. chem. Soc., 1953, **75**, 4078).

The base-catalysed hydrolysis of alkyl nitrates is a complex reaction.

Evidence has been produced for the occurrence of three simultaneous reactions (*Baker* and *Easty, loc. cit.*)

I 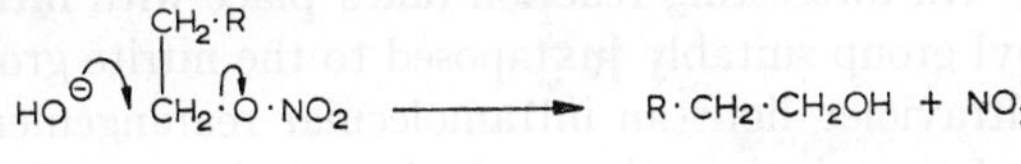

II 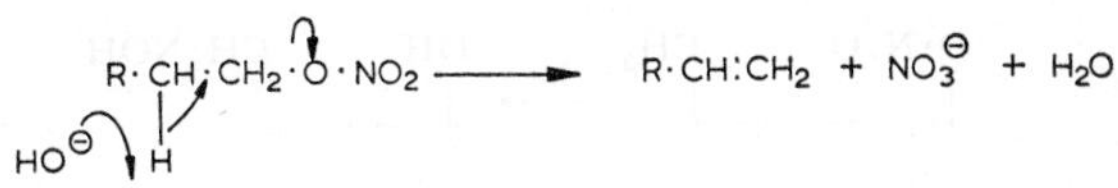

III 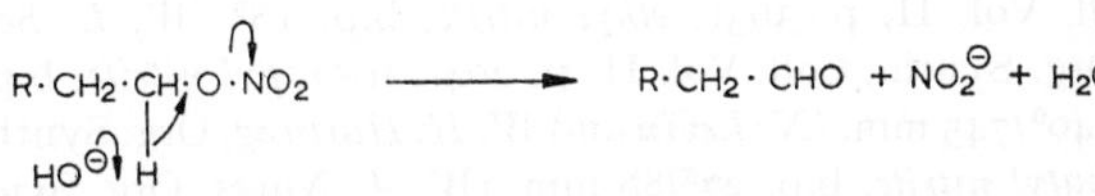

The first reaction, simple hydrolysis, is the predominating one. Dilute acids have little effect on the hydrolysis of nitrate esters. In concentrated sulphuric acid the alcohol and the nitronium ion are generated which accounts for the effectiveness of such solutions as nitrating agents:

$$RO\cdot NO_2 + H^{\oplus} \longrightarrow ROH + NO_2^{\oplus}$$

Compounds containing a reactive methylene group can be nitrated with alkyl nitrates in presence of alkoxide ions *i.e.* in a basic medium:

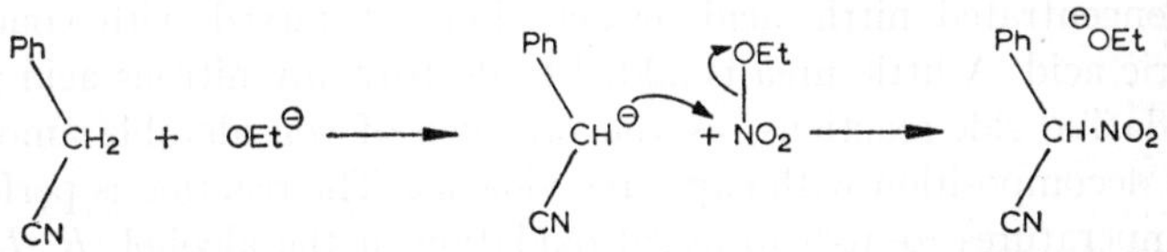

The product, present in the reaction mixture as an alkali metal salt, is isolated as the *aci*-form of the nitro compound (*W. Wislicenus* and *A. Endres*, Ber., 1902, 35, 1755).

Reduction of nitrates by a variety of methods gives the alcohol along with nitrogen or ammonia. The thermal decomposition of nitrates involves free radicals; ethyl nitrate gives mainly ethyl nitrite with some nitromethane and methyl nitrite. Explosive decomposition gives the aldehyde or ketone obtained by oxidation of the parent alcohol along with oxides of nitrogen and water.

For a fuller account of nitrate esters see *R. Boschan, R. T. Merrow,* and *R. W. van Dolah,* Chem. Reviews, 1955, 55, 485.

Methyl nitrate, $CH_3O \cdot NO_2$, b.p. 64·6° (*A. P. Black* and *F. H. Babers*, Org. Synth., Coll. Vol. II, p. 412) is the most readily prepared alkyl nitrate; it explodes violently when struck or heated at 150°.

Ethyl nitrate, b.p. 87°; n-*propyl nitrate*, b.p. 110°; *isopropyl nitrate*, b.p. 101°; n-*butyl nitrate*, b.p. 131°; sec-*butyl nitrate*, b.p. 124°; *isobutyl nitrate*, b.p. 123°; tert-*butyl nitrate*; b.p. 28°/4 mm.

2-*Chloroethyl nitrate*, b.p. 149–150°, d^{20} 1·388; 3-*chloro*-n-*propyl nitrate*, b.p. 173°; 2-*chloroisopropyl nitrate*, b.p. 157–158°.

2-*Bromoethyl nitrate*, b.p. 164–165°, d^{8} 1·735; 2-*iodoethyl nitrate*, b.p. 80°/22 mm.

2-*Nitroethyl nitrate*, b.p. 74°/1 mm.; n_D^{20} 1·455; 1-*nitro-isopropyl nitrate*, b.p. 78–80°/1 mm.; *nitro*-tert-*butyl nitrate*, b.p. 78°/1 mm., d_4^{20} 1·283, n_D^{20} 1·449; 1,1-*dimethyl-2-nitropropyl nitrate*, b.p. 135°/17 mm.

*(ii) Alkylene nitrosates, alkyl nitrosonitrates**

Among the products obtained by the action of dinitrogen tetroxide on olefins (see Vol. IA, p. 422) are the alkylene nitrosates, $R'R''C(NO) \cdot C(ONO_2)$-$R'''R^{iv}$, the nitrate esters of the unknown nitrosoalkanols. They are most conveniently prepared by the action of concentrated nitric acid on a cooled mixture of the olefin and pentyl nitrite (*N. Thorne*, J. chem. Soc., 1956, 4271), the reaction occurring most readily with highly alkylated olefins, such as 2,4,4-trimethylpent-2-ene. Monomeric nitrosates are blue or greenish-blue oils which very readily form colourless crystalline dimers. The alkylene nitrosites obtained by the addition of dinitrogen trioxide to olefins and formerly regarded as nitrosonitrites are best formulated as nitronitroso-compounds and are discussed in Chapter 6.

2-**Methylbut-2-ene nitrosate** is a blue oil which rapidly forms the *dimer* obtainable in two crystalline forms, cubes m.p. 93–98·5° (decomp.) and needles 86–90·4° (decomp.) (*C. A. Taylor* and *W. H. Rinkenbach*, J. Amer. chem. Soc., 1926, **48**, 1684). *Hex-2-ene nitrosate*, prepared by the action of concentrated nitric acid on a mixture of hex-2-ene and pentyl nitrite at 0°, rapidly forms the *dimer*, m.p. 110° (decomp.). Monomeric *hept-2-ene* and *oct-2-ene nitrosates* are similarly obtained as blue unstable oils. 2,4,4-*Trimethylpent-2-ene nitrosate, dimer*, m.p. 124°.

(e) Esters of phosphorous acid, alkyl phosphites and related compounds

Alkyl phosphites. Free phosphorous acid exists in the unsymmetrical form (I) in which phosphorus is pentavalent. Mono- and di-alkyl phosphites also

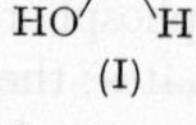

(I)

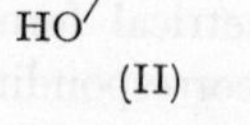

(II)

* Contributed by *E. S. Waight*.

have this structure. Trialkyl phosphites however are derived from the symmetrical form (II) in which phosphorus is trivalent, and are isomeric with dialkyl alkylphosphonates derived from form (I) in which all the hydrogen atoms are replaced by alkyl groups.

The reactions of alcohols with phosphorous acid are of little use for the preparation of phosphite esters which are more readily obtained from phosphorus trichloride. Reaction of an alcohol with an excess of this reagent gives the alkyl phosphorodichloridite in good yield though controlled hydrolysis of this to the monoalkyl phosphite is difficult. Using two equivalents of alcohol gives only a poor yield of dialkyl phosphorochloridite, but increasing the alcohol proportion still further results in the formation of dialkyl phosphite from the trialkyl phosphite and hydrogen chloride initially produced (*H. McCombie, B. C. Saunders,* and *G. J. Stacey,* J. chem. Soc., 1945, 380, 873):

$$ROH + PCl_3 \longrightarrow RO{\cdot}PCl_2 + HCl$$

$$3ROH + PCl_3 \longrightarrow (RO)_3P + 3HCl$$

$$(RO)_3P + HCl \longrightarrow (RO)_2HPO + RCl$$

Monoalkyl phosphites can then be obtained by alkaline hydrolysis of the dialkyl phosphites. If in the above preparation the liberation of free hydrogen chloride is avoided either by using sodium alkoxide in place of the alcohol or having a tertiary amine present, then good yields of the trialkyl phosphite can be obtained.

Trialkyl arsenites. The symmetrical esters are obtained by heating arsenious oxide in the alcohol in the presence of a suitable dehydrating agent (*P. Pascal* and *A. Dupine,* Compt. rend., 1932, 195, 14) or by the reaction of arsenic trichloride or tribromide with the alcohol in the presence of an organic base (*A. W. Reitz* and *R. Sabathy,* Z. physik. Chem., 1938, [41], 155):

$$As_2O_3 + 6ROH \longrightarrow 2(RO)_3As + 3H_2O$$

Trialkyl antimonites are made in a similar way (*J. F. MacKay,* J. chem. Soc., 1909, 95, 604).

Mono- and di-alkyl phosphites have properties in keeping with structure based on the unsymmetrical form (I) of phosphorous acid and are not readily oxidised to the corresponding phosphates; the monoalkyl phosphites are monobasic and the dialkyl phosphites give metal derivatives only under forcing conditions. Spectral studies have indicated the presence of P—H and P=O bonds (*C. F. Callis et al.*, J. Amer. chem. Soc., 1957, 79, 2719; and *L. W. Daasch, ibid.*, 1958, 80, 5301). Dialkyl phosphites react with halogens to give dialkyl phosphorohalidates (*McCombie et al., loc. cit.*; *F. R. Atherton, H. T. Openshaw*, and *A. R. Todd*, J. chem. Soc., 1945, 660):

$$(RO)_2HPO + Cl_2 \longrightarrow (RO)_2POCl + HCl$$

The trialkyl phosphites are rapidly hydrolysed by dilute acids but only slowly by dilute alkali and are oxidised readily to the corresponding phosphates. They undergo the Arbusov reaction on treatment with an alkyl iodide giving a dialkyl alkylphosphonate (*A. Arbusov*, Chem. Zentbl., 1906, II, 1639; *A. H. Ford-Moore* and *J. H. Williams*, J. chem. Soc., 1947, 1465):

$$(EtO)_3P + MeI \longrightarrow (EtO)_2Me\overset{\oplus}{P}\text{—}O\text{—}Et + I^{\ominus} \longrightarrow (EtO)_2MeP{:}O + EtI$$

Reaction of a trialkyl phosphite with a small amount of the corresponding alkyl iodide is sufficient to bring about the isomerisation of the phosphite ester to the dialkyl alkylphosphonate. A similar reaction takes place with halogens (*McCombie et al., loc. cit.*):

$$(EtO)_3P + Cl_2 \longrightarrow (EtO)_2POCl + EtCl$$

Phosphite and phosphate esters derived from *2-fluoroethanol* are highly toxic and attack the central nervous system (see also p. 64).

Dimethyl phosphite, $HPO(OCH_3)_2$, b.p. 170–171°; *diethyl*, b.p. 87°/20 mm.; *di-n-propyl*, b.p. 91·5°/11 mm.; *di-isopropyl*, b.p. 79°/14 mm.; *di-n-butyl*, b.p. 122°/13 mm.; *di-sec-butyl*, b.p. 111°/15 mm.; *di-isobutyl phosphite*, b.p. 106°/12 mm.

Bis-(2-fluoroethyl) phosphite, b.p. 109–110°/1·7 mm.; *bis-(2-chloroethyl) phosphite*, b.p. 129°/0·6 mm.

Methyl phosphorodichloridite, $MeO{\cdot}PCl_2$, b.p. 95°; *ethyl*, b.p. 117°; *n-propyl*, b.p. 144°; *n-butyl*, b.p. 157°; *isobutyl*, b.p. 155°; *2-chloroethyl phosphorodichloridite*, b.p. 172–175°/748 mm.; d_4^{20} 1·4765, n_D^{20} 1·5051.

Isobutyl phosphorodibromidite, $Me_2CH{\cdot}CH_2O{\cdot}PBr_2$, b.p. 86°/20 mm.; *2-bromoethyl phosphorodibromidite*, b.p. 79–80°/2 mm.; d_4^{20} 2·3758, n_D^{20} 1·6106.

Diethyl phosphorofluoridite, $(EtO)_2PF$, b.p. 81°/18 mm.

Diethyl phosphorochloridite, $(EtO)_2PCl$, b.p. 154°; *di-n-propyl*, b.p. 65°/8 mm.; *n-butyl ethyl*, b.p. 72–5°/9 mm.; *n-butyl n-propyl*, b.p. 85–6°/9 mm.; *di-n-butyl*, b.p. 92°/6 mm.; *bis-(2-chloroethyl) phosphorochloridite*, b.p. 101–103°/4·5 mm., d_4^{20} 1·4007, n_D^{20} 1·4950. *bis-(2-bromoethyl) phosphorobromidite*, b.p. 114–115·5°/1·7 mm., d_4^{20} 2·1109, n_D^{20} 1·5671.

Trimethyl phosphite, $P(OCH_3)_3$, b.p. 111–112°; *triethyl phosphite*, b.p. 57–58°/16 mm.; *tri-n-propyl phosphite*, b.p. 83°/10 mm.; *tri-isopropyl phosphite*, b.p.

43·5°/1 mm.; *tri-n-butyl phosphite*, b.p. 122°/12 mm.; *tri-isobutyl phosphite*, b.p. 135–136°/10 mm.

Tris-(2-fluoroethyl) phosphite, b.p. 114–116°/8 mm., d_{20}^{20} 1·285, n_D^{20} 1·417; *tris-(2-chloroethyl) phosphite*, b.p. 112–115°/2·5 mm., d_4^{26} 1·3443, n_D^{26} 1·4818.

Trimethyl arsenite, $As(OCH_3)_3$, b.p. 127°/730 mm.; *triethyl arsenite*, b.p. 162°/745 mm.; *tri-n-propyl arsenite*, b.p. 96°/10 mm.; *tri-isopropyl arsenite*, b.p. 72°/14 mm.; *tri-n-butyl arsenite*, b.p. 145°/15 mm.

Triethyl antimonite, $Sb(OEt)_3$, b.p. 115–120°.

(f) Esters of phosphoric acid, alkyl phosphates and related compounds

Alkyl phosphates. Phosphoryl chloride is a most useful starting material for preparing alkyl phosphates. Reaction of one, two or three equivalents of alcohol, the last in the presence of a tertiary base, with phosphoryl chloride gives respectively mono-alkyl phosphorodichloridate, dialkyl phosphorochloridate and trialkyl phosphate:

$$ROH + POCl_3 \longrightarrow RO\cdot POCl_2 \longrightarrow RO\cdot PO(OH)_2$$

$$2ROH + POCl_3 \longrightarrow (RO)_2POCl \longrightarrow (RO)_2PO(OH)$$

$$3ROH + POCl_3 \longrightarrow (RO)_3PO$$

The alkyl phosphorochloridates are readily hydrolysed by alkali to the mono- and di-alkyl phosphates as shown above.

Phosphorus pentoxide or polyphosphoric acid (phosphorus pentoxide in phosphoric acid) have been used as phosphorylating agents. The first reagent leads to mixtures of mono- and di-alkyl phosphates while the second may give pyro- and higher phosphates (*R. H. Hall* and *H. G. Khorana*, J. Amer. chem. Soc., 1955, 77, 1871). Reaction of alkyl halides with trisilver phosphate gives good yields of tertiary esters though substitution of the trisodium salt gives mainly the monoalkyl phosphate:

$$3\,RX + Ag_3PO_4 \longrightarrow (RO)_3PO + 3AgX$$

$$RX + Na_3PO_4 \longrightarrow RO\cdot PO(ONa)_2 + NaX$$

More recent methods of preparation include the use of diazoalkanes to esterify phosphates (*F. R. Atherton, H. T. Howard*, and *A. R. Todd*, J. chem. Soc., 1948, 1106), and the specific formation of monoalkyl phosphates by phosphorylating alcohols with dibenzyl phosphorochloridate giving a phosphate triester from which the two benzyl groups may be removed simultaneously or consecutively by hydrogenolysis or anionic displacement (*Atherton, Openshaw*, and *Todd*, *ibid.*, 1945, 382; and later papers):

$$ROH + (PhCH_2O)_2POCl \longrightarrow RO\cdot PO(OCH_2Ph)_2 \longrightarrow RO\cdot PO(OCH_2Ph)OH$$

$$RO\cdot PO(OCH_2Ph)_2 \longrightarrow RO\cdot PO(OH)_2 \longleftarrow RO\cdot PO(OCH_2Ph)OH$$

Dialkyl phosphates have been obtained from an alcohol and a monoalkyl phosphate in the presence of pyridine and dicyclohexylcarbodiimide (*Khorana, et al.*, J. Amer. chem. Soc., 1957, **79**, 1002; 1958, **80**, 6212):

$$C_6H_{11}N{:}C{:}NC_6H_{11} + R'\cdot OH + R''O\cdot PO(OH)_2 \rightarrow R'O\cdot PO(OR'')OH + (C_6H_{11}NH)_2CO$$

Cyclic trialkyl trimetaphosphates (see p. 66) are involved in this reaction. This synthesis has been further extended by using benzyl dihydrogen phosphate thus obtaining a diester from which the benzyl group may be removed by hydrogenolysis, effecting simple phosphorylation of the original alcohol; using 2-cyanoethyl phosphate gives a diester from which the cyanoethyl group may be removed by treatment with mild alkali (*P. T. Gilham* and *G. M. Tener*, Chem. and Ind., 1959, 542; *Tener*, J. Amer. chem. Soc., 1961, **83**, 159).

The mono- and di-alkyl phosphates are strong acids, the trialkyl esters have been used as alkylating agents.

Trimethyl phosphate, $(CH_3O)_3PO$, b.p. 197°; *triethyl phosphate*, b.p. 216°; *tri-*n*-propyl phosphate*, b.p. 252° (*G. R. Dutton* and *C. R. Noller*, Org. Synth., Coll. Vol. II, p. 110); *tri-isopropyl phosphate*, b.p. 83·5°/5 mm.; *tri-*n*-butyl phosphate*, b.p. 289°; *tri-*sec*-butyl phosphate*, b.p. 119–129°/8–12 mm.; *tri-isobutyl phosphate*, b.p. 117°/5·5 mm.

Tris-(2-fluoroethyl) phosphate, b.p. 169°/11 mm., d_{20}^{20} 1·365, n_D^{20} 1·4043; *tris-(2-chloroethyl) phosphate*, b.p. 214°/25 mm., d_4^{20} 1·4256, n_D^{20} 1·4731.

Alkyl phosphorohalidates, $(RO)_2POX$ and $(RO)POX_2$. Although some mention of compounds of this type has already been made the dialkyl phosphorofluoridates are particularly interesting in view of their high toxicity, especially di-isopropyl phosphorofluoridate. This toxicity is due to their anti-chlorine esterase activity. They are obtained by halide exchange from the corresponding phosphorochloridate (*B. C. Saunders et al.*, J. chem. Soc., 1948, 695, 699):

$$(RO)_2POCl + NaF \longrightarrow (RO)_2POF + NaCl$$

Ethyl phosphorodifluoridate, $EtO\cdot POF_2$, b.p. 85°.

Methyl phosphorodichloridate, $MeO\cdot POCl_2$, b.p. 63°/15 mm.; *ethyl*, b.p. 63°/19 mm.; n*-butyl*, b.p. 90°/17 mm.; *2-fluoroethyl*, b.p. 106–107°/30 mm., d_{20}^{20} 1·5637, n_D^{20} 1·4400; *2-chloroethyl phosphorodichloridate*, b.p. 71·5°/2 mm., d_4^{20} 1·5527, n_D^{20} 1·4690.

Dimethyl phosphorofluoridate, $(MeO)_2POF$, b.p. 150°, 43·5°/14 mm.; *diethyl*, b.p. 71°/18 mm.; *di-n-propyl*, b.p. 99°/20 mm.; *di-isopropyl*, b.p. 84°/25 mm.; n-*butyl ethyl*, b.p. 83°/13 mm.; *bis-(2-fluoroethyl)*, b.p. 125–127°/13 mm.; *bis-(2-chloroethyl) phosphorofluoridate*, b.p. 142°/15 mm.

Dimethyl phosphorochloridate, $(MeO)_2POCl$, b.p. 60°/4 mm.; *diethyl*, b.p. 94°/18 mm.; *di-n-propyl*, b.p. 65°/3–4 mm.; *di-isopropyl*, b.p. 95°/14 mm.; *di-n-butyl*, b.p. 133°/15 mm., *di-isobutyl*, b.p. 57°/0·1 mm.; *bis-(2-chloroethyl) phosphorochloridate*, b.p. 122–124°/0·6 mm., d_4^{20} 1·4623.

Alkyl pyrophosphates. The symmetrical tetra-alkyl pyrophosphates have been obtained by the controlled hydrolysis of dialkyl phosphorochloridates in aqueous pyridine (*A. D. F. Toy*, J. Amer. chem. Soc., 1948, **70**, 3882):

$$(RO)_2POCl + H_2O \longrightarrow (RO)_2PO\cdot OH + HCl$$

$$(RO)_2PO\cdot OH + (RO)_2POCl \longrightarrow (RO)_2\underset{\underset{O}{\|}}{P}\cdot O\cdot \underset{\underset{O}{\|}}{P}(OR)_2 + HCl$$

The latter stage of the above synthesis can be utilised to prepare unsymmetrical tetra-alkyl pyrophosphates:

$$(R'O)_2PO\cdot OH + (R''O)_2POCl \longrightarrow (R'O)_2\underset{\underset{O}{\|}}{P}\cdot O\cdot \underset{\underset{O}{\|}}{P}(OR'')_2 + HCl$$

Dicyclohexyl carbodiimide can be used to prepare symmetrical di- and tetra-alkyl pyrophosphates from mono- or di-alkyl phosphates (*H. G. Khorana* and *A. R. Todd*, J. chem. Soc., 1953, 2257), *e.g.*, for a dialkyl pyrophosphate:

$$2RO\cdot PO(OH)_2 + C_6H_{11}N{:}C{:}NC_6H_{11} \rightarrow RO\cdot\underset{\underset{O}{\|}}{P}(OH)\cdot O\cdot\underset{\underset{O}{\|}}{P}(OR)OH + (C_6H_{11}NH)_2CO$$

These pyrophosphate esters are reactive compounds since they are acid anhydrides and are readily hydrolysed to the corresponding phosphates. They are toxic and tetraethyl pyrophosphate especially has found much use as an insecticide (*S. A. Hall* and *M. Jacobson*, Ind. Eng. Chem., 1948, **40**, 694).

Tetramethyl pyrophosphate, b.p. 115°/0·5 mm.; *tetraethyl pyrophosphate*, b.p. 155°/5 mm.; *tetra-n-propyl pyrophosphate*, b.p. 179°/4 mm.; *tetraisopropyl pyrophosphate*, m.p. 14–15°, b.p. 92–95°/0·01 mm.; *tetra-n-butyl pyrophosphate*, b.p. 143–146°/0·01 mm.

Alkyl metaphosphates. The reaction of trialkyl phosphates or dialkyl ethers with phosphorus pentoxide leads to alkyl metaphosphates of undefined structure and generally represented as $(RO\cdot PO\cdot O)_x$ though it is fairly certain that both cyclic and linear polyphosphates are present. The first synthesis of cyclic trimeric trialkyl metaphosphates was reported in 1958, reaction of methyl or ethyl iodide with silver trimetaphosphate in acetonitrile giving respectively the cyclic methyl and ethyl esters characterised by their hydrolytic reactions (*F. Cramer* and *H. Hettler*, Ber., 1958, **91**, 1181):

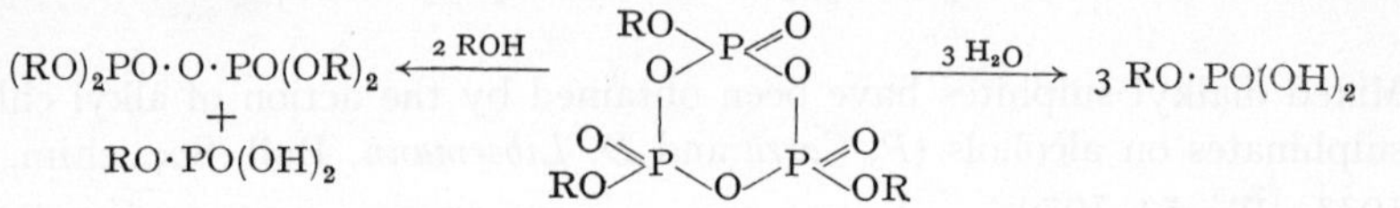

The cyclic triethyl metaphosphate has also been claimed to arise from the reaction of ethyl dihydrogen phosphate with dicyclohexyl carbodiimide in dioxan (*M. Smith* and *H. G. Khorana*, J. Amer. chem. Soc., 1958, **80**, 1141; *G. Weimann* and *Khorana*, Chem. and Ind., 1962, 271; J. Amer. chem. Soc., 1962, **84**, 4329). These authors also claim that cyclic trimetaphosphate esters are intermediates in phosphorylations involving carbodiimides.

For a more detailed account of phosphorus oxyacid esters see *G. M. Kosolapoff*, "Organophosphorus Compounds", Wiley, New York, 1950; *Cramer*, Angew. Chem., 1960, **72**, 236.

Trialkyl arsenates. These esters are formed by the reaction between silver arsenate and alkyl iodides (*M. J. Crafts*, Bull. Soc. chim. France, 1870, 14; 1870, 99; *G. Kamai* and *K. J. Kuzinin*, C.A., 1956, **50**, 6297 h):

$$Ag_3AsO_4 + 3RI \longrightarrow (RO)_3AsO + 3AgI$$

Trimethyl arsenate, b.p. 98°/13 mm.; *triethyl arsenate*, b.p. 119°/16 mm.; *tri-n-propyl arsenate*, b.p. 145°/14 mm.; *tri-n-butyl arsenate*, b.p. 170°/12 mm.

(g) Esters of sulphurous acid and related compounds

Alkyl sulphites are esters of sulphurous acid in its symmetrical form in which sulphur is tetravalent. They are isomeric with alkyl sulphonates in which there is a carbon-sulphur bond and sulphur is hexavalent.

RO·SO·OR	$R\cdot SO_2\cdot OR$
Dialkyl sulphite	Alkyl alkanesulphonate

The latter compounds are esters of alkanesulphonic acids which as oxidation products of the alkane thiols are described in the following chapter.

Dialkyl sulphites cannot be obtained from the reaction between alkyl halides and silver or alkali metal sulphites, instead the isomeric sulphonates are produced:

$$K_2SO_3 + 2C_2H_5I \longrightarrow C_2H_5\cdot SO_2\cdot OC_2H_5 + 2KI$$

They may be prepared from alcohols and sulphur chloride, or better, thionyl chloride (*W. E. Bessinger* and *F. E. Kung*, J. Amer. chem. Soc., 1947, 69, 2158):

$$S_2Cl_2 + 3C_2H_5OH \longrightarrow SO(OC_2H_5)_2 + C_2H_5SH + 2HCl$$

$$SOCl_2 + 2C_2H_5OH \longrightarrow SO(OC_2H_5)_2 + 2HCl$$

Mixed dialkyl sulphites have been obtained by the action of alkyl chlorosulphinates on alcohols (*P. Carré* and *D. Libermann*, Bull. Soc. chim. Fr., 1933, [iv], 53, 1070):

$$CH_3OH + C_2H_5{\cdot}O{\cdot}SO{\cdot}Cl \longrightarrow CH_3O{\cdot}SO{\cdot}OC_2H_5 + HCl$$

The dialkyl sulphites are readily hydrolysed by aqueous acid or alkali to alcohols and sulphur dioxide or sodium sulphite though the sodium alkyl hydrogen sulphite can be obtained by using sodium methoxide in methanol (*W. Voss* and *E. Blanke*, Ann., 1931, 485, 258). As with dialkyl sulphates, sodium iodide in acetone reacts with dialkyl sulphites to give alkyl iodides but no sodium alkyl sulphite is obtained, instead acetone sodium bisulphite is produced (*A. B. Foster et al.*, J. chem. Soc., 1956, 2589). Pyrolysis of methyl alkyl sulphites gives good yields of olefins (*G. Berti*, J. Amer. chem. Soc., 1954, 76, 1213):

$$>CH{\cdot}\overset{|}{\underset{|}{C}}{\cdot}O{\cdot}SO{\cdot}OCH_3 \xrightarrow{\Delta} >C{:}C< + SO_2 + CH_3OH$$

Under the influence of certain catalysts, especially tertiary amines, the lower members of the series rearrange to the corresponding sulphonate esters (*W. E. Bissinger, F. E. Kung*, and *C. W. Hamilton*, *ibid.*, 1948, 70, 3940). For a more complete account of interconversions and comparisons of sulphites and sulphonates see *A. Simon, R. Paetzold*, and *H. Kriegsmann*, Ber., 1956, 89, 883.

Dimethyl sulphite, $SO(OCH_3)_2$, b.p. 126°; *ethyl methyl sulphite*, b.p. 141°; *diethyl sulphite*, b.p. 157°; *di-*n*-propyl sulphite*, b.p. 192°; *di-isopropyl sulphite*, b.p. 170°; *di-*n*-butyl sulphite*, b.p. 90·5°/5 mm. (*C. M. Suter* and *H. L. Gerhart*, Org. Synth., Coll. Vol. II, 112); *di-isobutyl sulphite*, b.p. 209°/741 mm.

Bis-(*2-chloroethyl*) *sulphite*, b.p. 133°/12 mm., d_0^{20} 1·422, n_D^{20} 1·481; *bis-*(*2-chloropropyl*) *sulphite*, b.p. 142–143°/6–7 mm., d^{23} 1·27; *bis-*(*3-chloropropyl*) *sulphite*, b.p. 161–162°/13 mm., d_0^{19} 1·313, n_D^{13} 1·480. *Bis-*(*2-bromoethyl*) *sulphite*, b.p. 158–159°/13 mm., d_0^{0} 1·959, $n_D^{20.5}$ 1·525.

Chlorosulphinates. Although the reaction between thionyl chloride and an alcohol in the presence of a tertiary amine is used to prepare alkyl chlorides, the intermediate chlorosulphinate can be isolated when equimolar quantities of thionyl chloride and the alcohol are allowed to react in the cold:

$$ROH + SOCl_2 \longrightarrow RO{\cdot}SO{\cdot}Cl + HCl$$

They have also been obtained by the action of thionyl chloride or phosphorus trichloride on dialkyl sulphites.

They are liquids readily decomposed by water into hydrochloric acid, sulphur dioxide, and alcohols and are decomposed by heating into alkyl chlorides and sulphur dioxide (*Bissinger* and *Kung*, *loc. cit.*).

Methyl chlorosulphinate, b.p. 35°/60 mm.; *ethyl chlorosulphinate*, b.p. 32°/16 mm.; n-*propyl chlorosulphinate*, b.p. 46°/16 mm.; *isopropyl chlorosulphinate*, b.p. 55°/40 mm.; n-*butyl chlorosulphinate*, b.p. 73°/28 mm.; *isobutyl chlorosulphinate*, b.p. 67°/20 mm. 3-*Chloropropyl chlorosulphinate*, b.p. 111–112°/21 mm.

(h) Esters of sulphuric acid and related compounds

Dialkyl sulphates. The reaction between sulphuric acid and alcohols gives rise to a variety of products and ethers, olefins, mono- and di-alkyl sulphates can be obtained by controlling the reaction conditions. Dialkyl sulphates are sometimes more conveniently obtained from the corresponding sulphites by reaction with sulphuryl chloride:

$$(CH_3O)_2SO + SO_2Cl_2 \longrightarrow CH_3Cl + CH_3O{\cdot}SO_2Cl + SO_2$$

$$(CH_3O)_2SO + CH_3O{\cdot}SO_2Cl \longrightarrow (CH_3O)_2SO_2 + SO_2 + CH_3Cl$$

the methyl chlorosulphonate reacting with another molecule of dimethyl sulphite when the temperature is raised (*C. M. Suter* and *H. L. Gerhart*, Org. Synth., Coll. Vol. II, p. 111). Mixed dialkyl sulphates may be prepared by reaction of chlorosulphonate esters with sodium alkoxides (*R. Levaillant*, Ann. Chim., 1936, 6, 560):

$$C_3H_7O{\cdot}SO_2Cl + NaOCH_3 \longrightarrow C_3H_7O{\cdot}SO_2{\cdot}OCH_3$$

and by the reaction of an alkyl hydrogen sulphate with a chloroformic ester (*M. J. Kratt* and *F. W. Ljutkina*, Zhur. Obschei. Khim., 1931, 1, 190):

$$C_2H_5O{\cdot}SO_2OH + Cl{\cdot}COOCH_3 \longrightarrow C_2H_5O{\cdot}SO_2{\cdot}OCH_3 + CO_2 + HCl$$

The reaction between alkyl iodides and silver sulphate has been used to prepare dialkyl sulphates. Technically the action of sulphuric acid on alcohols or olefins is the most important method of manufacture of these esters (*S. G. P. Plant* and *N. V. Sidgwick*, J. Soc. chem. Ind., 1921, **40**, 14).

The dialkyl sulphates are heavy, distillable liquids, soluble in organic solvents and sparingly so in water in which they slowly decompose to the alcohol and the alkyl hydrogen sulphate. They are useful alkylating agents and are generally used in the presence of a basic catalyst, *e.g.*, sodium hydroxide or potassium carbonate. Sodium iodide in acetone reacts readily with dialkyl sulphates to give alkyl iodides and sodium alkyl sulphates (*Foster et al., loc. cit.*).

Dimethyl sulphate, $SO_2(OCH_3)_2$, b.p. 189°, 72·6°/13 mm., m.p. —31·4° is a highly toxic substance (*S. Weber*, Chem. Ztbl., 1902, I, 364). Technically it is prepared from dimethyl ether and sulphur trioxide or methanol and sulphuric acid (*J. Guygot* and *L. J. Simon*, Compt. rend., 1919, **164**, 795).

Diethyl sulphate, b.p. 208° (93·1°/13 mm.), m.p. —25·2°, is non-toxic; *ethyl methyl sulphate,* b.p. 101°/45 mm.; *methyl* n-*propyl sulphate,* b.p. 80°/5 mm.; *di-*n-*propyl sulphate,* b.p. 108–109°/11 mm. (*Suter* and *Gerhart, loc. cit.*); *di-isopropyl sulphate,* b.p. 94°/7 mm.; *di-*n-*butyl sulphate,* b.p. 109·5°/4 mm. (*Suter* and *Gerhart, loc. cit.*).

Bis-(2-*fluoroethyl*) *sulphate,* b.p. 145°, d_{20}^{20} 1·3191, n_D^{20} 1·4080; *bis-*(2-*chloroethyl*) *sulphate,* m.p. 11°, b.p. 154°/8 mm., d_4^{20} 1·4801, n_D^{20} 1·4622; *bis-*(3-*chloropropyl*) *sulphate,* b.p. 163–165°/4 mm., d_0^{25} 1·363, n_D^{20} 1·460.

Alkyl hydrogen sulphates. These acid esters arise when the dialkyl sulphates are hydrolysed in alkaline media in which the monoesters are relatively stable, in acid media hydrolysis proceeds further to give sulphuric acid and alcohols. They are also formed by reaction of alcohols with sulphuric acid, pyrosulphates, chlorosulphonic acid or its sodium salt; sulphur trioxide has been used in the form of its complex with dioxan (*C. M. Suter, P. B. Evans,* and *J. M. Kiefer,* J. Amer. chem. Soc., 1938, **60**, 538). The reaction of sulphuric acid with olefins to give alkyl hydrogen sulphates is used industrially in the separation of olefins from paraffins, the hydration of olefins (by subsequent hydrolysis of the ester) and the preparation of wetting agents.

The alkyl hydrogen sulphates are difficult to obtain in anhydrous form; some crystalline hydrates are known. They are generally isolated as their barium salts from which the free acid may be obtained by the addition of the calculated amount of sulphuric acid. Their aqueous solutions are strongly acid and decompose into the alcohol and sulphuric acid. When heated alone they give olefins and in the presence of alcohols ethers are obtained:

$$CH_3 \cdot CH_2O \cdot SO_2 \cdot OH \longrightarrow H_2SO_4 + CH_2{:}CH_2$$

$$C_2H_5O \cdot SO_2 \cdot OH + C_2H_5OH \longrightarrow H_2SO_4 + (C_2H_5)_2O$$

Their alkali salts are frequently used in synthetic reactions, with alkali hydrosulphides and sulphides they give mercaptans and thioethers respectively, with carboxylic acid salts they give esters and with potassium cyanide alkyl cyanides are obtained.

Methyl hydrogen sulphate, *methyl sulphuric acid*, $CH_3O \cdot SO_3H$, when anhydrous is soluble in ether.

Ethyl hydrogen sulphate (ammonium salt, m.p. 96–97°) cannot be obtained in a pure state because of the equilibrium:

$$2C_2H_5O \cdot SO_3H \rightleftharpoons (C_2H_5O)_2SO_2 + H_2SO_4$$

The final mixture consists of 45 % ethyl hydrogen sulphate, 25 % diethyl sulphate and 30 % sulphuric acid (*D. S. Breslow, R. R. Hough,* and *J. T. Fairclough,* J. Amer. chem. Soc., 1954, **76**, 5361). A similar equilibrium for methyl hydrogen sulphate has been postulated but it is attained much more slowly (*Guygot* and *Simon, loc. cit.*).

Higher alkyl hydrogen sulphates. A number of these esters are produced commercially as wetting agents (*e.g.*, dodecyl) and detergents (hexadecyl, octadecyl). Their sodium salts form soapy solutions which are unaffected by acids or by hard water, since the calcium and magnesium salts are soluble in water. The higher alcohols from which these esters are prepared are obtained by reduction of fatty acid esters, from saponification of waxes, and from olefins (obtained from petroleum) by way of the Oxo reaction with carbon monoxide and hydrogen to give an aldehyde which is further reduced to an alcohol. Sulphuric acid, chlorsulphonic acid, sulphamic acid and sulphur trioxide have all been used to convert the alcohols to alkyl sulphuric acids. See *A. M. Schwartz, J. W. Perry,* and *J. Berch,* "Surface Active Agents and Detergents", Vol. II, Interscience, New York, 1958, p. 40 *et seq.*, for a more detailed account.

For properties of individual alkyl sodium sulphates, see *E. E. Dreger et al.*, Ind. Eng. Chem., 1944, **36**, 610.

Chlorosulphonic esters. These are the acid chlorides of the alkyl hydrogen sulphates. Most conveniently prepared by the action of sulphuryl chloride on alcohols (*W. W. Binkley* and *E. F. Degering*, J. Amer. chem. Soc., 1938, **60**, 2810):

$$ROH + SO_2Cl_2 \longrightarrow RO \cdot SO_2Cl + HCl$$

they have also been obtained by the action of phosphorus pentachloride on salts of the alkyl sulphuric acids; by the addition of chlorosulphonic acid to olefins; by the reaction of sulphur trioxide with alkyl chlorides and by the reaction of sulphur dioxide on alkyl hypochlorites (*J. Sandmeyer*, Ber., 1888, **19**, 857).

They are liquids soluble in organic solvents and insoluble in water; the lower members are lachrymators. Water decomposes them into hydrochloric acid and alkyl hydrogen sulphates; the kinetics of this reaction have been studied (*H. K. Hall,* J. Amer. chem. Soc., 1956, **78,** 1450). Their reaction with alcohols or sodium alkoxides is complex, ethers and alkyl halides being among the products obtained; amines are generally alkylated (*F. W. Bushong,* Amer. chem. J., 1903, **30,** 212; *O. W. Willcox, ibid.,* 1904, **32,** 446).

Methyl chlorosulphonate, b.p. 48·1°/29 mm.; *ethyl chlorosulphonate,* b.p. 42·3°/10 mm.; n-*propyl chlorosulphonate,* b.p. 53·2°/10 mm.; *isopropyl chlorosulphonate,* b.p. 50°/5 mm.; n-*butyl chlorosulphonate,* b.p. 59°/3 mm.; *isobutyl chlorosulphonate,* b.p. 46°/4 mm.

2-*Fluoroethyl chlorosulphonate,* b.p. 79–80°/18 mm., d_{20}^{20} 1·4970, n_D^{20} 1·4198; 2-*bromoethyl chlorosulphonate,* b.p. 100–105°/18 mm., d^{24} 1·863, n_D^{24} 1·484.

(i) Esters of hypochlorous and perchloric acids

Alkyl hypochlorites. These are most conveniently prepared by passing chlorine into an alkaline solution of the alcohol (*F. D. Chattaway* and *O. G. Backeberg,* J. chem. Soc., 1923, 2999) or from hypochlorous acid and alcohols in a suitable solvent such as carbon disulphide or carbon tetrachloride to facilitate the isolation of the ester (*R. Fort* and *L. Denivelle,* Bull. Soc. chim. Fr., 1954, 1109; *M. C. Taylor, R. B. MacMullin* and *C. A. Gammal,* J. Amer. chem. Soc., 1925, **47,** 395).

The lower members of the series are unstable though stability increases with molecular weight and degree of substitution. Primary and secondary hypochlorites are decomposed by light or bases to hydrogen chloride and aldehydes or ketones, respectively. Tertiary hypochlorites on heating decompose to ketones and alkyl halides:

$$(CH_3)_3C{\cdot}O{\cdot}Cl \longrightarrow (CH_3)_2CO + CH_3Cl$$

Alkyl hypochlorites are a convenient organic source of positive chlorine and can be used to prepare chlorohydrins or β-chloroethers from olefins (*C. F. Irwin* and *G. F. Hennion,* J. Amer. chem. Soc., 1941, **63,** 858):

$$>C{:}C< + R{\cdot}OCl + R'OH \longrightarrow >CCl{\cdot}\overset{|}{\underset{|}{C}}(OR')$$

They oxidise alcohols to aldehydes or ketones, presumably via the hypochlorite ester; excessive reagent can lead to further chlorination of the aldehyde or ketone (*J. J. Beereboom, C. Djerassi, D. Ginsburg,* and *L. F. Fieser, ibid.,* 1953, **75,** 3500). Similarly, primary amines are oxidised to carbonyl compounds (*W. E. Bachmann, M. P. Cava,* and *A. S. Dreiding, ibid.,* 1954, **76,** 5554):

$$>CH{\cdot}NH_2 \longrightarrow >CH{\cdot}NHCl \longrightarrow >C{:}NH \longrightarrow >CO$$

Methyl hypochlorite, $ClOCH_3$, b.p. 12°/763 mm. is an explosive gas.

tert-*Butyl hypochlorite*, b.p. 77–78° (*H. M. Teeter* and *E. W. Bell*, Org. Synth., 1952, 32, 20), can be stored in darkness without decomposition.

Alkyl perchlorates. Although the products obtained from the reaction between alkyl iodides and silver perchlorate have been regarded as organic esters (*J. Meyer* and *W. Sporman*, Z. anorg. Chem., 1936, **228**, 341), there is evidence that they are ionic in character, since alkylation of the aromatic ring occurs when they are prepared in anisole (*H. Burton* and *P. F. G. Praill*, Chem. and Ind., 1951, 939; *Burton, D. A. Munday*, and *Praill*, J. chem. Soc., 1956, 3933):

$$CH_3I + AgClO_4 \longrightarrow CH_3{\cdot}OClO_3$$

$$CH_3{\cdot}OClO_3 \rightleftharpoons CH_3^{\oplus} + ClO_4^{\ominus}$$

$$CH_3^{\oplus} + C_6H_5{\cdot}OCH_3 \longrightarrow p\text{-}CH_3{\cdot}C_6H_4{\cdot}OCH_3 + H^{\oplus}$$

The alkyl perchlorates explode readily and are too unstable to allow characterisation.

Chapter 5

Sulphur Analogues of the Alcohols and their Derivatives

H. GOLDWHITE

Although a number of classes of sulphur compounds, including the thiols, the sulphides, and the disulphides, are formally analogous to oxygenated compounds, the chemistry of the two series shows more differences than similarities. The ease with which sulphur, an element in the second row of the Periodic Table, can attain the higher oxidation numbers +4 and +6, means that series of stable sulphur compounds exist which have no oxygen analogues.

The hypothetical relationships which exist between these various types of compounds may be indicated as follows:

$$R{\cdot}SH \xrightarrow{[O]} R{\cdot}S{\cdot}OH \xrightarrow{[O]} R{\cdot}SO{\cdot}OH \xrightarrow{[O]} R{\cdot}SO_2{\cdot}OH$$

Thiol	Sulphenic acid	Sulphinic acid	Sulphonic acid

$$R{\cdot}S{\cdot}R' \xrightarrow{[O]} R{\cdot}SO{\cdot}R' \xrightarrow{[O]} R{\cdot}SO_2{\cdot}R'$$

Sulphide	Sulphoxide	Sulphone

This chapter deals with the chemistry of these and related sulphur compounds, *viz.*: disulphides, polysulphides, sulphonium compounds, alkylthiolsulphonic acids and alkanethiosulphonic acids. Those selenium and tellurium analogues which exist are dealt with in Chapter VII.

A recent general review of preparative sulphur chemistry appears in *Houben-Weyl*, "Methoden der Organischen Chemie", Vol. IX, Georg Thieme Verlag, Stuttgart, 1955. A number of chapters in "Organic Sulphur Compounds", Vol. I, Ed. *N. Kharasch*, Pergamon Press, Oxford, 1961, are of general interest for the whole area of sulphur chemistry. These include chapters on the thermodynamics of organic sulphur compounds; on the bonding characteristics of the sulphur atom; and on infrared and ultraviolet absorption spectra of organic sulphur compounds.

Other theoretical aspects of organic sulphur chemistry are discussed by *C. C. Price* and *S. Oae* in "Sulphur Bonding", Ronald Press Co., New York, 1962, and by *W. A. Pryer* in "Mechanisms of Sulphur Reactions", McGraw-Hill, New York, 1962. A comprehensive discussion of thiols, sulphides, sulphenic acid derivatives, etc. is given by *E. E. Reid* in "Organic Chemistry of Bivalent Sulphur", Chemical Publishing Co., New York, 1960. The natural occurrence of sulphur compounds is reviewed by *A. I. Virtanen*, Angew. Chem. Internat. Edn., 1962, **1**, 199.

1. Thiols, mercaptans, or alkyl hydrosulphides

The thiols, RSH, are the sulphur analogues of the alcohols. Their structure follows from the mode of synthesis from alkyl halides and alkali-metal hydrosulphides, and is confirmed by the reactions of the class. Ethanethiol (ethyl mercaptan) was the first member of the class to be discovered and investigated (*W. C. Zeise*, Ann., 1834, **11**, 1), and a wide variety of thiols has been prepared.

These compounds were originally called mercaptans (Lat. *mercurium aptans*) because of their ready reaction with mercuric oxide to form crystalline mercury derivatives:

$$2C_2H_5SH + HgO \longrightarrow (C_2H_5S)_2Hg + H_2O$$

The I.U.P.A.C. system of nomenclature uses the suffix thiol appended to the name of the hydrocarbon from which the compound is derived. The metal derivatives of the thiols are usually called mercaptides.

Occurrence. Free thiols rarely occur in natural products. Methanethiol is found in the radish (*N. Nakamura*, Biochem. Z., 1925, **164**, 31) and the skunk employs 1-butanethiol as a defensive weapon (see *S. Blackburn* and *F. Challenger*, J. chem. Soc., 1938, 1872, for reference to this and other natural sources of thiols). Some natural gases contain considerable amounts of methanethiol and ethanethiol, and large amounts of the lower alkanethiols are produced in the thermal cracking of crude petroleum oils.

Preparation. (1) By the action of alkyl halides or sodium alkyl sulphates on an alkali-metal hydrosulphide:

$$RX + NaSH \longrightarrow RSH + NaX$$

(R can be a primary, secondary, or tertiary alkyl group; X can be Cl, Br, I, $—OSO_3Na$ etc.).

(2) By alkaline hydrolysis of *S*-alkyl isothiuronium salts. These are prepared by reaction between thiourea and an alkyl halide or alkyl sulphate, and therefore this route is a convenient alternative to (1):

$$RX + (NH_2)_2C{=}S \longrightarrow [RSC({=}NH_2)NH_2]^{\oplus}X^{\ominus} \xrightarrow{NaOH}$$

$$RSH + NaX + H_2O + NH_2CN$$

The thiol is produced together with polymers of cyanamide (*G. G. Urquhart, J. W. Gates* and *R. Connor*, Org. Synth. Coll. Vol. III, 1955, p. 363).

(3) By addition of hydrogen sulphide to olefins. This reaction in its catalytic form is carried out under rather vigorous conditions and the products are often mixtures:

$$C_3H_6 + H_2S \xrightarrow[200\text{–}300°]{NiS} \underset{33\%}{CH_3\cdot CH_2\cdot CH_2\cdot SH} + \underset{67\%}{CH_3\cdot CH(SH)\cdot CH_3}$$

(*F. T. Barr* and *D. B. Keyes*, Ind. Eng. Chem., 1935, 26, 1111). However, the ultraviolet light-induced addition of hydrogen sulphide to olefins is a smooth reaction in which, presumably by a free-radical mechanism, the anti-Markownikoff product is formed. A dialkyl sulphide, produced by a secondary addition of the thiol to excess olefin, is a frequent by-product:

$$C_3H_6 + H_2S \xrightarrow[0°]{h\nu} \underset{65\%}{CH_3\cdot CH_2\cdot CH_2\cdot SH} + \underset{35\%}{(CH_3\cdot CH_2\cdot CH_2)_2S}$$

(*W. E. Vaughan* and *F. F. Rust*, J. org. Chem., 1942, 7, 472; *F. W. Stacey* and *J. F. Harris*, Org. Reactions, 1963, 13, 150).

Nitroalkanethiols are obtained by addition of hydrogen sulphide to the unsaturated bond in nitro-olefins (*F. D. Chattaway* and *J. M. Harrison*, J. chem. Soc., 1961, 109, 171).

(4) On a technical scale the "reductive thiolation" of aldehydes and ketones can be carried out using a catalyst, *e.g.* cobalt polysulphide:

$$R\cdot CO\cdot R' + H_2S + H_2 \xrightarrow{Cat.} R\cdot CH(SH)\cdot R' + H_2O$$

(*M. W. Farlow, W. A. Lazier*, and *F. K. Signaigo*, Ind. Eng. Chem., 1950, 42, 2547).

(5) The passage of a mixture of hydrogen sulphide and the vapour of an alcohol over a thoria catalyst yields a thiol:

$$C_2H_5OH + H_2S \xrightarrow[350\text{–}380°]{ThO_2} C_2H_5SH + H_2O$$

(*R. L. Kramer* and *E. E. Reid*, J. Amer. chem. Soc., 1921, 43, 880).

Properties and reactions. The thiols are colourless liquids which are only slightly soluble in water. Their boiling points (see Table 1) are lower than those of the corresponding alcohols, despite their higher molecular weights, because of diminished intermolecular attractions. Thiols are much more acidic than alcohols (EtSH, pKa ~ 12; EtOH, pKa ~ 17) and are soluble in aqueous alkalis.

The most noticeable feature of the lower-molecular-weight thiols is their intense and disagreeable smell; 1 part in 10^{10} of 1-butanethiol in air can be detected by smell. This property diminishes with increasing length of the hydrocarbon chain and 1-nonanethiol, n-$C_9H_{19}SH$, does not have an unpleasant smell.

(1) Thiols react with heavy metal oxides and salts to give stable and usually crystalline derivatives (*mercaptides*) which can be used to characterize the thiols:

$$2EtSH + HgO \longrightarrow (EtS)_2Hg + H_2O$$

$$EtSH + HgCl_2 \longrightarrow EtSHgCl + HCl$$

$$2MeSH + Pb(OAc)_2 \longrightarrow (MeS)_2Pb + 2AcOH$$

(2) Thiols are easily oxidised. A mild oxidant, *e.g.* iodine, converts a thiol into the corresponding *disulphide*:

$$2RSH + I_2 \longrightarrow RS{\cdot}SR + 2HI$$

This reaction is quantitative and provides a convenient technique for estimating thiols. (For a discussion of its mechanism see *D. S. Tarbell* in "Organic Sulphur Compounds", Vol. I, p. 97).

More vigorous oxidation, by nitric acid, hydrogen peroxide, or potassium permanganate, leads to a sulphonic acid:

$$MeSH + 3H_2O_2 \longrightarrow MeSO_3H + 3H_2O$$

(3) Thiols can be added to olefins to give *sulphides* (*q.v.*).

(4) The alkali-metal mercaptides are rapidly alkylated by alkyl halides to give *sulphides*:

$$C_2H_5SNa + CH_3I \longrightarrow C_2H_5{\cdot}S{\cdot}CH_3 + NaI$$

(5) Thiols are esterified by acids to give *alkyl thio-esters* (*acyl thiols*) but the equilibrium does not favour the ester and only low yields can be obtained:

$$CH_3{\cdot}COOH + C_2H_5{\cdot}SH \rightleftharpoons CH_3{\cdot}COSC_2H_5 + H_2O$$

The reaction of an acid halide with a thiol, however, gives good yields of the alkyl thioester (acyl thiol):

$$CH_3{\cdot}COCl + C_2H_5{\cdot}SH \longrightarrow CH_3{\cdot}COSC_2H_5 + HCl$$

(A review of the preparation and properties of acyl thiols is given by *T. C. Bruice, ibid.*, p. 421).

(6) Thiols give stable derivatives with carbonyl compounds: *mercaptals*, $RCH(SR')_2$, with aldehydes, *mercaptols*, $R_2C(SR')_2$, with ketones (see *E. Campaigne, ibid.*, p. 134).

TABLE 1

PHYSICAL PROPERTIES OF THIOLS, RSH

R	B.p. °C	d^{25}	n_D^{20}	Solubility in water at 25° C g./l.
Me	6	0·860	—	23·3
Et	35	0·832	1·431	6·8
n-Pr	68	0·836	1·435	2·0
iso-Pr	58	0·809	1·422	—
n-Bu	99	0·841	1·443	0·57
tert-Bu	64	0·800	1·423	—
n-C_9H_{19}	220	0·840	—	0·0012
n-$C_{12}H_{25}$	169/39 mm.	0·845	1·459	—

Methanethiol, *methyl mercaptan*, CH_3SH, b.p. 6°, is available commercially from petroleum cracking. It is used in the synthetic preparation of DL-methionine $CH_3SCH_2CH_2CH(NH_2)CO_2H$, which has therapeutic uses.

Ethanethiol, C_2H_5SH, b.p. 35°, is also available from petroleum cracking, and has been made on a technical scale from ethyl chloride and potassium hydrosulphide. It is an intermediate in the preparation of the hypnotic drug sulphonal $(CH_3)_2C(SO_2C_2H_5)_2$.

1-*Propanethiol*, b.p. 68°, is found in small amounts in onions, (*F. Challenger* and *D. Greenwood*, Biochem. J., 1949, **44**, 87).

1-*Dodecanethiol* (*dodecyl mercaptan*), *n*-$C_{12}H_{25}SH$, b.p. 169°/39 mm., is used as a chain-length regulator in the production of butadiene-styrene synthetic rubbers.

2-*Chloroethane*-1-*thiol*, $ClCH_2 \cdot CH_2SH$, b.p. 113–113·5°/760 mm., n_D^{20} 1·4929, is obtained together with 2,2′-dichlorodiethyl sulphide by the addition of hydrogen sulphide to vinyl chloride, or by hydrolysis of 2-*chloroethyl thiolacetate*, b.p. 176–178°, n_D^{23} 1·4950, from vinyl chloride and thiolacetic acid (*E. K. Ellingboe*, U.S.P., 2,423,641/1947).

2-*Nitroethanethiol*, $CH_2(NO_2) \cdot CH_2SH$, b.p. 86–87°/14 mm., from potassium hydrogen sulphide and 2-nitroethyl nitrate.

2-*Nitropropane*-1-*thiol*, $CH_3 \cdot CH(NO_2) \cdot CH_2SH$, b.p. 83°/12 mm., from nitropropene and hydrogen sulphide.

2. Sulphides or thio-ethers

(a) Dialkyl sulphides, alkylthioalkanes

The sulphides are the sulphur analogues of the ethers. They are usually named in a similar way to the ethers *e.g.* $CH_3 \cdot S \cdot C_2H_5$, ethyl methyl sulphide or ethyl methyl thioether. The I.U.P.A.C. system, however, is to name sulphides as substitution products of hydrocarbons and on this system this same sulphide would be methylthioethane.

A wide variety of sulphides occurs in crude petroleum oil including both open-chain and cyclic compounds. Dimethyl sulphide is present in some essential oils and is produced by certain algae and plants (*F. Challenger*, "Aspects of the Organic Chemistry of Sulphur", Butterworths, London, 1959, Chaps. I and II). The sulphide link is frequently found in biologically important compounds *e.g.* the vitamin biotin, and the amino-acids methionine, $CH_3 \cdot S \cdot CH_2 \cdot CH_2 \cdot CH(NH_2)CO_2H$, and lanthionine, $S[CH_2 \cdot CH(NH_2)CO_2H]_2$.

Preparation. (1) By alkylation of alkali-metal sulphides or mercaptides (these latter can yield mixed sulphides) with alkyl halides or sodium alkyl sulphates:

$$2MeBr + Na_2S \longrightarrow Me_2S + 2NaBr$$

$$C_2H_5 \cdot O \cdot SO_3Na + CH_3 \cdot SNa \longrightarrow C_2H_5 \cdot S \cdot CH_3 + Na_2SO_4$$

(2) By the addition of thiols to unsaturated compounds.

(a) Very pure reagents give the product expected from Markownikoff's rule (cf. Vol. IA, p. 301). The reaction is very slow unless the double-bond is activated by, *e.g.*, a carbonyl group:

$$CH_3CH{=}CH_2 + CH_3SH \xrightarrow[\text{absence of peroxides}]{\text{Dark}} CH_3 \cdot CH(SCH_3) \cdot CH_3$$

Nitro-olefins give nitroalkyl thioethers with alkanethiols (*R. L. Heath* and *A. Lambert*, J. chem. Soc., 1947, 1477).

(b) In the presence of peroxides or ultra-violet light, thiols add rapidly to olefins to yield the anti-Markownikoff product:

$$R \cdot CH{=}CH_2 + R'SH \xrightarrow{h\nu} R \cdot CH_2 \cdot CH_2 \cdot SR'$$

(see *F. R. Mayo* and *C. Walling*, Chem. Reviews, 1940, **27**, 351 and *Stacey* and *Harris*, *loc. cit.* for reviews of this reaction).

(3) Heating lead or mercury mercaptides yields sulphides:

$$(RS)_2Hg \longrightarrow R_2S + HgS$$

(4) Sulphur chloride and sulphur dichloride react with olefins to give β,β'-dichloroalkyl sulphides:

$$2C_2H_4 + S_2Cl_2 \longrightarrow (ClCH_2\cdot CH_2)_2S + S$$

(5) Sulphenyl halides (*q.v.*) react with unsaturated compounds to give sulphides:

$$ClCH_2\cdot CH_2\cdot SCl + CH_3\cdot CH{=}CH_2 \longrightarrow ClCH_2\cdot CH_2\cdot S\cdot CH_2\cdot CHCl\cdot CH_3$$

(*R. C. Fuson, C. C. Price* and *D. M. Burness*, J. org. Chem., 1946, **11**, 475).

Properties and reactions. The sulphides are colourless liquids, insoluble in water. If they are carefully purified their smells are not unpleasant. They are, like the ethers, very weak bases and will dissolve in pure sulphuric acid to give solutions which presumably contain sulphonium salts, *e.g.* $[(C_2H_5)_2SH]^{\oplus}HSO_4^{\ominus}$, and which regenerate the sulphide on dilution.

Although sulphides are formally analogous to ethers they are much more reactive. The unshared pairs of electrons on the sulphur atom in a sulphide are readily donated and many types of addition compounds are formed by sulphides.

(1) Sulphides form stable addition compounds with halogens, *e.g.* $(CH_3)_2SBr_2$, and with many metal salts, *e.g.* $(C_2H_5)_2S\cdot 2HgCl_2$.

(2) Sulphides form crystalline sulphonium salts (p. 85) with alkyl halides.

(3) Oxidation of sulphides gives first sulphoxides (p. 86) and then sulphones (p. 88). (See *D. Barnard, L. Bateman* and *J. I. Cunneen*, in "Organic Sulphur Compounds", Vol. I, p. 229).

(4) Sulphides are cleaved by strong acids, though more slowly than ethers. Simple alkyl sulphides can be cleaved quantitatively by cyanogen bromide (von Braun method):

$$(CH_3)_2S + CNBr \longrightarrow CH_3\cdot SCN + CH_3Br$$

Unsymmetrical sulphides usually give mixtures of products, but most methyl sulphides give methyl bromide (*D. S. Tarbell* and *D. P. Harnish*, Chem. Reviews, 1951, 49, 1).

(5) Sulphides are "desulphurized" by Raney nickel:

$$RSR' + H_2(Ni) \longrightarrow RH + R'H + NiS$$

This reaction has proved to be useful both as a method for determining the structures of sulphides and in general organic synthesis (see *Challenger*, *loc. cit.*, pp. 90–111; *H. Hauptmann* and *W. F. Walter*, Chem. Reviews, 1962, **62**, 347; *G. R. Pettit* and *E. E. van Tamelen*, Org. Reactions, 1962, **12**, 356, for a full discussion).

Dimethyl sulphide, *methylthiomethane*, $CH_3 \cdot S \cdot CH_3$, b.p. 37·5°; *diethyl sulphide*, b.p. 92°; *di-n-propyl sulphide*, b.p. 143°; *di-n-butyl sulphide*, b.p. 182°; *dicetyl sulphide*, $(n\text{-}C_{16}H_{33})_2S$, m.p. 57–58°. (For the physical constants of many of the lower alkyl sulphides see *D. T. Mc.Allan et al.*, J. Amer. chem. Soc., 1951, **73** 3627).

α,α′-Dichlorodiethyl sulphide, $(CH_3 \cdot CHCl)_2S$, b.p. 52°/12 mm., is obtained by the action of hydrogen chloride and hydrogen sulphide on paraldehyde at 0°:

$$2CH_3CHO + 2HCl + H_2S \longrightarrow CH_3CHClSCHClCH_3$$

bromination and subsequent dehydrohalogenation with diethylaniline affords 2-bromovinylthioethene, $CHBr{=}CHSCH{=}CH_2$ (*L. Brandsma* and *J. F. Arens*, Rec. trav. chim., 1961, **80**, 237).

β,β′-Dichlorodiethyl sulphide, $(ClCH_2 \cdot CH_2)_2S$, m.p. 13–13·5°, b.p. 215–217° (dec.), is a vesicant compound with a toxic vapour, which was used as a poison gas (mustard gas) during the war of 1914–18. It was prepared by the addition of sulphur chloride to ethylene (see *R. C. Fuson et al.*, J. org. Chem., 1946, **11**, 469–517) or by the action of hydrogen chloride on the alcohol prepared from sodium sulphide and ethylene chlorohydrin:

$$Na_2S + 2ClCH_2 \cdot CH_2OH \longrightarrow S(CH_2 \cdot CH_2OH)_2 \xrightarrow{2HCl} S(CH_2 \cdot CH_2Cl)_2$$

The physiological action of mustard gas has been the subject of much investigation (see, for example, Biochem. Soc. Symposia, 1948, No. 1, 2, 8, 16; *R. M. Herriott*, J. gen. Physiol., 1948, **32**, 221) Substitution reactions of mustard gas and its analogues have been thoroughly investigated. Solvolytic reactions are much more rapid than for the oxygen analogues and involve as a likely intermediate an ethylenesulphonium ion which reacts rapidly with any nucleophilic species present in solution (*P. D. Bartlett* and *C. G. Swain*, J. Amer. chem. Soc., 1949, **71**, 1406).

Heating with alcoholic potash gives 2-chloroethyl vinyl sulphide and divinyl sulphide.

2-Chloroethyl methyl sulphide, b.p. 55–56°/30 mm, from 2-mercaptoethanol and thionyl chloride (Org. Synth., Coll. Vol. II, 136).

Methyl 2-nitroethyl sulphide, b.p. 105°/20 mm.; *methyl 2-nitropropyl sulphide*, b.p. 98–100°/16 mm., from methanethiol and the appropriate nitroalkene.

(b) Olefinic sulphides, dialkenyl sulphides or thioethers, alkenylthioalkenes

These compounds, in modes of preparation and in general properties, are similar to the dialkyl sulphides (see *L. Brandsma* and *J. F. Arens*, Rec. Trav. chim., 1962, **81**, 33).

Divinyl sulphide, *divinyl thioether*, *ethenylthioethene* $(CH_2{=}CH)_2S$, b.p. 84–86°, has been prepared by dehydrohalogenation of αα′-dichloro- and β,β′-dibromodiethyl sulphide (*Brandsma* and *Arens, loc. cit.*). It was claimed that this unsaturated sulphide was a main ingredient of the oil of the leek, *allium ursinum* L. (*F. W. Semmler,* Ann., 1887, **241,** 92) but the chemical properties of pure vinyl sulphide are at variance with those reported for the compound extracted from leek oil (see *W. L. Ruigh* and *A. E. Erickson,* J. Amer. chem. Soc., 1939, **61,** 915).

Di(propen-1-yl) sulphide, $(CH_3 \cdot CH{=}CH)_2S$, b.p. 38°/10 mm.; *diallyl sulphide* [*di(propen-2-yl) sulphide*], $(CH_2{=}CH \cdot CH_2)_2S$, b.p. 139°, has been isolated from various plants of the mustard family (*Cruciferae*) and has been synthesised from allyl iodide and potassium sulphide.

(c) *Acetylenic sulphides, ethynyl thioethers or alkylthioalkynes*

The preparation and reactions of these interesting compounds, in which an acetylenic link is adjacent to the sulphur atom: $R \cdot C{\vdots}C \cdot SR'$, have been studied only recently. (For a full review see *J. F. Arens* in "Advances in Organic Chemistry, Methods and Results", Vol. II, Ed. *R. A. Raphael, E. C. Taylor* and *H. Wynberg,* Interscience Publishers Inc., New York, 1960, p. 117).

They are generally made from β-halogeno-α,β-unsaturated sulphides which, in turn, can be prepared in a number of ways, *e.g.* from vinyl sulphides. A typical route to an acetylenic sulphide is as follows:

$$R \cdot C{\vdots}CH + R'SH \longrightarrow R \cdot CH{:}CH \cdot SR' \xrightarrow{Br_2} R \cdot CHBr \cdot CHBr \cdot SR'$$

$$\xrightarrow{KOH} R \cdot CBr{:}CH \cdot SR' \xrightarrow{KOH} R \cdot C{\vdots}C \cdot SR'$$

A direct nucleophilic substitution of a thiolate ion on a halogenoalkyne in dimethylformamide also gives an acetylenic sulphide (*S. I. Miller et al.,* J. Amer. chem. Soc., 1962, **84,** 2020):

$$RC{\vdots}CBr + NaSR' \xrightarrow{(DMF)} RC{\vdots}CSR' + NaBr$$

The compounds are usually colourless liquids with characteristic smells. The ethynyl compounds (R=H) are not very stable, though they can be kept for several days at 0°. Higher homologues ($R{=}CH_3$, C_2H_5) are stable at room temperature in the absence of air.

The acetylenic sulphides show an interesting duality in addition reactions to the triple bond. Electrophilic reagents invoke electron release from the sulphur atom and induce a polarization of the type $R \cdot C{\equiv}C{-}\ddot{S}R'$. Thus addition of hydrogen chloride to (ethylthio)ethyne proceeds as follows:

$$HC{\vdots}C \cdot SC_2H_5 + HCl \longrightarrow H_2C{:}CCl \cdot SC_2H_5$$

On the other hand nucleophilic reagents induce the reverse polarization $R{\cdot}C{\equiv}\overset{\curvearrowright}{C}{-}SR'$ and amines and alcohols add to (ethylthio)ethyne in the following way:

$$HC{:}C{\cdot}SC_2H_5 + RNH_2 \longrightarrow RNH{\cdot}CH{:}CH{\cdot}SC_2H_5$$

$$HC{:}C{\cdot}SC_2H_5 + ROH \xrightarrow{RO^{\ominus}} ROCH{:}CH{\cdot}SC_2H_5$$

Free radical additions to the triple bond have also been effected.

A number of *ethynylene bis-thioethers*, $RS{\cdot}C{:}C{\cdot}SR$, have also been prepared, *e.g.*

$$(CH_3)_3\,C{\cdot}SNa + CCl_2{:}CHCl \longrightarrow (CH_3)_3\,CS{\cdot}C{:}C{\cdot}SC(CH_3)_3$$

Their reactions have not yet been very fully explored.

Ethynyl methyl sulphide, *methylthioethyne,* $HC{:}C{\cdot}SCH_3$, b.p. 71°; *ethyl ethynyl sulphide,* b.p. 92°; n-*butyl ethynyl sulphide,* b.p. 83°/93 mm.; *methyl propynyl sulphide,* $CH_3C{:}C{\cdot}SCH_3$, b.p. 114–116°; *ethyl propynyl sulphide,* b.p. 136°; *vinylthioethyne,* $CH_2{:}C{\cdot}S{\cdot}C{:}CH$, b.p. 84°; *bis-(ethylthio)ethyne,* $C_2H_5S{\cdot}C{:}C{\cdot}SC_2H_5$, b.p. 97°/12 mm.; *bis-*(tert-*butylthio)ethyne,* m.p. 40°.

3. Derivatives of alkanesulphenic acids

No free alkanesulphenic acid, $R{\cdot}S{\cdot}OH$, has yet been prepared but the corresponding sulphenyl chlorides and bromides, $R{\cdot}SCl$ and $R{\cdot}SBr$, are well known and from them many sulphenamides and sulphenic esters have been made (see *I. B. Douglass* in "Organic Sulphur Compounds", Vol. I, p. 350, for a full review of these compounds).

The *sulphenyl halides* can be prepared by the cleavage of disulphides with halogens:

$$MeS{\cdot}SMe + Cl_2 \longrightarrow 2MeSCl$$

They react with amines and with metal alkoxides to form corresponding *amides* and *esters*. Hydrolysis of sulphenyl chlorides leads, presumably via the unstable sulphenic acid, to sulphenic *anhydrides*, $(RS)_2O$. Sulphenyl halides add to unsaturated compounds yielding sulphides (*Stacey* and *Harris*, *loc. cit.*):

$$ClCH_2CH_2SCl + CH_3CH{=}CH_2 \longrightarrow ClCH_2CH_2{\cdot}S{\cdot}CH_2{\cdot}CHClCH_3$$

$$HC{\equiv}CH + RSCl \longrightarrow RSCH{=}CHCl \xrightarrow{RSCl} RS{\cdot}CHCl{\cdot}CHCl{\cdot}SR$$

Trichloromethanesulphenyl chloride, $CCl_3{\cdot}SCl$, b.p. 147–149°, is easily prepared by the chlorination of carbon disulphide in the presence of iodine as a catalyst:

$$2CS_2 + 5Cl_2 \longrightarrow 2CCl_3SCl + S_2Cl_2$$

Some of its derivatives have been suggested as additives for lubricants and fuels (see *F. A. Drahowzal* in "Organic Sulphur Compounds", Vol. I, p. 361).

4. Dialkyl disulphides or alkyldithioalkanes

These compounds are usually formulated RS·SR. The alternative formulation, $R_2S{=}S$, is incorrect for the simple alkyl disulphides (*F. Challenger* and *D. Greenwood*, J. chem. Soc., 1950, 26).

The disulphide link is a common one in natural products. The amino-acid cystine, $[S{\cdot}CH_2{\cdot}CH(NH_2)CO_2H]_2$, provides a cross link in many polypeptides and proteins (*e.g.* insulin). Simpler disulphides, *e.g.* allyl disulphide, have also been isolated from natural sources.

The common name for these compounds is disulphide, but the name preferred by I.U.P.A.C. is the substitutive name with the —S·S— linkage termed "dithio". On this system $CH_3S{\cdot}SC_2H_5$ (ethyl methyl disulphide) is called methyldithioethane.

The disulphides often occur as by-products in the synthesis of sulphides, but the most important method of preparing them is by mild oxidation of thiols *e.g.* by iodine:

$$2RSH + I_2 \longrightarrow RS{\cdot}SR + 2HI \quad \text{(see p. 76)}$$

The disulphides can easily be reduced back to thiols with hydrogen. The oxidation-reduction system of disulphide-thiol is of great biological significance.

Disulphides react with thiols in an equilibrium of the following type:

$$RS{\cdot}SR + R'SH \rightleftharpoons RS{\cdot}SR' + RSH$$

and mixed disulphides can be prepared by distilling out the more volatile thiol (*D. T. Mc. Allan et al.*, J. Amer. chem. Soc., 1951, 73, 3627).

Cleavage of disulphides with halogens yields sulphenyl halides; oxidation of disulphides leads to thio-esters of alkanesulphonic acids (alkanethiosulphonates) (p. 92) and eventually to alkanesulphonic acids:

$$RS{\cdot}SR + Br_2 \longrightarrow 2RSBr$$

$$RS{\cdot}SR \xrightarrow{HNO_3} RSO_2{\cdot}SR \longrightarrow 2\ RSO_3H$$

Dimethyl disulphide, *methyldithiomethane,* $CH_3S{\cdot}SCH_3$, b.p. 112°, and *diethyl disulphide,* b.p. 153°, are colourless liquids with strong, garlic-like, smells. (For the physical constants of many of the lower dialkyl disulphides see *McAllan et al., loc. cit.*). 2,2'-*Dichlorodiethyl disulphide,* b.p. 155°/30 mm. (*J. Bennett,* J. chem. Soc., 1921, **119**, 418).

Di-allyl disulphide, b.p. 117°/16 mm., is obtained from garlic, *Allium sativum* (*F. W. Semmler*, Arch. Pharm., 1892, **230**, 434) but is probably formed from other precursors during the extraction (*A. Stoll* and *E. Seebeck*, Adv. in Enzymology, 1951, 11, 377).

5. Alkyl polysulphides

Compounds of the general formula $R\cdot(S)_n\cdot R$, when n is 3, 4, or 5 have been prepared. They are generally assumed to contain linear chains of sulphur atoms and the methods by which the compounds are prepared, and physical measurements on them, support this (see *O. Foss* in "Organic Sulphur Compounds", Vol. I, p. 75).

Dialkyl trisulphides have been prepared by the interaction of thiols and sulphur dichloride:

$$2RSH + SCl_2 \longrightarrow RS\cdot S\cdot SR + 2HCl$$

tetra- and *penta-sulphides* by the reaction between alkyl halides and alkali-metal polysulphides:

$$2EtI + K_2S_5 \longrightarrow Et\cdot S_5\cdot Et + 2KI$$

(*R. W. Riding* and *J. S. Thomas*, J. chem. Soc., 1923, 3271; 1924, 2214).

The preparation of a number of *alkyl hydropolysulphides*, $R\cdot(S)_n\cdot H$ has been described (*H. Böhme* and *G. Zinner*, Ann., 1954, **585**, 142).

Thiokols. This interesting and commercially important class of polymeric polysulphides is prepared by reactions between alkyl dihalides and sodium polysulphides. The simplest example, prepared from sodium tetrasulphide and ethylene dichloride, can be given the formula $\text{—}(\text{–}CH_2\cdot CH_2\cdot S_4\text{–})_n\text{–}$. Polymers of this type are rubbery and have good resistance to oils and organic solvents. They have found a wide range of industrial uses (see *E. M. Fettes* and *J. S. Jorczak*, Ind. Eng. Chem., 1950, **42**, 2217; *Fettes* in "Organic Sulphur Compounds" Vol. I, p. 266). They will be discussed further in the subsequent chapter on synthetic macromolecules.

6. Alkylthiosulphuric acids

The free acids of this series, $RS\cdot SO_3H$, are not known but their salts (usually known as *Bunte salts*) can be prepared by reaction between:

(1) an alkyl halide and a metal thiosulphate:

$$RBr + Na_2S_2O_3 \longrightarrow RS\cdot SO_3Na + NaBr$$

(*H. E. Westlake Jr.* and *G. Dougherty*, J. Amer. chem. Soc., 1941, 63, 658); sodium thiosulphate is usually used but it may be advantageous to use thallous thiosulphate with an alkyl chloride since thallous chloride is sparingly soluble in water and may be separated from the soluble Bunte salt (*H. Z. Lecher* and *E. M. Hardy*, J. org. Chem., 1955, **20**, 475);

(2) a thiolate, sodium sulphite, and iodine:

$$C_2H_5SNa + Na_2SO_3 + I_2 \longrightarrow C_2H_5S\cdot SO_3Na + 2NaI$$

Bunte salts are usually crystalline and fairly water-soluble. On treatment with acids they break down and give thiols. They give disulphides when they are heated alone or oxidised in solution. Bunte salts are reviewed by *B. Milligan* and *J. M. Swan* (Reviews, pure and appl. Chem., [Australia], 1962, **12**, 72).

7. Sulphonium compounds

The lone pairs of electrons on the sulphur atom of the sulphides make these compounds nucleophilic, and dialkyl sulphides react rapidly, in general, with alkyl halides to produce the salt-like sulphonium compounds:

$$R\text{—}\ddot{\underset{..}{S}}\text{—}R' + R''Cl \longrightarrow \left[\begin{array}{c} R\text{—}\ddot{S}\text{—}R' \\ | \\ R'' \end{array}\right]^{\oplus} Cl^{\ominus}$$

The structure of the sulphonium compounds follows from the following observations. The compounds are often water-soluble and are strong electrolytes in solution. Their aqueous solutions show the usual reactions of the halide anion in solution. By whatever route the ion $RR'R''S^{\oplus}$ is prepared the same cation results; this demonstrates the equivalence of all three bonds to sulphur:

$$\begin{array}{lcl} Me\cdot S\cdot Et + n\text{-}PrBr & \longrightarrow & MeEt\,n\text{-}Pr\overset{\oplus}{S}\overset{\ominus}{Br} \\ & & \uparrow \\ & & Me\cdot S\cdot n\text{-}Pr + EtBr \end{array}$$

When all three groups attached to sulphur are different, *e.g.* in carboxymethyl-ethyl-methylsulphonium bromide, the salt can be resolved into two optical isomers (*W. J. Pope* and *S. J. Peachey*, J. chem. Soc., 1900, **77**, 1072), thus showing the asymmetry of the cation; the sulphonium ion is tetrahedral with sulphur at the apex of the tetrahedron and the three attached groups at the corners of the base.

The *sulphonium halides*, prepared from alkyl halides and alkyl sulphides, can be converted into the parent strong base, the *sulphonium hydroxide*, by the action of moist silver oxide. Sulphonium hydroxides are usually hygroscopic solids, strongly basic, (they absorb carbon dioxide from the air), which can be neutralised by acids to form sulphonium salts:

$$\overset{\oplus\ominus}{Et_3SBr} \xrightarrow[H_2O]{Ag_2O} \overset{\oplus\ \ominus}{Et_3SOH} \xrightarrow{HClO_4} \overset{\oplus\ \ominus}{Et_3SClO_4} + H_2O$$

Sulphonium compounds are thermally unstable. The halides revert to alkyl halide and sulphide on heating, but the hydroxides undergo an elimination reaction and give an olefin:

$$[CH_3CH_2SR_2]^{\oplus}\overset{\ominus}{O}H \longrightarrow CH_2{=}CH_2 + R_2S + H_2O$$

(see *C. K. Ingold*, "Structure and Mechanism in Organic Chemistry", Cornell University Press, Ithaca, N.Y., 1953, Ch. VIII, for a discussion of this reaction and its mechanism.)

8. Dialkyl sulphoxides or alkylsulphinylalkanes

The sulphoxides, (alkylsulphinylalkanes in I.U.P.A.C. nomenclature), R·SO·R′, result from mild oxidation of the sulphides. The best reagent for this purpose is hydrogen peroxide used in equivalent amount but many other oxidants (per-acids, nitric acid etc.) have been used. (For a review of the preparation and properties of sulphoxides see *H. Szmant* in "Organic Sulphur Compounds", Vol. I, Pergamon Press, Oxford, 1961, p. 154).

Unsymmetrical sulphoxides, R·SO·R′, can be resolved (*P. W. B. Harrison, J. Kenyon* and *H. Phillips*, J. chem. Soc., 1926, 2079) showing that the sulphoxide molecule is not planar but pyramidal. The mode of synthesis of the sulphoxides, and the fact that they can be reduced to sulphides by zinc dust, confirms their structure. The exact nature of the sulphur to oxygen bond in sulphoxides is not certain, but it can be described in resonance terms as a hybrid of two contributing structures:

$$R{-}\underset{\cdot\cdot}{\overset{\overset{O}{\|}}{S}}{-}R \longleftrightarrow R{-}\underset{\cdot\cdot}{\overset{\overset{O^{\ominus}}{|}}{S^{\oplus}}}{-}R$$

No simple sulphoxide has been found in nature but the sulphoxide group is present in a number of naturally occurring compounds including alliin $CH_2{:}CH{\cdot}CH_2{\cdot}SO{\cdot}CH_2{\cdot}CH(NH_2)CO_2H$, obtained from garlic (*A. Stoll* and *E. Seebeck*, Helv., 1948, 31, 189), and the optically active sulphoraphen, $CH_3{\cdot}SO{\cdot}CH{:}CH{\cdot}CH_2NCS$ (*P. Karrer, ibid.*, 1948, 31, 1017).

The aliphatic sulphoxides are low-melting crystalline solids, somewhat soluble in water. They can be distilled at reduced pressures but decompose on being heated at normal pressure. They are reactive compounds, easily oxidised to sulphones, and reducible to sulphides. The sulphoxides are weakly basic and form salts with strong acids; they also form complexes with ferric chloride, chloroplatinic acid, and many other metallic salts.

The sulphinyl group, —SO—, has a formal resemblance to the carbonyl group, —CO—, but it does not exert a comparable activating effect on adjacent methylene groups. Sulphoxides do not undergo condensation reactions at the α-carbon atom as readily as do ketones, but the presence of two adjacent sulphoxide groups does activate a methylene group very considerably.

Dimethyl sulphoxide, *methylsulphinylmethane,* $CH_3 \cdot SO \cdot CH_3$, m.p. 8°, has recently found use as a solvent for polar compounds and as a specific oxidant in *e.g.* the preparation of aldehydes from alkyl halides (*N. Kornblum et al.,* J. Amer. chem. Soc., 1957, **79,** 6562; 1959, **81,** 4113):

$$R \cdot CH_2Br + Me_2SO \longrightarrow R \cdot CHO + Me_2S + HBr$$

It is available commercially and is the subject of a recent review (*W. O. Ranky* and *D. C. Nelson* in "Organic Sulphur Compounds, Vol. I, Pergamon Press, Oxford, 1961, p. 170).

Diethyl sulphoxide, m.p. 4–6°; *diisoamyl sulphoxide,* m.p. 35°; *divinyl sulphoxide* (*vinylsulphinylethene*) b.p. 70°/13 mm.; *diallyl sulphoxide* m.p. 23–23·5°.

2,2′-*Dichlorodiethyl sulphoxide,* m.p. 109·5°.

9. Alkanesulphinic acids

Alkanesulphinic acids, RSO_2H, can be regarded as derivatives of sulphurous acid in which a hydroxyl group has been replaced by an alkyl radical. In formula they are analogous to carboxylic acids; they are stronger acids than these (though considerably weaker than sulphonic acids).

Sulphinic acids can be prepared by the following reactions.

(1) Reduction of alkanesulphonyl chlorides with zinc dust or alkaline sulphite solutions:

$$2RSO_2Cl + 2Zn \longrightarrow (RSO_2)_2Zn + ZnCl_2$$

(2) Reaction between Grignard reagents and sulphur dioxide or sulphuryl chloride:

$$RMgBr + SO_2 \longrightarrow RSO_2MgBr \xrightarrow{H_2O} RSO_2H$$

$$2RMgBr + SO_2Cl_2 \longrightarrow RSO_2MgBr + RCl + MgBrCl$$

(3) Alkaline cleavage of sulphones (*q.v.*). (For a review of methods of preparation see *W. E. Truce* and *A. M. Murphy,* Chem. Reviews, 1951, 48, 69).

Alkanesulphinic acids are viscous oils which are thermally unstable. On being heated they disproportionate to sulphonic acids and thiosulphonic esters:

$$3RSO_2H \longrightarrow RSO_3H + R \cdot SO_2 \cdot SR + H_2O$$

Salts of sulphinic acids are much more stable than the free acids. The acids are easily oxidised to sulphonic acids and with halogens form sulphonyl halides:

$$RSO_2H + Br_2 \longrightarrow RSO_2Br + HBr$$

Salts of sulphinic acids can be alkylated with alkyl halides to give sulphones:

$$RSO_2Na + R'X \longrightarrow RSO_2R' + NaX$$

The free sulphinic acids behave in many ways like substituted hydrogen sulphites (compare $R{\cdot}SO_2H$ and $NaO{\cdot}SO_2H$) and will add to activated double bonds, for example in $\alpha\beta$-unsaturated ketones, to give sulphones:

$$RSO_2H + CH_3{\cdot}CH{=}CH{\cdot}CO{\cdot}CH_3 \longrightarrow CH_3{\cdot}CH(SO_2R){\cdot}CH_2{\cdot}CO{\cdot}CH_3$$

They also give unstable derivatives with aldehydes:

$$RSO_2H + R'CHO \longrightarrow RSO_2CH(OH)R'$$

Thionyl chloride converts sulphinic acids into *sulphinyl chlorides* which in turn can be converted into *sulphinamides* and *sulphinic esters*. Unsymmetrical sulphinic esters, $R{\cdot}SO{\cdot}OR'$, have been resolved, showing the non-planarity of this type of molecule (*H. Phillips*, J. chem. Soc., 1925, **127**, 2552).

Because of their low stabilities few alkanesulphinic acids have been prepared in a pure state. They are usually made only as intermediates, or stored as salts and not isolated. However 1-*dodecanesulphinic acid* has been prepared as a pure crystalline solid, m.p. 29–30° (*C. S. Marvel* and *R. S. Johnson*, J. org. Chem., 1948, **13**, 822).

Sodium hydroxymethanesulphinate, $HO{\cdot}CH_2SO_2Na$, commonly known as "sodium formaldehyde sulphoxylate", is an industrially important reducing agent for vat dyeing. It is prepared by the reduction, either with hydrogen, catalytically, or with zinc dust and acid, of the addition product of formaldehyde and sodium hydrogen sulphite:

$$HCHO + NaHSO_3 \longrightarrow HO{\cdot}CH_2SO_3Na \longrightarrow HO{\cdot}CH_2SO_2Na$$

10. Sulphones

Further oxidation of sulphoxides or vigorous oxidation of sulphides yields sulphones, (alkylsulphonylalkanes), $R{\cdot}SO_2{\cdot}R'$. The sulphones do not have planar molecules; the bonds are arranged in an approximately tetrahedral way around sulphur.

In addition to oxidative methods of preparation sulphones are also prepared by alkylating salts of sulphinic acids:

$$RSO_2Na + R'X \longrightarrow RSO_2R' + NaX$$

(*P. Allen et al.*, J. org. Chem., 1951, 16, 767).

Sulphones are stable compounds. They resist further oxidation and are reduced with difficulty, although lithium aluminium hydride does reduce them to sulphides (*F. G. Bordwell* and *W. H. McKellin*, J. Amer. chem. Soc., 1951, 73, 2251). They are stable to acids but are cleaved by vigorous alkaline treatment to give salts of sulphinic acids:

$$(C_2H_5)_2SO_2 + KOH \longrightarrow C_2H_5SO_2K + CH_2{=}CH_2 + H_2O$$

(*G. W. Fenton* and *C. K. Ingold*, J. chem. Soc., 1928, 3127; 1929, 2338).

The sulphonyl group, —SO_2—, activates adjacent methylene groups, though rather weakly, and in most of the condensation reactions which sulphones have been reported as undergoing, other activating groups were also present (see *J. Strating* in "Organic Sulphur Compounds", Vol. I, p. 146).

The presence of a sulphonyl group adjacent to a double bond greatly enhances the reactivity of the latter. Thus vinyl sulphones have become important technically as reactive linking groups in dyestuffs and textile finishing agents (see *Tessoro et al.*, Textile Res. J., 1961, 31, 283; *J. Heyna*, Angew. Chem. internat., 1963, **2**, 20; *G. Kränzlein, Heyna* and *W. Schumacher*, G.P., 842,198/1942). The reactive vinyl sulphonyl group, which is part of the dyestuff or finishing agent reacts either with the reactive groups in the textile fibre (*e.g.* the hydroxyl groups in cotton or the amino or thiol groups in wool) to form a substituted cellulose ethyl ether, etc., or with itself to form a polymeric coating which is difficult to remove by washing. The simple vinyl sulphones are obtained, *e.g.*, by dehydrohalogenation of β-halogenated sulphones prepared by oxidation of the appropriate sulphides (*J. R. Alexander* and *H. McCombie*, J. chem. Soc., 1931, 1913).

Dimethyl sulphone, *methylsulphonylmethane*, $CH_3\cdot SO_2\cdot CH_3$, m.p. 109°; *diethyl sulphone* m.p. 73°; *di-n-butyl sulphone*, m.p. 43·5°; *di-n-hexadecyl sulphone*, m.p. 103°; *methyl ethyl sulphone*, m.p. 36°; *ethyl n-propyl sulphone*, m.p. 25°, b.p. 142–143°/23 mm.

Simple alkyl sulphones have found few applications. Two disulphones, sulphonal, $Me_2C(SO_2Et)_2$, and trional, $MeEtC(SO_2Et)_2$, have been used as hypnotics.

Divinyl sulphone, *ethenylsulphonylethene*, b.p. 90–92°/8 mm, (*Alexander* and *McCombie, loc. cit.*; *Overberger et al.*, J. org. Chem., 1954, **19**, 1486). *Dichlorodiethyl sulphone*, m.p. 56°; *2-nitro-1-methylsulphonylethane*, m.p. 78°. *2-Nitro-1-methylsulphonylpropane*, m.p. 69–70° (*J. D. Rose et al.*, J. chem. Soc., 1947, 1514).

11. Alkanesulphonic acids and derivatives

Alkanesulphonic acids, of general formula $R{\cdot}SO_3H$, may be regarded as derivatives of sulphuric acid in which an alkyl radical has replaced a hydroxyl group. Their formation by oxidation of thiols indicates that they contain a carbon-sulphur bond.

Preparation. Simple alkanesulphonic acids are not found in nature. They can be prepared in a number of ways.

(1) Direct introduction of a sulphonic acid group into an alkane is possible by reaction with sulphuric acid, oleum, chlorosulphonic acid, or fluorosulphonic acid. The yields are generally low, oxidation products are also formed, and the method is not a useful one (see *E. E. Gilbert*, Chem. Reviews, 1962, **62**, 549).

(2) Thiols and disulphides are oxidised by potassium permanganate, hydrogen peroxide, sodium hypobromite etc., to give excellent yields of sulphonic acids:

$$EtSH + 3H_2O_2 \longrightarrow EtSO_3H + 3H_2O$$

$$MeS{\cdot}SMe \xrightarrow{KMnO_4} 2MeSO_3H$$

(3) Alkylation of alkali-metal sulphites with alkyl halides gives salts of sulphonic acids:

$$RX + Na_2SO_3 \longrightarrow RSO_3Na + NaX$$

(4) Addition of alkali metal hydrogen sulphites (bisulphites) to unsaturated compounds occurs at moderate temperatures and pressures to give excellent yields of sulphonic acid salts:

$$R{\cdot}CH{=}CH_2 + NaHSO_3 \longrightarrow R{\cdot}CH_2{\cdot}CH_2SO_3Na$$

Nitroalkane sulphonic acids can be made from nitro-olefins in this way (*R. L. Heath* and *H. A. Piggott*, J. chem. Soc., 1947, 1481).

The addition is catalysed by peroxides and gives the anti-Markownikoff product; it is presumably free-radical in type (*D. Harman*, U.S.P., 2,504,411; B.P., 661,771/1950; *Stacey* and *Harris, loc. cit.*).

(5) The reaction between alkanes and sulphuryl chloride, or a mixture of chlorine and sulphur dioxide (Reed reaction), which is catalysed by peroxides or ultra-violet light, gives alkanesulphonyl chlorides (Vol. IA, p. 386):

$$R{\cdot}H + SO_2 + Cl_2 \longrightarrow R{\cdot}SO_2Cl + HCl$$

The chlorides can be hydrolysed to the free acids.

Properties and reactions: The alkanesulphonic acids are hygroscopic liquids or low-melting solids. They are strong acids and are soluble in water and in a variety of organic solvents. Their solubility in organic solvents, plus the fact that they are not oxidising acids, makes them useful as catalysts for esterification, hydrolysis, alkylation etc., and these properties have been applied on an industrial scale (see *W. A. Proell, C. E. Adams* and *B. H. Shoemaker*, Ind. Eng. Chem., 1948, **40**, 1129).

Alkanesulphonic acids form well-defined salts with inorganic and organic bases; those with phenylhydrazine have been used as derivatives for characterisation of the acids. Phosphorus pentachloride and pentabromide convert the acids and their salts into the corresponding *sulphonyl halides* and these react with ammonia and amines to give *sulphonamides*:

$$RSO_2Cl + R'_2NH \longrightarrow R\cdot SO_2NR'_2 + HCl$$

and with alkoxides to give *esters*:

$$RSO_2Cl + EtONa \longrightarrow RSO_3Et + NaCl$$

Primary alkanesulphonic acids are cleaved by heating them with alkali under pressure and yield *alcohols*:

$$RSO_3Na + NaOH \longrightarrow ROH + Na_2SO_3$$

Under similar conditions secondary alkane sulphonates give *olefins*.

Methanesulphonic acid, CH_3SO_3H, m.p. 20°, b.p. 167°/10 mm., [*phenylhydrazine salt,* m.p. 193–194° (dec.)], was originally prepared by the reduction with sodium amalgam of trichloromethanesulphonic acid produced, via the chloride, from the aqueous chlorination of carbon disulphide (*H. Kolbe,* Ann., 1845, **54**, 145):

$$CS_2 \longrightarrow CCl_3SO_2Cl \longrightarrow CCl_3SO_3H \longrightarrow CH_3SO_3H$$

Methanesulphonyl chloride, b.p. 160°; *amide,* m.p. 88°; *ethyl* ester, b.p. 86°/10 mm.; *anhydride,* $(CH_3SO_2)_2O$, m.p. 71°, b.p. 138°/10 mm. (*O. C. Billeter,* Ber., 1905, **38**, 2018). Sodium iodomethanesulphonate (*Abrodil*) is used in X-ray therapy.

Chloroiodomethanesulphonic acid, $CHClISO_3H$, is of historical interest as the first compound containing only one carbon atom to be resolved into optical isomers (*W. J. Pope* and *J. Read,* J. chem. Soc., 1914, **105**, 811; *Read* and *A. M. McMath, ibid.,* 1932, 2723).

Ethanesulphonyl chloride, b.p. 171°; *amide,* m.p. 60°; *methyl* ester, b.p. 198°; *ethyl* ester, b.p. 213°.

Taurine, $NH_2\cdot CH_2\cdot CH_2SO_3H$, m.p. 328–9° (dec.), a simple derivative of ethanesulphonic acid, is found in bile salts. It is probably formed in the body from cysteine. Derivatives prepared from long-chain fatty acids and taurine, or the related isethionic acid, $HO\cdot CH_2\cdot CH_2SO_3H$, *e.g.* sodium *N*-methyl-*N*-oleyl taurate, and sodium oleyl isethionate, have been used commercially as surface-active agents (*Igepons*).

1-*Propanesulphonyl chloride*, b.p. 77–78°/13 mm.; *amide*, m.p. 52°; 2-*propanesulphonyl chloride*, b.p. 79°/18 mm.; *amide*, m.p. 60°.

Alkene- and alkyne-sulphonic acids are also known but have not been studied in great detail (cf. *E. P. Kohler*, Amer. chem. J., 1898, **20**, 680; 1899, **21**, 349; *A. Lambert* and *J. D. Rose*, J. chem. Soc., 1949, 48).

12. Alkanethiosulphonic acids

In this series the free acids, $R\cdot SO_2\cdot SH$, are not known, but the salts and esters are readily available. The *salts* are made by treating alkanesulphonyl chlorides with potassium sulphide or hydrosulphide:

$$RSO_2Cl + KSH \longrightarrow RSO_2SK + HCl$$

The *esters* are prepared by:

(1) Alkylating the salts with alkyl halides:

$$RSO_2SK + R'Br \longrightarrow RSO_2SR'$$

(see *B. G. Boldyrev et al.*, Doklady Akad. Nauk S.S.S.R., 1954, **94**, 877: 1956, **107**, 697; *Zhur. Obshchei Khim.*, 1956, **26**, 3360; 1957, **27**, 1006).

(2) Thermal decomposition of sulphinic acids:

$$3\ RSO_2H \longrightarrow RSO_2SR + RSO_3H + H_2O$$

(3) Oxidation of disulphides with nitric acid or ozone (*L. Horner*, *H. Schaefer* and *W. Ludwig*, Ber., 1958, **91**, 75):

$$RS\cdot SR \xrightarrow{O_3} RSO_2\cdot SR$$

The esters are liquids, insoluble in and heavier than water, with intense garlic- or onion-like smells. They have pronounced anti-bacterial activity.

At one time there was some doubt as to whether the esters had the thiosulphonic structure $—SO_2\cdot S—$, or whether they were disulphoxides $—SO\cdot SO—$. This point has been resolved in favour of the thiosulphonic structure by physical and chemical methods (see, *e.g.*, *N. I. Grishko* and *E. N. Gur'yanova*, Zhur. Obshchei Khim., 1959, **29**, 878).

Methanethiosulphonic acid, CH_3SO_2SH, *ethyl* ester, b.p. 95°/1–2 mm.; n-*propyl* ester, b.p. 106–7°/1–2 mm. *Ethanethiosulphonic acid, ethyl* ester, b.p. 102°/1–2 mm.; n-*propyl* ester, b.p. 109°/1–2 mm. For physical properties of a large number of these esters see the papers by *B. G. Boldyrev et al.*, mentioned above.

Chapter 6

Nitrogen Derivatives of the Aliphatic Hydrocarbons

H. GOLDWHITE

1. Nitro- and nitroso-compounds

(a) Nitroalkanes and related compounds

(i) Nitroalkanes or nitroparaffins

Nitroalkanes, (nitroparaffins), $R \cdot NO_2$, are those compounds in which the nitro group, $—NO_2$, has replaced a hydrogen atom in an alkane. The nitro group is attached to carbon through the nitrogen atom, as is shown by reduction of nitroalkanes to primary amines:

$$R \cdot NO_2 + 6H \longrightarrow R \cdot NH_2 + 2H_2O$$

and nitroalkanes are isomeric with alkyl nitrites, R·ONO. This section deals with nitroalkanes, their halogenated derivatives, nitro-olefins, and di- and poly-nitroalkanes. (For a review of nitroalkanes see *H. B. Hass, E. F. Riley* and *H. Shechter*, "The Science of Petroleum", Vol. V Part II, Oxford University Press, London, 1953, p. 70 and for one on nitro-aliphatic compounds, *N. Levy* and *J. D. Rose*, Quart. Reviews, 1947, 1, 358).

Nitroalkanes are named as substitution products of the parent hydrocarbon; thus $CH_3CH(NO_2)CH_3$ is 2-nitropropane. Primary and secondary nitroalkanes can assume a tautomeric form (see below):

$$CH_3 \cdot CH_2 \cdot NO_2 \rightleftharpoons CH_3 \cdot CH{:}NO_2H$$

and the tautomeric form has several alternative names. In the above example the tautomeric form of nitroethane is usually referred to as acinitroethane; isonitroethane and ethanenitronic acid are alternative names.

The nitro group may be considered as a resonant hybrid of the following two equivalent structures:

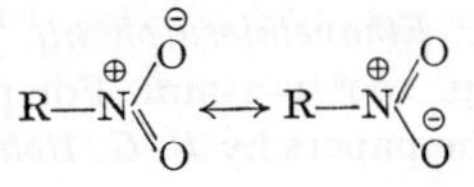

Electron diffraction measurements of nitroethane show that the nitro group and the carbon atom to which it is attached are coplanar; the O—N—O bond angle is 127 ± 3°, the N—O bond length is 1·21 ± 0·02 Å and the C—N bond length is 1·46 ± 0·02 Å. Nitroalkanes are polar substances and have high dipole moments of about 3·5 D.

The first preparation of a simple nitroalkane was reported by *Victor Meyer* and *O. Stuber* in 1872; 3-methyl-1-nitrobutane was one of the products from the reaction between isoamyl iodide and silver nitrite (Ber., 1872, 5, 203). In the same year the preparation of nitromethane was also reported (*H. Kolbe*, J. pr. Chem., 1872, 5, 427).

Nitroalkanes have not appeared as natural products but β-nitropropionic acid, $NO_2 \cdot CH_2 \cdot CH_2 \cdot CO_2H$, has been isolated recently both from plants (*C. L. Carter* and *W. J. McChesney*, Nature, 1949, **164**, 575) and from cultures of micro-organisms (*M. T. Bush, O. Touster*, and *J. E. Brockman*, J. biol. Chem., 1951, 188, 685).

Preparation (see *N. Kornblum*, Org. Reactions, 1962, **12**, 101). (1) *Direct nitration of alkanes*. The most important industrial method for the production of nitroalkanes is the vapour phase nitration of alkanes introduced in 1934 (*H. B. Hass, E. B. Hodge* and *B. M. Vanderbilt*, U.S.P., 1,967,667; Ind. Eng. Chem., 1936, **28**, 339). Moderately concentrated nitric acid (35–70%) is used in a continuous process at 375–450°. The products are exclusively mononitroalkanes and the nitro group can replace any hydrogen atom or any alkyl group in the starting alkane; no rearrangement of the carbon skeleton occurs.

The nitrating reagent shows very little selectivity; *e.g.* the nitration of 2-methylbutane, $CH_3 \cdot CH(CH_3) \cdot CH_2 \cdot CH_3$, at 420° yields nitromethane, $CH_3 \cdot NO_2$, (4%), nitroethane, $CH_3 \cdot CH_2 \cdot NO_2$, (9%), 2-nitropropane, $CH_3 \cdot CH(NO_2) \cdot CH_3$ (16%), both possible nitrobutanes, $CH_3 \cdot CH(NO_2) \cdot CH_2 \cdot CH_3$ and $(CH_3)_2CH \cdot CH_2 \cdot NO_2$, (10%), and the four possible nitropentanes, $CH_3 \cdot CH(CH_3) \cdot CH_2 \cdot CH_2 \cdot NO_2$, $CH_3 \cdot CH(CH_3) \cdot CH(NO_2) \cdot CH_3$, $(CH_3)C(NO_2) \cdot CH_2CH_3$, and $(NO_2CH_2 \cdot CH(CH_3) \cdot CH_2 \cdot CH_3$ (61% total, the highest individual yield being 24% of 2-methyl-1-nitrobutane).

In practice, the vapour phase process is run for the production of C_1, C_2, and C_3 nitroalkanes as the products of commercial interest. The addition of a small amount of air or oxygen to the reaction mixture is sometimes employed to increase the efficiency of the nitrating agent. Vapour phase nitration seems to involve a radical-chain reaction.

Liquid phase nitration of alkanes was used long before the vapour phase process (*F. Beilstein* and *A. Kurbatov*, Ber., 1880, **13**, 1818). In this process tertiary hydrogen atoms are replaced more readily than secondary which, in turn, are more readily replaced than primary ones. Reaction is slow and is accompanied by much oxidation and the production of large amounts of polynitroalkanes.

(2) *Replacement of halogen by a nitro group* (Victor Meyer reaction) remains the best method for laboratory preparation of primary and secondary nitroalkanes. In its original form, reaction between alkyl halides and silver nitrite, fair yields of primary nitroalkanes could be obtained but the reaction was useless for the preparation of secondary (or tertiary) compounds and considerable amounts of by-products, mainly alkyl nitrites, were produced also. The mechanism of the reaction has been elucidated and a modified Victor Meyer synthesis evolved between alkyl halides and sodium nitrite in either dimethylformamide or dimethyl sulphoxide as solvent, by which means good yields of both primary and secondary nitroalkanes are obtained:

$$n\text{-}C_6H_{13}\cdot CHBr\cdot CH_3 + NaNO_2 \longrightarrow NaBr + n\text{-}C_6H_{13}\cdot CH(NO_2)CH_3 \quad (58\%)$$

(*N. Kornblum et al.*, J. Amer. chem. Soc., 1955, 77, 5528, 6269; 1956, 78, 1497).

(3) *Oxidation of aldoximes or ketoximes with peroxytrifluoroacetic acid* gives good yields of primary and secondary nitroalkanes:

$$(CH_3\cdot CH_2\cdot CH_2)_2C{:}N\cdot OH \xrightarrow{CF_3\cdot CO_3H} (CH_3\cdot CH_2\cdot CH_2)_2\cdot CH\cdot NO_2 \quad (64\%)$$

(*W. D. Emmons* and *A. S. Pagano, ibid.*, 1955, 77, 4557).

(4) Tertiary nitroalkanes result from *controlled oxidation of the corresponding amines with permonosulphuric acid* (*E. Bamberger* and *R. Seligman*, Ber., 1903, 36, 685) or, more conveniently, with potassium permanganate:

$$(CH_3)_3C\cdot NH_2 \longrightarrow (CH_3)_3C\cdot NO_2 \quad (83\%)$$

(*Kornblum, R. J. Clutter* and *W. J. Jones*, J. Amer. chem. Soc., 1956, 78, 4003).

(5) Nitromethane results from *the spontaneous decarboxylation of nitroacetic acid* and was first prepared by reacting an alkali metal chloroacetate and aqueous potassium nitrite (cf. Org. Synth. Coll. Vol. I, p. 401):

$$KNO_2 + Cl\cdot CH_2\cdot CO_2K \longrightarrow [NO_2\cdot CH_2\cdot CO_2K] \longrightarrow CH_3\cdot NO_2 + K_2CO_3$$

The yield of nitromethane is only moderate and yields of higher nitroalkanes, from analogous reactions between potassium nitrite and α-halogenocarboxylic acids, are very low (*W. Treibs* and *H. Reinheckel*, Ber., 1954, 87, 341).

(6) *Nitryl chloride, NO_2Cl, adds to olefins* to give β-chloronitroalkanes. With terminal olefins the primary nitro compound is formed exclusively (*J. Ville* and *G. Dupont*, Bull. Soc. chim. Fr., 1956, 804); *F. W. Stacey* and *J. F. Harris*, Org. Reactions, 1963, 13, 150):

$$CH_3\cdot CH{:}CH_2 + NO_2Cl \longrightarrow CH_3\cdot CHCl\cdot CH_2\cdot NO_2$$

Properties and reactions. Nitroalkanes are colourless liquids of low water solubility. Some physical constants are recorded in Table 1. They are miscible with most organic liquids and are excellent solvents for many organic substances including certain synthetic rubbers and resins; this property has inspired much of the commercial interest in these compounds.

TABLE 1

PHYSICAL PROPERTIES OF NITROALKANES

Compound	M.p. °C	B.p. °C	d_{20}^{20}	n_D^{20}	Dipole moment, D (Vapour)
Nitromethane	—20	102	1·139	1·3818	3·50
Nitroethane	—90	114	1·052	1·3916	3·70
1-Nitropropane	—108	132	1·003	1·4015	3·72
2-Nitropropane	—93	120	0·992	1·3941	3·73
1-Nitrobutane		153	0·975	1·4112	
2-Methyl-2-nitropropane	25·5	127			
1-Nitropentane		173	0·974 (d_4^{20})	1·4218	
1-Nitrohexane		193	0·949 (d_4^{20})		

Nitroalkanes are, in general, stable compounds but nitromethane can decompose explosively under certain conditions. All nitroalkanes are moderately toxic and should be handled carefully.

(1) *Tautomerism and acidity of nitroalkanes; nitronic acids.* Primary and secondary nitroalkanes exist in tautomeric equilibrium with aci-forms (*nitronic acids*):

$$RR'CH\cdot NO_2 \rightleftharpoons RR'C{:}\overset{\oplus}{N}\begin{matrix}\nearrow \overset{\ominus}{O} \\ \searrow OH\end{matrix}$$

Nitro- Aci-form (nitronic acid)

The position of equilibrium under normal conditions greatly favours the nitro form. Both forms are acidic, but the aci-form is much the stronger acid of the two; *e.g.* for nitromethane, $K_{Nitro} = 6\cdot1 \times 10^{-11}$, $K_{Aci} = 5\cdot6 \times 10^{-4}$. Neutralization of a nitroalkane by a strong base is a slow bimolecular reaction. The resulting salt can usually be isolated. Its anion can be represented as a resonance hybrid of the following important contributory forms:

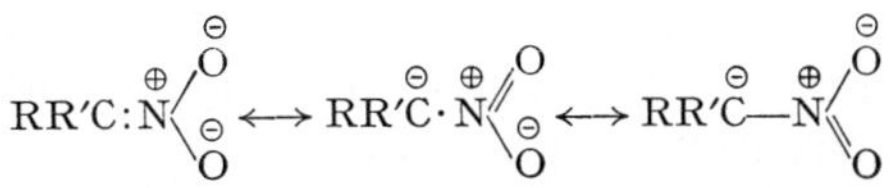

When the salts are carefully acidified the free aci-form is liberated, and this, more or less rapidly, reverts to the nitro form. In a few cases (*e.g.* *p*-bromophenylmethanenitronic acid) the aci-form is stable enough to be stored for some time (*A. Hantzsch* and *O. W. Schultze*, Ber., 1896, **29**, 2251).

Salts of nitronic acids give esters when they are treated with alkyl halides or sulphates:

$$CH_3\cdot CH{:}\overset{\ominus}{N}O_2\overset{\oplus}{Na} + CH_3\cdot I \longrightarrow CH_3\cdot CH{:}N(O)O\cdot CH_3 + NaI$$

Methyl nitronates are also formed when either the nitro or aci-form of a nitroalkane is treated with diazomethane.

(2) *Action of bases.* Prolonged action of bases on nitroalkanes usually leads to extensive decomposition. A major product from primary nitroalkanes with, *e.g.* isopropylamine or diethylamine in presence of a small amount of water, is a β-dioxime, which gives an isoxazole on dehydration with acids (*S. B. Lippincott*, J. Amer. chem. Soc., 1940, **62**, 2604):

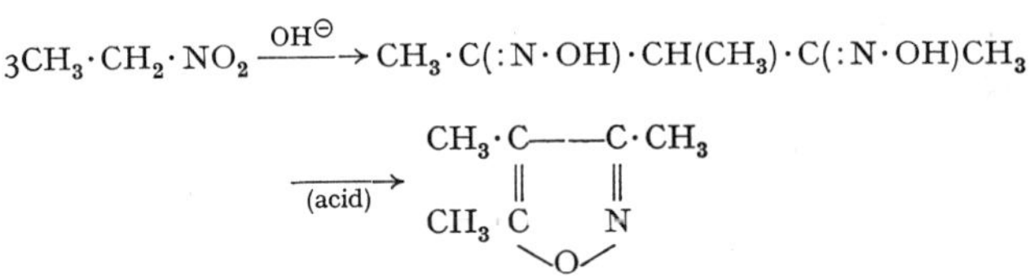

Nitromethane, however, with aqueous alkali hydroxide gives a salt of methazonic acid (the oxime of nitroacetaldehyde) which on warming with excess of 50% reagent gives an alkali metal salt of nitroacetic acid:

$$2CH_3\cdot NO_2 + KOH \rightarrow HO\cdot N{:}CH\cdot CH{:}NO_2K + 2H_2O \rightarrow CO_2K\cdot CH{:}NO_2K + NH_3$$

(*W. Steinkopf*, Ber., 1909, **42**, 2026, 3925; *H. Feuer et al.*, J. Amer. chem. Soc., 1949, **71**, 3078).

(3) *Action of acids.* Aqueous acids hydrolyse nitroalkanes under controlled conditions giving, initially, hydroxamic acids, and, finally, carboxylic acids and hydroxylammonium salts:

$$R{\cdot}CH_2{\cdot}NO_2 \longrightarrow R{\cdot}C(:N{\cdot}OH)OH \longrightarrow R{\cdot}CO_2H + H_2N{\cdot}OH$$

This reaction has been used commercially to produce hydroxylamine from nitromethane (*Lippincott* and *H. B. Hass*, Ind. Eng. Chem., 1939, 31, 118).

Treatment of salts of nitronic acids with concentrated mineral acids at moderate temperatures gives good yields of carbonyl compounds (Nef reaction):

$$RR'C{:}NO_2Na + H_2SO_4 \longrightarrow RR'C{:}O + N_2O + H_2O + NaHSO_4$$

(*K. Johnson* and *E. F. Degering*, J. org. Chem., 1943, 8, 10).

(4) *Action of nitrous acid.* The various classes of nitroalkanes can be distinguished by the different reactions they undergo with nitrous acid. Primary nitroalkanes yield *nitrolic acids* (see p. 110); these are colourless but give intensely coloured, usually red, anions:

$$CH_3{\cdot}CH_2{\cdot}NO_2 + HONO \longrightarrow CH_3{\cdot}C(:NOH)NO_2 + H_2O$$

$$CH_3{\cdot}C(:N{\cdot}OH)NO_2 + KOH \longrightarrow [CH_3C(:NO)NO_2]^{\ominus}\overset{\oplus}{K} + H_2O$$

Nitrolic acid salts can be isolated as coloured crystalline solids; they are explosive.

Secondary nitroalkanes on treatment with nitrous acid give *pseudonitroles* (see p. 110), which are blue in solution and can be extracted into chloroform from the reaction mixture:

$$(CH_3)_2CH{\cdot}NO_2 + HONO \longrightarrow (CH_3)_2C(NO)NO_2 + H_2O$$

In the crystalline state pseudonitroles are colourless and dimeric, while in solution or in a melt they are blue and monomeric—typical behaviour of a true *C*-nitroso compound (see p. 107). Tertiary nitroalkanes are unaffected by nitrous acid.

(5) *Halogenation.* Nitroalkanes are easily halogenated in the presence of bases with stepwise replacement of the hydrogen atoms on the carbon atom bearing the nitro group:

$$CH_3{\cdot}CH_2{\cdot}NO_2 \xrightarrow[OH^{\ominus}]{Cl_2} CH_3{\cdot}CHCl{\cdot}NO_2 \xrightarrow[OH^{\ominus}]{Cl_2} CH_3{\cdot}CCl_2{\cdot}NO_2$$

Photochemical chlorination, which is presumably a free-radical reaction, gives chlorinated products in which the chlorine atoms are not attached to the carbon atom bearing the nitro group (see *H. B. Hass* and *E. F. Riley*, Chem. Reviews, 1943, 32, 373).

(6) *Condensation reactions with carbonyl compounds.* Primary and secondary nitroalkanes react with aldehydes and ketones in the presence of bases to give, initially, β-nitroalcohols (see Chap. 4, p. 37):

$$CH_3 \cdot CHO + CH_3 \cdot NO_2 \longrightarrow CH_3 \cdot CH(OH) \cdot CH_2 \cdot NO_2$$

In some cases, especially with aromatic aldehydes, the β-nitroalcohol dehydrates spontaneously to a nitro-olefin, and a second molecule of nitroalkane may add to this (see next paragraph).

(7) *Addition to activated unsaturated compounds.* Primary and secondary nitroalkanes undergo Michael addition reactions, in the presence of catalytic amounts of base, with αβ- unsaturated esters, aldehydes, ketones, cyanides, nitro-compounds, sulphones etc.:

$$(CH_3)_2CH \cdot NO_2 + CH_2{:}CH \cdot CO_2CH_3 \longrightarrow (CH_3)_2C(NO_2) \cdot CH_2 \cdot CH_2 \cdot CO_2CH_3$$

$$CH_3NO_2 + CH_3CH{:}CH \cdot NO_2 \longrightarrow CH_3 \cdot CH(CH_2 \cdot NO_2)_2$$

The adduct formed initially may still be an acidic nitro-compound and may undergo further reaction:

$$CH_3 \cdot NO_2 + CH_2{:}CH \cdot CN \longrightarrow CH_2(NO_2) \cdot CH_2 \cdot CH_2CN \xrightarrow{CH_2:CH \cdot CN}$$

$$CH(NO_2)(CH_2 \cdot CH_2 \cdot CN)_2 \xrightarrow{CH_2:CH \cdot CN} C(NO_2)(CH_2 \cdot CH_2 \cdot CN)_3$$

By controlling the ratios of reactants the intermediate products can be isolated. (Michael additions are fully reviewed by *E. D. Bergmann, D. Ginsburg* and *R. Pappo*, Org. Reactions, 1959, 10, 179).

(8) *Primary and secondary nitroalkanes undergo Mannich reactions with secondary amines and formaldehyde* (or with carbinolamines):

$$R_2NH + HCHO + R'_2CH \cdot NO_2 \longrightarrow R'_2C(NO_2) \cdot CH_2 \cdot NR_2 + H_2O$$

Nitromethane and nitroethane may undergo condensation with either one or two molecules of carbinolamine (*M. Zief* and *J. P. Mason*, J. org. Chem., 1943, 8, 1).

(9) *Reduction.* Two stages can be observed in the reduction of nitroalkanes; initial formation of an *N*-alkylhydroxylamine and then complete reduction to a primary amine:

$$R \cdot NO_2 \longrightarrow RNH \cdot OH \longrightarrow R \cdot NH_2$$

Alkylhydroxylamines (p. 139) are the major products when zinc dust and ammonium chloride solution is the reducing agent, or when electrolytic reduction is employed. More vigorous reducing agents (iron and dilute acid, hydrogen and a Raney nickel catalyst, lithium aluminium hydride) give the primary amine directly.

Nitromethane is available commercially from vapour phase nitration plants and is used as a solvent or co-solvent for resins and polymers, especially cellulose esters. Its potentialities as an explosive and as a rocket fuel have been explored (*A. Makovky* and *L. Lenji*, Chem. Reviews, 1958, **58**, 627).

The sodium and potassium salts of methanenitronic acid are also explosive, and the mercuric salt decomposes readily into mercury fulminate (*J. V. Nef*, Ann., 1894, **280**, 275).

(ii) Halogenonitroalkanes or halogenonitroparaffins

Chlorination or bromination of nitroalkanes in alkaline solution leads to the progressive substitution of the halogen for all α-hydrogen atoms:

$$R{\cdot}CH_2{\cdot}NO_2 \xrightarrow[OH^{\ominus}]{Br_2} R{\cdot}CHBr{\cdot}NO_2 \xrightarrow[OH^{\ominus}]{Br_2} R{\cdot}CBr_2{\cdot}NO_2$$

By the use of the calculated quantity of halogen it is possible to control this reaction and obtain mono-, di-, or, for nitromethane, tri-halogeno-nitroalkanes (*R. Scholl*, Ber., 1896, **29**, 1823). This is the most general method for preparing these compounds but other methods have been used in special cases; *e.g.*, hydrolysis of bromonitroketones gives bromonitro-alkanes (*C. D. Hurd* and *M. E. Nilson*, J. org. Chem., 1955, **20**, 927):

$$CH_3{\cdot}CO{\cdot}CBr_2{\cdot}NO_2 \xrightarrow{H_2O} CH_3CO_2H + CHBr_2{\cdot}NO_2$$

Addition of halogen to α-nitro-olefins gives stable nitro-1,2-dihalogeno-alkanes:

$$CH_3CH{:}CH{\cdot}NO_2 + Cl_2 \longrightarrow CH_3{\cdot}CHCl{\cdot}CHCl{\cdot}NO_2$$

Trihalogenonitromethanes are frequently prepared by special methods.

Chloronitromethane, CH_2ClNO_2, b.p. 122–123°; *bromonitromethane*, b.p. 145–147°/750 mm.; 1-*chloronitroethane*, $CH_3CHClNO_2$, b.p. 124–125°; 1-*bromonitroethane*, b.p. 55°/22 mm.; 2-*chloronitroethane*, $CH_2ClCH_2NO_2$, b.p. 173°; 1-*chloro-1-nitropropane*, b.p. 141°; 2-*chloro-1-nitropropane*, b.p. 172°; 3-*chloro-1-nitropropane*, b.p. 197°; 1-*chloro-2-nitropropane*, b.p. 170°; 2-*chloro-2-nitropropane*, b.p. 131°.

Dichloronitromethane, b.p. 106–107°; *dibromonitromethane*, b.p. 58–60°/

13 mm.; 1,1-*dichloronitroethane*, $CH_3 \cdot CCl_2NO_2$, b.p. 124°, has been used as an insecticide; 1,1-*dibromonitroethane*, b.p. 165°.

Fluoropicrin, *trifluoronitromethane*, CF_3NO_2, b.p. —20°, is made by the oxidation of trifluoronitrosomethane (*R. N. Haszeldine*, J. chem. Soc., 1953, 2075).

Chloropicrin, *trichloronitromethane*, CCl_3NO_2, b.p. 112°, d^0 1·692, derives its name from the fact that it is produced when chlorine or a hypochlorite reacts with picric acid or its salts (see *H. Wieland* and *H. Jung*, Ann., 1925, **445**, 82, for a discussion of this reaction). This method is still used in its manufacture; a more recent alternative is to react hypochlorites with nitromethane (*J. B. Tindall*, U.S.P., 2,365,981/194). Chloropicrin is also obtained when concentrated nitric acid reacts with highly chlorinated compounds, *e.g.* chloral or chloroform, and when *aqua regia* reacts with acetone.

It is a colourless liquid with a sweetish odour, a powerful lachrymator and irritant to the respiratory tract, hence its use as a poison gas in World War I. Later, its insecticidal activity was observed, and it is now used on a large scale as a fumigating agent, particularly in the milling and grain industry (*B. G. Peters*, Ann. Report appl. Chem., 1948, **33**, 479). It is stable at ordinary temperatures but at its boiling point it slowly decomposes into phosgene and nitrosyl chloride; this change is powerfully accelerated by fuming sulphuric acid. Alcoholic ammonia on chloropicrin produces guanidine and alkali sulphites nitromethanedisulphonates:

$$CCl_3 \cdot NO_2 \xrightarrow{Na_2SO_3} NO_2 \cdot CNa(SO_3Na)_2 \qquad CCl_3 \cdot NO_2 \xrightarrow{NH_3} NH{:}C(NH_2)_2$$

Reduction of chloropicrin with iron and acetic acid gives methylamine, but reduction with tin and hydrochloric acid gives cyanogen chloride. Electrolytic reduction produces *dichloroformoxime*, $Cl_2C{:}NOH$, m.p. 39°, b.p. 52–54°/28 mm.

Bromopicrin, *tribromonitromethane*, CBr_3NO_2, m.p. 10°, b.p. 73°/13 mm., is produced by methods similar to those used for chloropicrin, *e.g.* by the action of hypobromites on picric acid. Many of its reactions parallel those of chloropicrin. It differs from chloropicrin, however, in its addition to olefins, which is induced by peroxides or ultra-violet light:

$$>C{=}C< + CBr_3NO_2 \xrightarrow{h\nu} >CBr \cdot \overset{|}{\underset{|}{C}} \cdot CBr_2NO_2$$

A variety of α,α-dibromonitro compounds can be made in this way (*Yu A. Baskakov* and *N. N. Mel'nikov*, Zhur. Obshchei Khim., 1959, **29**, 1233). By the action of 50% potassium hydroxide, it yields the yellow, explosive dipotassium derivative of *s*-tetranitroethane (*L. Hunter*, J. chem. Soc., 1923, **123**, 543). With liquid ammonia, the ammonium salt of dibromonitromethane is formed (*E. Schultz*, *S. Meisel* and *F. C. Schmidt*, J. Amer. chem. Soc., 1951, **73**, 2962):

$$3CBr_3(NO_2) + 8NH_3 \longrightarrow 3CBr_2(NO_2)NH_4 + 3NH_4Br + N_2.$$

(b) Nitro-olefins or nitroalkenes

The conjugated (α-β unsaturated) nitro-olefins are of particular interest because of the ease with which they are prepared and the many easy addition reactions they undergo (cf. *Levy* and *Rose*, Quart. Reviews, 1947, **1**, 383 *et seq.*). The strongly electron-attracting nitro group promotes reactions of nucleophiles at the β-carbon atom:

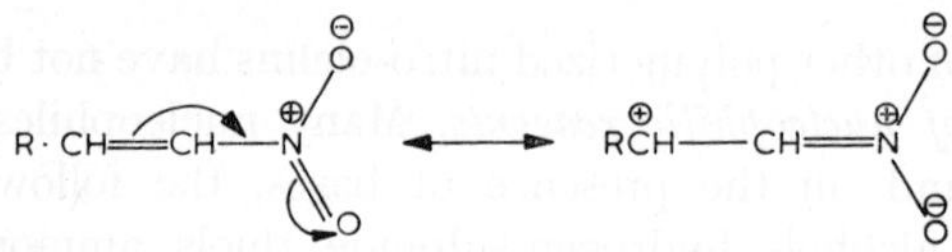

Preparation. (1) *From nitroalkanols.* β-Nitroalcohols (see p. 37), readily obtained from primary or secondary nitroalkanes and aldehydes or ketones, may dehydrate spontaneously, or can be dehydrated under mild conditions, to give nitroolefins:

$$RR'C(OH)\cdot CHR''NO_2 \longrightarrow RR'C{:}CR''\cdot NO_2 + H_2O$$

The best dehydrating agent is phthalic anhydride (*G. D. Buckley* and *C. W. Scaife*, J. chem. Soc., 1947, 1471) but frequently a better procedure is to make the acetate of the nitro-alcohol and remove the elements of acetic acid from it with a weak base:

$$CH_3\cdot CH(OH)\cdot CH_2\cdot NO_2 \xrightarrow{CH_3COCl} CH_3\cdot CH(O\cdot CO\cdot CH_3)\cdot CH_2\cdot NO_2 \xrightarrow{KHCO_3}$$
$$CH_3\cdot CH{:}CH\cdot NO_2$$

(*E. Schmidt* and *G. Rutz*, Ber., 1928, **61**, 2142; *H. B. Hass, A. G. Susie* and *R. L. Heider*, J. org. Chem., 1950, **15**, 8).

(2) *Addition of pure dinitrogen tetroxide to olefins* gives dinitroparaffins, nitro-nitrites, and nitro-nitrates. The dinitroparaffins yield nitro-olefins when they are boiled with ethanol; the nitro-nitrites and -nitrates can be hydrolysed to the nitro-alcohol which can be dehydrated to the nitro-olefin (*N. Levy* and *Scaife*, J. chem. Soc., 1946, 1093 *et seq.*):

$$CH_3\cdot CH{:}CH_2 \xrightarrow{N_2O_4} CH_3\cdot CH(NO_2)\cdot CH_2\cdot NO_2 + CH_3\cdot CH(ONO)\cdot CH_2\cdot NO_2$$

EtOH N_2O_4

$$CH_3\cdot CH{:}CH\cdot NO_2 \leftarrow CH_3\cdot CH(OH)\cdot CH_2\cdot NO_2 \leftarrow CH_3\cdot CH(ONO_2)\cdot CH_2\cdot NO_2$$

Properties and reactions. α-Nitro-olefins are colourless or pale-yellow liquids which are extremely lachrymatory. They polymerize readily in the presence of bases to give short-chain polymers which are essentially poly-nitroalkanes, when the polymer structure is that of a tertiary nitro-compound (*A. T. Blomquist, W. J. Tapp* and *J. R. Johnson*, J. Amer. chem. Soc., 1945, 67, 1519):

$$n\ CH_3\cdot C(NO_2):CH_2 \longrightarrow [-C(CH_3)(NO_2)\cdot CH_2-]_n \quad (n \sim 20\text{–}30)$$

The structures of other polymerized nitro-olefins have not been determined.

(1) *Addition of nucleophilic reagents.* Many nucleophiles add readily to α-nitro-olefins and, in the presence of bases, the following form stable adducts: water, alcohols, hydrogen sulphide, thiols, ammonia, primary and secondary amines, hydrogen cyanide, nitriles, primary and secondary nitro-alkanes, sodium hydrogen sulphite:

$$CH_3\cdot CH:CH\cdot NO_2 + CH_3\cdot CH_2\cdot OH \longrightarrow CH_3\cdot CH(OCH_2\cdot CH_3)\cdot CH_2NO_2$$

$$CH_2:CH\cdot NO_2 + H_2S \longrightarrow HS\cdot CH_2\cdot CH_2\cdot NO_2 + S(CH_2\cdot CH_2\cdot NO_2)_2$$

(see *J. D. Rose et al.*, J. chem. Soc., 1947, 1474, 1477, 1481, 1486, 1489, 1500, 1504).

(2) *Addition of electrophilic reagents.* Chlorine and bromine add to α-nitro-olefins to give stable 1,2-dihalonitroalkanes. Hydrogen chloride in ether gives chloronitrosoalkanes which rearrange, if free α-hydrogen is present, to oximes. These oximes on hydrolysis give α-chloro- or α-hydroxy-carboxylic acids plus hydroxylamine:

$$CH_3\cdot CH:CH\cdot NO_2 \xrightarrow{HCl} CH_3\cdot CHCl\cdot CCl:NOH \xrightarrow{H_2O}$$

$$CH_3\cdot CH(OH)\cdot CO_2H + H_2NOH$$

(*R. L. Heath* and *Rose*, *ibid.*, 1947, 1485).

(3) *Grignard reagents and zinc alkyls* undergo a 1,4 addition reaction to α-nitro-olefins, giving higher nitroalkanes:

$$CH_3C(NO_2):CH\cdot CH_3 + CH_3\cdot CH_2\cdot MgBr \longrightarrow CH_3\cdot CH(NO_2)\cdot CH(CH_3)\cdot CH_2\cdot CH_3$$

(*Buckley* and *Buckley* and *E. Ellery*, J. chem. Soc., 1947, 1494, 1497).

(4) *α-Nitro-olefins act as dienophiles in Diels-Alder addition reactions with conjugated dienes*:

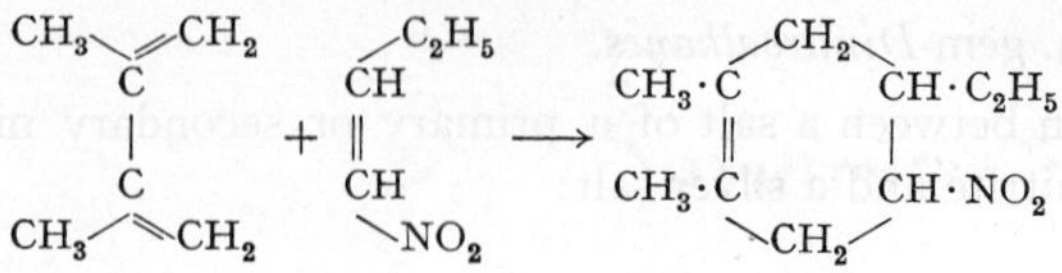

(*H. L. Holmes*, Org. Reactions, 1948, **4**, 60: *N. L. Drake* and *A. B. Ross*, J. org. Chem., 1958, **23**, 717).

(5) *Hydrolysis* of α-nitro-olefins with dilute acids or bases gives a nitroalkane plus an aldehyde or ketone:

$$R_2C{:}CR'NO_2 + H_2O \longrightarrow R_2CO + R'CH_2NO_2$$

(6) *Reduction* of α-nitro-olefins can lead to various products. Hydrogen with a Raney nickel catalyst, or lithium aluminium hydride at ordinary temperatures gives the saturated amine.

Moderate yields of nitroalkanes can be obtained with lithium aluminium hydride at low temperatures:

$$C_6H_5\cdot CH{:}C(CH_3)\cdot NO_2 \xrightarrow{LiAlH_4} C_6H_5\cdot CH_2\cdot CH(CH_3)\cdot NO_2$$

(*R. T. Gilsdorf* and *F. F. Nord*, J. Amer. chem. Soc., 1952, **74**, 1837). Reduction with hydrogen and a palladised charcoal catalyst may give oximes:

$$R\cdot CH{:}CH\cdot NO_2 \xrightarrow{H_2,\ Pd/C} R\cdot CH_2\cdot CH{:}N\cdot OH$$

(*B. Reichert* and *W. Koch*, Arch. Pharm., 1935, **273**, 265).

Nitroethylene, *nitroethene*, $CH_2{:}CHNO_2$, b.p. 38–39°/80 mm.; *1-nitropropene*, $CH_3\cdot CH{:}CHNO_2$, b.p. 54°/28 mm.; *2-nitropropene*, $CH_3\cdot C(NO_2){:}CH_2$, b.p. 58°/90 mm.; *3-nitroprop-1-ene*, $CH_2{:}CH\cdot CH_2NO_2$, b.p. 88°/180 mm.; *1-nitrobut-1-ene*, b.p. 55°/12 mm.; *2-nitrobut-1-ene*, b.p. 60·5°/50 mm.; *2-nitrobut-2-ene*, b.p. 70°/30 mm.; *2-methyl-1-nitropropene*, b.p. 56°/11 mm.; *2-nitropent-1-ene*, b.p. 58°/20 mm.; *3-nitropent-2-ene*, b.p. 58°/10 mm.; *1-nitro-oct-1-ene*, b.p. 112°/9 mm.

(c) Dinitroalkanes or dinitroparaffins

Only those dinitroalkanes with both nitro groups on the same carbon atom (*gem*-dinitroalkanes) or on adjacent carbon atoms (*vic*-dinitroalkanes) show chemical behaviour different in kind from that of mononitroalkanes, and only these two classes of compounds will be considered here.

Preparation. gem-*Dinitroalkanes.*

(1) Reaction between a salt of a primary or secondary nitroalkane, an alkali-metal nitrite and a silver salt:

$$R_2C{:}\overset{\ominus}{NO_2} + \overset{\ominus}{NO_2} + 2\overset{\oplus}{Ag} \longrightarrow R_2C(NO_2)_2 + 2Ag$$

(*R. B. Kaplan* and *H. Shechter*, J. Amer. chem. Soc., 1961, **83**, 3535).

(2) Reaction between α-halonitroalkanes and potassium nitrite:

$$CH_3{\cdot}CHCl{\cdot}NO_2 + KNO_2 \longrightarrow CH_3{\cdot}CH(NO_2)_2 + KCl$$

(*E. ter Meer*, Ann., 1876, **181**, 1).

(3) Degradation of various classes of compounds, particularly secondary alcohols and ketones, with concentrated nitric acid:

$$(CH_3{\cdot}CH_2)_2CO \xrightarrow{HNO_3} 2CH_3{\cdot}CH(NO_2)_2$$

(*G. Chancel*, Compt. rend., 1878, **87**, 1405; 1882, **94**, 399).

(4) By oxidation of pseudonitroles (p. 110):

$$R_2CH{\cdot}NO_2 \xrightarrow{HONO} R_2C(NO)NO_2 \xrightarrow{CrO_3} R_2C(NO_2)_2$$

(*G. Born*, Ber., 1896, **29**, 90).

(5) Oxidation of nitrolic acids:

$$R{\cdot}CH_2{\cdot}NO_2 \xrightarrow{HONO} R{\cdot}C({:}NOH)NO_2 \xrightarrow{HNO_3} R{\cdot}CH(NO_2)_2$$

(*R. Behrend* and *H. Tryller*, Ann., 1894, **283**, 209).

vic-*Dinitroalkanes.*

(6) Addition of pure dinitrogen tetroxide to olefins:

$$CH_3{\cdot}CH{:}CH_2 + N_2O_4 \longrightarrow CH_3{\cdot}CH(NO_2){\cdot}CH_2{\cdot}NO_2$$

(*H. Baldock, N. Levy* and *C. W. Scaife*, J. chem. Soc., 1949, 2627 and earlier papers; *F. W. Stacey* and *J. F. Harris, loc. cit.*).

(7) Reaction between an α-halonitroalkane and the salt of a nitronic acid; this reaction is successful only for derivatives of secondary nitroalkanes:

$$(CH_3)_2C{\cdot}(Br)NO_2 + (CH_3)_2C{:}NO_2Na \longrightarrow (CH_3)_2C(NO_2){\cdot}C(NO_2)(CH_3)_2$$

(*L. W. Seigle* and *H. B. Hass*, J. org. Chem., 1940, **5**, 100).

(8) Reacting two molecules of an α-halonitroalkane with silver:

$$2R_2C\cdot(Br)NO_2 + 2Ag \longrightarrow R_2C(NO_2)\cdot C(NO_2)R_2 + 2AgBr$$

(*J. Bewad*, Ber., 1906, 39, 1231).

(9) Oxidative coupling, *e.g.* by persulphates, of secondary alkanenitronates:

$$2(CH_3)_2C{:}NO_2Na \xrightarrow{(NH_4)_2S_2O_8} (CH_3)_2C(NO_2)\cdot C(NO_2)(CH_3)_2$$

(*H. Shechter* and *R. B. Kaplan*, J. Amer. chem. Soc., 1953, 75, 3980).

Properties and reactions. gem-Dinitroalkanes with a free hydrogen atom on the carbon atom bearing the nitro groups are strong acids; dinitromethane has $pK_a \sim 4$. These compounds also add readily to activated double bonds:

$$CH_3\cdot CH(NO_2)_2 + CH_2{:}CH\cdot CO_2CH_3 \longrightarrow CH_3\cdot C(NO_2)_2\cdot CH_2\cdot CH_2\cdot CO_2CH_3$$

and a variety of substituted *gem*-dinitro-compounds can be prepared by similar reactions (*Shechter* and *L. Zeldin*, *ibid.*, 1951, 73, 1276).

Reduction of *gem*-dinitroalkanes gives oximes, which can be further reduced to primary amines:

$$R\cdot CH(NO_2)_2 \longrightarrow H_2N\cdot OH + R\cdot CH{:}N\cdot OH \longrightarrow R\cdot CH_2\cdot NH_2$$

vic-Dinitroalkanes, in which at least one of the carbon atoms carrying a nitro group also carries a hydrogen atom, lose the elements of nitrous acid either spontaneously or on being boiled in ethanol and form nitro-olefins:

$$CH_3\cdot CH(NO_2)CH_2\cdot NO_2 + C_2H_5\cdot OH \longrightarrow CH_3CH{:}CH\cdot NO_2 + C_2H_5\cdot ONO + H_2O$$

(*Levy, Scaife*, and *A. E. W. Smith*, J. chem. Soc., 1946, 1096).

Dinitromethane, $CH_2(NO_2)_2$, is an unstable liquid which has not been fully characterized; 1,1-*dinitroethane*, $CH_3\cdot CH(NO_2)_2$, b.p. 185–186°; 1,1-*dinitropropane*, b.p. 189°; 2,2-*dinitropropane*, m.p. 54°, b.p. 185·5°; 2,2-*dinitrobutane*, b.p. 199°; 1,1-*dinitrohexane*, b.p. 212°.

1,2-**Dinitroethane** $CH_2NO_2\cdot CH_2NO_2$, m.p. 39–40°, b.p. 88°/< 1 mm.; 1,2-*dinitropropane*, b.p. 88°/< 1 mm.; 1,2-*dinitrobutane*, b.p. 90–92°/< 1 mm.; 2,3-*dinitrobutane*, m.p. 48–49°, b.p. 76–77°/< 1 mm.; 2,3-*dimethyl*-2,3-*dinitrobutane*, m.p. 211°; 3,4-*dimethyl*-3,4-*dinitrohexane*, m.p. 79·5–80°.

(d) Polynitroalkanes

The only general method applicable to the preparation of polynitroalkanes is further nitration of nitroalkanes. Nitration by nitric acid leads to extensive fragmentation and is not generally useful. The recent method of *Kaplan* and *Shechter* (*loc. cit.*), *i.e.* the reaction of their salts with silver salts and alkali-metal nitrites, seems to be applicable to dinitroalkanes which are not *gem*- or *vic*-compounds. In this way further nitro groups can be introduced on to carbon atoms already bearing one nitro-group:

$$CH_3C(:NO_2Na)\cdot CH_2\cdot C(:NO_2Na)CH_3 \xrightarrow{AgNO_3,\ NaNO_2} CH_3\cdot C(NO_2)_2\cdot CH_2\cdot C(NO_2)_2\cdot CH_3$$

For review see *P. Noble, F. G. Borgardt* and *W. L. Reed*, Chem. Reviews, 1964, 64, 19.

Trinitromethane, nitroform, $CH(NO_2)_3$, m.p. 23°, is most conveniently prepared in the laboratory by the action of aqueous potassium hydroxide and glycerol on tetranitromethane (*A. K. Macbeth* and *W. B. Orr*, J. chem. Soc., 1932, 534). It can also be prepared in good yield by reaction between acetylene and nitric acid (*K. F. Hager*, Ind. Eng. Chem., 1949, **41**, 2168). It is an explosive, low-melting solid and is a strong monobasic acid ($pK_a \sim 1$) forming salts, which are often explosive. With halogens the salts of trinitromethane yield halogenotrinitromethanes (halogenonitroforms); *chloro-*, $CCl(NO_2)_3$, m.p. 4·5°; *bromo-*, $CBr(NO_2)_3$, m.p. 12° and *iodo-trinitromethane*, $CI(NO_2)_3$, m.p. 58°; nitration affords tetranitromethane.

Tetranitromethane, $C(NO_2)_4$, m.p. 13°, b.p. 126°, is obtained by nitration of acetylene, trinitromethane being an intermediate (*Hager, loc. cit.*) or by passing keten into nitric acid at 0° (*G. Darzens* and *G. Levy*, Compt. rend., 1949, **229**, 1081). A convenient laboratory preparation is nitration of acetic anhydride (*F. D. Chattaway*, J. chem. Soc., 1910, **97**, 2099; Org. Synth. Coll. Vol. III, p. 803). Tetranitromethane is a colourless liquid which is quite stable when it is pure; mixtures of it with hydrocarbons are explosive. Although the structure of tetranitromethane is the symmetrical $C(NO_2)_4$, one nitro group is easily removed, for example, by bases, yielding trinitromethane. Tetranitromethane can also be used as a mild nitrating agent, for instance to nitrate aromatic amines. With unsaturated compounds and aromatic compounds tetranitromethane gives a yellow or red-yellow colour; this reaction has been widely used as a test for unsaturation (see, *e.g.*, *E. Heilbronner*, Helv., 1953, **36**, 1121).

1,1,1-*Trinitroethane*, m.p. 56°, is prepared by methylating the silver salt of trinitromethane (*A. Hantzsch* and *A. Rinckenberger*, Ber., 1899, **32**, 636).

1,1,2,2-*Tetranitroethane* is known only in the form of its explosive *dipotassium* salt (*F. Allsop* and *J. Kenner*, J. chem. Soc., 1923, **123**, 2312).

Hexanitroethane, m.p. 142° (dec.); 2,2-*dimethyl*-1,1,3-*trinitropropane*, m.p. 122°; 2,3-*dimethyl*-2,4,4-*trinitropentane*, m.p. 83°; 2,2,4,4-*tetranitropentane*, m.p. 87·5°.

(e) *Nitrosoalkanes and derivatives*

Nitrosoalkanes, R·NO, are derived from alkanes by replacement of a hydrogen atom by the monovalent —N:O group. Although it was assumed for a long period that primary, $R{\cdot}CH_2{\cdot}NO$, and secondary nitrosoalkanes, $R_2{\cdot}CH{\cdot}NO$, could not be isolated because of their rapid isomerisation to oximes, more recent work has succeeded in preparing these compounds and they have proved to be relatively stable (for a recent review of *C*-nitroso compounds see *B. G. Gowenlock* and *W. Lüttke*, Quart. Reviews, 1958, **12**, 321).

Preparation. The nitrosoalkanes generally exist as dimers (see below) but the equations written in this section will, for simplicity, show the monomeric form.

(1) *Oxidative methods.* Older procedures, *e.g.* the oxidation of amines with permonosulphuric acid, which were applied to the preparation of tertiary nitrosoalkanes, gave very poor yields:

$$(CH_3)_3C{\cdot}NH_2 \longrightarrow (CH_3)_3C{\cdot}NO \quad (4\%)$$

A far better method is the oxidation of a ketimine with an aliphatic peracid; an oxaziran is produced as an intermediate:

$$(CH_3)_2CH{\cdot}N{:}C(CH_3)_2 \xrightarrow{CH_3CO_3H} (CH_3)_2{\cdot}CH{\cdot}\underset{\diagdown O \diagup}{N\text{——}C}(CH_3)_2 \longrightarrow (CH_3)_2CH{\cdot}NO$$

(*W. D. Emmons*, J. Amer. chem. Soc., 1957, **79**, 6522).

(2) *Decomposition of alkyl nitrites.* Nitrosomethane was first prepared by photochemical decomposition of *tert*-butyl nitrite (*C. S. Coe* and *T. F. Doumani*, *ibid.*, 1948, **70**, 1516):

$$(CH_3)_3C{\cdot}ONO \longrightarrow (CH_3)_2CO + CH_3{\cdot}NO$$

This reaction, and the pyrolysis of *tert*-alkyl nitrites at low pressures for short reaction times, is a general way of preparing nitrosoalkanes (*Gowenlock* and *J. Trotman*, J. chem. Soc., 1956, 1670).

(3) *From free alkyl radicals.* Free radicals produced thermally or photochemically, for instance from alkylmercury- or iodo-compounds, may combine with nitric oxide to give nitrosoalkanes. The interesting monomeric trifluoronitrosomethane was prepared in this way:

$$CF_3I \xrightarrow{h\nu,\ Hg} CF_3{\cdot} \xrightarrow{NO} CF_3{\cdot}NO$$

(*R. N. Haszeldine*, *ibid.*, 1953, 2075).

(4) *Controlled pyrolysis of highly fluorinated acyl nitrites* gives fluoronitrosoalkanes in excellent yield:

$$C_3F_7{\cdot}CO_2Ag \xrightarrow{NOCl} C_3F_7{\cdot}CO_2{\cdot}NO \longrightarrow C_3F_7{\cdot}NO + CO_2$$

(*R. E. Banks, Haszeldine,* and *M. K. McCreath,* Proc. chem. Soc., 1961, 64).

(5) *Halogenation of oximes* gives halogenonitrosoalkanes:

$$\text{Ketoximes: } R_2C\!:\!N\cdot OH + Cl_2 \longrightarrow R_2CCl\cdot NO + HCl$$

$$\text{Aldoximes: } R\cdot CH\!:\!N\cdot OH + Cl_2 \longrightarrow HCl + R\cdot CHCl\cdot NO \text{ (isolable)}$$

$$R\cdot CHCl\cdot NO \longrightarrow R\cdot CCl\!:\!N\cdot OH \xrightarrow{Cl_2} R\cdot CCl_2\cdot NO + HCl$$

These compounds are sometimes referred to as *meso*-halogenonitrosoalkanes.

(6) *Addition of nitrosyl chloride to olefins* gives chloronitrosoalkanes:

$$R_2C\!:\!CHR' + NOCl \longrightarrow R_2CCl\cdot CH(NO)R'$$

(*W. A. Tilden* and *J. J. Sudborough,* J. chem. Soc., 1893, 63, 479), a reaction widely used by early workers for the separation and characterization of terpenes.

Properties and reactions. Halogenonitrosoalkanes, *e.g.* $CCl_3\cdot NO$, $CF_3\cdot NO$, exist as monomers in the pure state and are deep blue liquids or gases. Most simple nitrosoalkanes are dimeric in the pure state and these dimers can sometimes be obtained in two forms, which are *cis*- and *trans*-isomers of the following types:

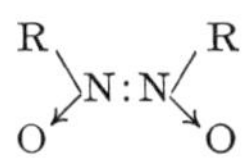

cis-form

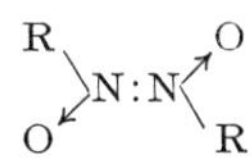

trans-form

(*Gowenlock* and *Trotman, ibid.,* 1955, 4190).

The *trans*-dimer is probably the more stable. The dimers generally dissociate into the monomer (as shown by the production of a blue colour) in the melt or in solution. With primary and secondary nitrosoalkanes this dissociation may be complicated by irreversible rearrangement of the monomeric nitrosoalkane to the isomeric oxime:

$$(RR'CH\cdot NO)_2 \rightleftharpoons 2RR'CH\cdot NO \longrightarrow 2RR'C\!:\!N\cdot OH$$

(1) Nitrosoalkanes are easily oxidised to nitroalkanes by a variety of oxidising agents, *e.g.* hydrogen peroxide, nitric acid.

(2) Reduction of nitrosoalkane dimers leads to hydrazine derivatives in the few cases which have been examined.

(3) Addition of nitrosoalkanes to olefins has been most fully explored with perfluoronitrosoalkanes. Two products are formed; an oxazetidine and a rubbery polymer:

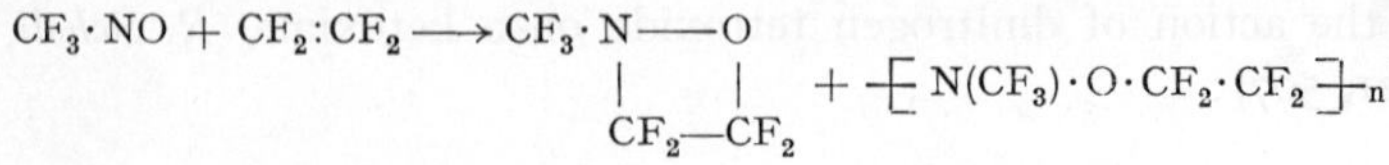

(*D. A. Barr* and *R. N. Haszeldine*, *ibid.*, 1955, 1881).

(4) Negatively substituted nitrosoalkanes (but not, apparently, simple nitrosoalkanes) act as dienophiles in Diels-Alder reactions; 1,2-dihydro-oxazines are formed:

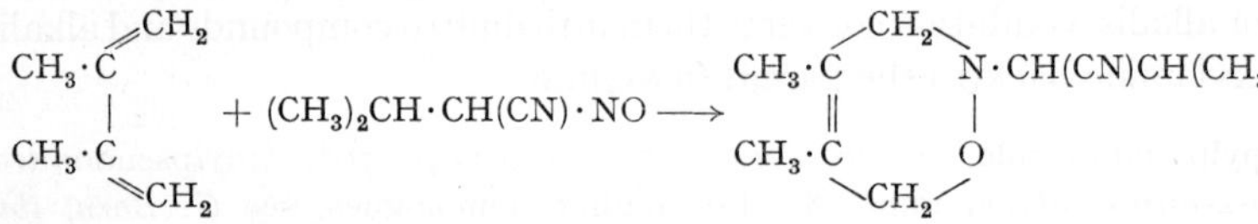

(*O. Wichterle et al.*, Chem. Listy, 1957, **51**, 605; *S. B. Needleman* and *M. C. Chang Kuo*, Chem.Reviews, 1962, **12**, 405).

Monomeric nitrosoalkanes. **Trifluoronitrosomethane,** $CF_3{\cdot}NO$, b.p. —84°, obtained by irradiation of trifluoronitromethane (*J. Jander* and *Haszeldine*, J. chem. Soc., 1954, 912) or by the action of nitric acid on trifluoroacetic anhydride (*R. Boschau*, J. org. Chem., 1960, **25**, 1450); *trichloronitrosomethane*, b.p. 5°/70 mm.; *tribromonitrosomethane*, b.p. 36–38°/14 mm.; 1,1-*dichloro*-1-*nitrosoethane*, $CCl_2(NO){\cdot}CH_3$, b.p. 68°; *pentafluoronitrosoethane*, b.p. —42°; 2-*chloro*-2-*nitrosopropane*, $CH_3{\cdot}CCl(NO){\cdot}CH_3$, b.p. 7°/18 mm.; 2-*bromo*-2-*nitrosopropane*, b.p. 41·5°/161 mm.; 2-*bromo*-2-*nitrosobutane*, b.p. 28°/15 mm.

Dimeric nitrosoalkanes. **Nitrosomethane dimer,** $(CH_3{\cdot}NO)_2$, *cis*-form., m.p. 97·5°, *trans*-form., m.p. 122°; *nitrosoethane dimer*, *cis*-form, m.p. 84°; 1-*nitrosopropane dimer*, *cis*-form, m.p. 76°; 2-*nitrosopropane dimer*, *cis*-form, m.p. 60°; *trans*-form (?), m.p. 53–55°; 1-*nitrosobutane dimer*, *cis*-form, m.p. 72°; 2-*nitrosobutane dimer*, *cis*-form, m.p. 61°; 2-*methyl*-2-*nitrosopropane dimer*, *trans*-form (?), m.p. 83–84°; 1-*nitrosododecane dimer*, *trans*-form (?), m.p. 76–78°; 1-*nitroso-octadecane dimer*, *trans*-form (?), m.p. 83–84°; 1-*chloro*-1-*nitrosoethane*, m.p. 65°.

(*f*) *Nitronitrosoalkanes*, pseudo*nitroles and nitrolic acids*

(*i*) pseudo-*Nitroles or* meso-*nitronitrosoparaffins*

Preparation. These compounds are obtained: (1) by the action of nitrous acid on *sec*-nitroalkanes (see p. 98) (*V. Meyer*, Ann., 1875, **175**, 120); (2) from *meso*halogenonitrosoalkanes and silver nitrite (*O. Piloty* and *A. Stock*, Ber., 1902, **35**, 3093):

$$(CH_3)_2C\begin{matrix} Br \\ NO \end{matrix} + AgNO_2 \longrightarrow (CH_3)_2C\begin{matrix} NO_2 \\ NO \end{matrix} + AgBr$$

(3) by the action of dinitrogen tetroxide on a ketoxime (*R. Scholl, ibid.*, 1888, **21**, 507):

$$4(CH_3)_2C{:}NOH + 3N_2O_4 \longrightarrow 4(CH_3)_2C(NO_2)NO + 2H_2O + 2NO$$

*Properties and reactions. pseudo*Nitroles are pungent, crystalline substances, colourless and bimolecular in the solid state but passing into deep blue monomolecular forms on melting or in solution. They are insoluble in aqueous acids or alkalis. Oxidation converts them into dinitro compounds and alkaline hydroxylamine back to the parent ketoxime.

Propylpseudo**nitrole,** *2-nitro-2-nitrosopropane*, m.p. 76°; *butyl*pseudo*nitrole*, *2-nitro-2-nitrosobutane*, m.p. 58°. For higher homologues, see *G. Born, ibid.*, 1896, **29**, 90.

(ii) Nitrolic acids

These compounds are weak acids, colourless to pale yellow in the solid state but giving blood-red anions in solution and are usually formulated (and described in I.U.P.A.C. rules) as oximes of nitroaldehydes, $R{\cdot}C(NO_2){:}NOH$.

They are prepared from nitrous acid and a primary nitroalkane (p. 98) or an α-oximinoketone (α-diketone monoxime); from dinitrogen tetroxide and an α-oximinocarboxylic acid, which is decarboxylated in the process; from hydroxamyl chlorides and silver nitrite:

$$CH_3CCl{:}NOH \xrightarrow{AgNO_2} CH_3C(NO_2){:}NOH$$

or by reducing a dibromonitroalkane with hydroxylamine.

Methanenitrolic acid (*nitroformaldehyde oxime,*) $CH(NO_2){:}NOH$, m.p. 68° (decomp.); *ethanenitrolic acid*, m.p. 88° (decomp.); *propanenitrolic* acid, m.p. 60° (decomp.).

2. Amines and quaternary ammonium compounds

(a) Alkylamines

When the hydrogen atoms in ammonia are replaced by alkyl groups alkylamines are the result; introduction of one alkyl group produces a primary amine, RNH_2, two alkyl groups a secondary amine, $RR'NH$, and replacement of all three hydrogen atoms by alkyl groups gives a tertiary amine, RR′R″N. Amines are differentiated into simple, in which all the alkyl groups attached to nitrogen are the same, or mixed, in which more than one kind of alkyl group is attached to the same nitrogen atom. Amines are named (I.U.P.A.C.) by stating the radical names in alphabetical order (disregarding numerical prefixes) and following them by the suffix-amine:

$CH_3 \cdot NH_2$	$(C_2H_5)_2NH$	$(CH_3CH_2CH_2)_3N$
Methylamine	Diethylamine	Tri-*n*-propylamine

$C_2H_5 \cdot NH \cdot CH_3$	$C_2H_5 \cdot N(CH_3)_2$
Ethylmethylamine	Ethyldimethylamine

Nitrogen bonded to three other groups should have a tetrahedral configuration and when the groups are all different the compound should be asymmetric and capable of resolution; the enantiomers of ethylmethylpropylamine would be:

No simple amine whose stereoisomerism would depend on this feature alone has, in fact, been resolved; this is presumably because of the very easy racemisation of this type of compound at ordinary temperatures. The nitrogen atom is oscillating and can pass through the plane of the other three groups thus inverting the configuration of the amine. The energy of activation for this reaction in ammonia is about 6 kcals./mole and it is probably of the same order for simple amines. Since it has been estimated that to separate enantiomers successfully under ordinary conditions the energy barrier to racemisation must be at least 12 kcals./mole, attempts to separate enantiomers of simple amines under these conditions would appear to be fruitless. One compound containing two tervalent nitrogen atoms incorporated into a cage structure which prevents inversion of the type described above has been resolved (*V. Prelog* and *P. Wieland*, Helv., 1944, 27, 1127).

The geometry of trimethylamine, which may be taken as a typical alkylamine, is as follows: the molecule is tetrahedral with the nitrogen atom at one vertex of the tetrahedron and the carbon atoms at the other vertices; the C—N—C angles are 108 ± 4^0 and the C—N distance is $1 \cdot 47 \pm 0 \cdot 02$ Å.

(i) *Occurrence and methods of preparation of amines*

Many compounds containing amino groups are found in nature (alkaloids, amino-acids etc.) but few simple alkylamines have been isolated from natural sources. The three methylamines have been isolated from plants, and trimethylamine occurs in the tissues of marine fish. A number of amines are formed in plants and animals (*e.g.* isoamylamine in ergot of rye) as the result of enzymatic decarboxylation of amino acids.

The more important and interesting methods for preparing alkylamines are considered below under four headings:

(1) alkylation of ammonia or its derivatives; (2) reductive methods; (3) degradative methods; (4) miscellaneous methods. (For a comprehensive review of methods of preparing amines see *Houben-Weyl*, "Methoden der Organischen Chemie", 4th edn. Vol. XI/I, Georg Thieme Verlag, Stuttgart, 1957).

(1) *Alkylation reactions.* (a) By heating an *alkylating agent*, *e.g.* alkyl halide, an alkyl sulphate or nitrate, with alcoholic or aqueous *ammonia* in a sealed tube, the whole spectrum of alkylamines and quaternary ammonium compounds is produced by the following series of reactions (*A. W. Hofmann*, Ann., 1850, 73, 91):

$$R{\cdot}X + NH_3 \longrightarrow R{\cdot}\overset{\oplus}{N}H_3\overset{\ominus}{X}\ ;\ R{\cdot}\overset{\oplus}{N}H_3\overset{\ominus}{X} + NH_3 \rightleftharpoons R{\cdot}NH_2 + \overset{\oplus}{N}H_4\overset{\ominus}{X}$$

$$R{\cdot}X + R{\cdot}NH_2 \longrightarrow R_2{\cdot}\overset{\oplus}{N}H_2\overset{\ominus}{X}\ ;\ R_2{\cdot}\overset{\oplus}{N}H_2\overset{\ominus}{X} + NH_3 \rightleftharpoons R_2NH + \overset{\oplus}{N}H_4\overset{\ominus}{X}$$

$$R{\cdot}X + R_2NH \longrightarrow R_3\overset{\oplus}{N}H\overset{\ominus}{X}\ ;\ R_3\overset{\oplus}{N}H\overset{\ominus}{X} + NH_3 \rightleftharpoons R_3N + \overset{\oplus}{N}H_4\overset{\ominus}{X}$$

$$RX + R_3N \longrightarrow R_4\overset{\oplus}{N}\overset{\ominus}{X}$$

The relative amounts of each amine produced depend on many factors including the proportions of reagents, relative rates of the various reactions, and relative base strengths of the intermediate amines. In general, alkyl iodides react more readily than bromides, which react more rapidly than chlorides; alkyl chlorides are, however, usually used in industrial preparation of alkylamines because of their lower cost. Primary halides give satisfactory results in direct alkylation but secondary halides give considerable amounts of olefinic by-products, and tertiary halides give olefins almost exclusively.

A large excess of ammonia favours the production of the primary amine; this can be achieved by using liquid ammonia as the reaction medium:

$$n\text{-}C_{12}H_{25}Br \xrightarrow[\text{18-fold excess}]{NH_3} n\text{-}C_{12}H_{25}{\cdot}NH_2\ (54\%) + (n\text{-}C_{12}H_{25})_2NH\ (7\%$$

(*J. von Braun*, Ber., 1937, 70, 979).

Primary amines react with alkyl halides to give mixtures of secondary and tertiary amines and quaternary ammonium halides; a large excess of primary amine results in fair yields of the secondary amine. Alkylation of a secondary amine with one equivalent of an alkyl halide is a reasonable route to a tertiary amine:

$$C_2H_5\cdot NH\cdot CH_3 + n\text{-}C_3H_7\cdot Br \longrightarrow CH_3(C_2H_5)(n\text{-}C_3H_7)N \ (30\%)$$

(*J. Meisenheimer* and *H. Bernhard*, Ann., 1922, **428**, 254).

An alternative route to primary amines is the reaction between an alkyl halide and sodium amide; some elimination of hydrogen halide occurs, but the method is fairly good for converting primary alkyl halides into the corresponding amine:

$$n\text{-}C_6H_{13}Br + NaNH_2 \longrightarrow n\text{-}C_6H_{13}\cdot NH_2 + NaBr$$

(*R. N. Shreve* and *L. W. Rothenberger*, Ind. Eng. Chem., 1937, **29**, 1361).

(b) Reaction between *potassium phthalimide* and an *alkyl halide* gives an *N*-substituted phthalimide which can be cleaved to give a primary amine:

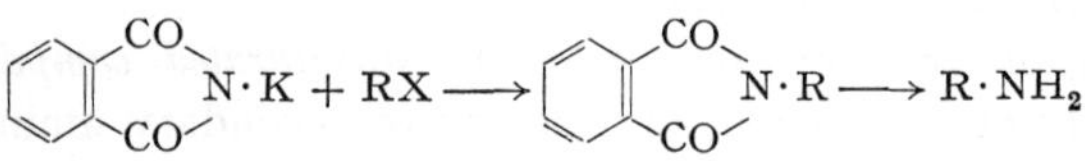

(*S. Gabriel*, Ber., 1887, **20**, 2224). The initial reaction between potassium phthalimide and alkyl halide is most rapid if dimethylformamide is used as solvent (*J. C. Sheehan* and *W. A. Bolhofer*, J. Amer. chem. Soc., 1950, **72**, 2786). The *N*-alkylphthalimide is cleaved most readily by hydrazine (*H. R. Ing* and *R. H. F. Manske*, J. chem. Soc., 1926, 2348). The Gabriel reaction is successful with primary and secondary alkyl halides, but tertiary alkyl halides lose hydrogen halide when they are treated with potassium phthalimide.

(c) *Alkylation* of *arene sulphonamides* or *N*-alkylsulphonamides gives *N,N*-dialkylsulphonamides which can be hydrolysed by acids or bases, or cleaved reductively, to give secondary amines:

$$Ar\cdot SO_2\cdot NH_2 + 2RX \xrightarrow{NaOEt} Ar\cdot SO_2\cdot NR_2 \xrightarrow{HCl} R_2NH + Ar\cdot SO_3H$$

$$Ar\cdot SO_2Cl + R'NH_2 \longrightarrow Ar\cdot SO_2\cdot NHR' \xrightarrow[NaOEt]{RX} Ar\cdot SO_2\cdot NRR'$$

$$\xrightarrow{HCl} RR'NH + Ar\cdot SO_3H$$

(*O. Hinsberg*, Ann., 1891, **265**, 178; *H. Lecher* and *F. Graf*, *ibid.*, 1925, **445**, 67).

(d) *Catalytic ammonolysis.* Alcohols react with ammonia at high temperatures and in the presence of catalysts to give mainly primary amines:

$$R \cdot OH + NH_3 \xrightarrow[300\text{–}500^0]{ThO_2} R \cdot NH_2 + H_2O$$

(*P. Sabatier* and *A. Mailhe,* Compt. rend., 1910, **150,** 823).

A great deal of more recent work, described in patents, has examined the optimum conditions of temperature, (usually about 400°), pressure, (about 100 atmospheres), and reaction time (two to three hours). Many catalysts are claimed to be active including alumina, magnesia, silica, chromic oxide, aluminium phosphate, etc. By variation of catalysts and conditions production of either primary or secondary amines can be favoured. (*R. H. Goshorn,* U.S.P., 2,349,222/1944; 2,394,515–6/1946; *W. Whitehead,* B.P., 649,980/1948; *U. A. Krishnamurthy* and *M. R. A. Rao,* J. Indian Inst. Sci., 1957, **39,** 138).

Ammonolysis of ethers also yields amines, and ethylamine is prepared industrially in this way:

$$(C_2H_5)_2O + 2NH_3 \longrightarrow 2C_2H_5 \cdot NH_2 + H_2O$$

(*Goshorn,* U.S.P., 2,389,500/1945).

(e) *Addition* of *ammonia* or *amines* to *unsaturated compounds.* Under ordinary conditions ammonia, or primary or secondary amines, add only to the double bond of activated unsaturated compounds:

$$(C_2H_5)_2NH + CH_2{:}CH \cdot CH_2OH \xrightarrow{NaOEt} (C_2H_5)_2N \cdot CH_2 \cdot CH_2 \cdot CH_2OH$$

(*O. Hromatka,* Ber., 1942, **75,** 131, 379);

$$RR'NH + C_6H_5CH{:}CH_2 \xrightarrow{Na} C_6H_5 \cdot CH_2 \cdot CH_2 \cdot NRR'$$

(*R. Wegler* and *G. Pieper, ibid.,* 1950, **83,** 1).

However at high temperatures and pressures, ammonia and primary and secondary amines add to simple olefins in the presence of catalytic amounts of sodium:

$$CH_2{:}CH_2 + NH_3 \xrightarrow[800\text{–}1000 \text{ ats.}]{Na,\ 200^0} C_2H_5 \cdot NH_2,\ (C_2H_5)_2NH,\ (C_2H_5)_3N$$

(*B. W. Howk et al.,* J. Amer. chem. Soc., 1954, **76,** 1899).

Amines add to acetylene derivatives to give, initially, vinylamines; these compounds are not isolated because they react rapidly with more acetylene to give acetylenic amines:

$$(C_2H_5)_2NH + CH{\vdots}CH \longrightarrow [(C_2H_5)_2N \cdot CH{:}CH_2] \xrightarrow{CH{\vdots}CH}$$

$$(C_2H_5)_2N \cdot CH(CH_3) \cdot C{\vdots}CH$$

(*W. Reppe et al.,* Ann., 1955, **596,** 14).

(2) *Amines by reduction.* Many classes of compounds containing carbon-nitrogen bonds yield amines by reduction (*Houben-Weyl, loc. cit.,* lists 22 different classes). Only the more important methods are discussed here.

(f) *Nitroalkanes* are best reduced to amines with hydrogen and a Raney nickel catalyst, or with iron powder and dilute hydrochloric acid.

(g) *Nitriles,* R·CN, can be reduced catalytically with hydrogen and Raney nickel. Under these conditions considerable amounts of secondary amine are formed, presumably by reaction between primary amine and the imine which is the initial product of reduction:

$$R\cdot C{\vdots}N \xrightarrow{H_2} R\cdot CH{:}NH \xrightarrow{H_2} R\cdot CH_2\cdot NH_2$$

$$R\cdot CH_2\cdot NH_2 + R\cdot CH{:}NH \longrightarrow R\cdot CH_2\cdot NH\cdot CH(NH_2)R$$

$$\xrightarrow{H_2} (R\cdot CH_2)_2NH + NH_3$$

Formation of secondary amine can be suppressed either by carrying out the reduction in presence of ammonia (*E. J. Schwoegler* and *H. Adkins,* J. Amer. chem. Soc., 1939, 61, 3499), on in acetic anhydride solution, when the primary amine forms an amide which does not react with the imine; the amide can be hydrolysed to the amine subsequently:

$$R\cdot C{\vdots}N \xrightarrow{H_2,\,(CH_3CO)_2O} R\cdot CH_2\cdot NH\cdot CO\cdot CH_3 \xrightarrow{HCl} R\cdot CH_2\cdot NH_2$$

(*K. Kindler* and *K. Schrader,* Ann., 1949, 564, 49).

Reduction of nitriles with lithium aluminium hydride gives very good yields of primary amines; this reagent can be used when a variety of other functional groups is present in the nitrile (*N. Gaylord,* "Reduction with Complex Metal Hydrides", Interscience Publishers Inc., New York, 1956, p. 731).

(h) *Oximes* can be reduced catalytically, or with sodium and ethanol, to give good yields of amines:

$$n\text{-}C_6H_{13}\cdot CH{:}N\cdot OH \xrightarrow{Na,\ EtOH} n\text{-}C_6H_{13}\cdot CH_2\cdot NH_2\ (70\%)$$

(*W. H. Lycan, S. S. Puntambeker,* and *C. S. Marvel,* Org. Synth. Coll., Vol. II, 1950, p. 318).

(i) *Amides* can be reduced to amines with lithium aluminium hydride (*Gaylord, op. cit.,* p. 544). A mixture of sodium borohydride and aluminium chloride has also been recommended for this reduction:

$$n\text{-}C_5H_{11}\cdot CONH_2 \xrightarrow{NaBH_4,\ AlCl_3} n\text{-}C_5H_{11}\cdot CH_2\cdot NH_2\ (65\%)$$

(*H. C. Brown* and *B. C. Subba Rao*, J. Amer. chem. Soc., 1956, 78, 2582). The use of complex metal hydrides has superseded older methods of reduction (*e.g.* catalytic or electrolytic) of amides.

(j) *Reductive alkylation* by reaction between an *aldehyde* or *ketone* and *ammonia* or a primary or secondary *amine* in the presence of hydrogen and a hydrogenation catalyst gives good yields of amines. The reaction probably proceeds by reduction of the initially formed condensation product of the carbonyl compound with the amine:

$$(n\text{-}C_3H_7)_2CO + NH_2 \xrightarrow{H_2\ \text{Raney Ni}} (n\text{-}C_3H_7)_2CH\cdot NH_2\ (80\%)$$

$$(CH_3)_2CH\cdot CH_2\cdot NH_2 + n\text{-}C_3H_7\cdot CHO \xrightarrow{H_2/\text{Raney Ni}}$$

$$n\text{-}C_3H_7\cdot CH_2\cdot NH\cdot CH_2\cdot CH(CH_3)_2\ (86\%)$$

(This method is reviewed by *W. S. Emerson*, Org. Reactions, 1948, 4, 174.)

(k) Alternatively the condensation products of aldehydes or ketones with ammonia or primary or secondary amines can be reduced to amines with formic acid (Leuckart reaction):

$$R'R''CO + R_2NH + HCO_2H \longrightarrow R'R''CH\cdot NR_2 + CO_2 + H_2O$$

(The Leuckart reaction is reviewed by *M. L. Moore*, *ibid.*, 1949, 5, 301.)

(3) *Amines by degradation.* (l) Hofmann degradation (cf. *E. S. Wallis* and *J. F. Lane*, *ibid.*, 1946, 3, 267).

Carboxyamides are converted into primary amines containing one fewer carbon atom by the action of *chlorine* or *bromine* and *alkali* (*A. W. Hofmann*, Ber., 1881, 14, 2725).

The probable course of reaction is:

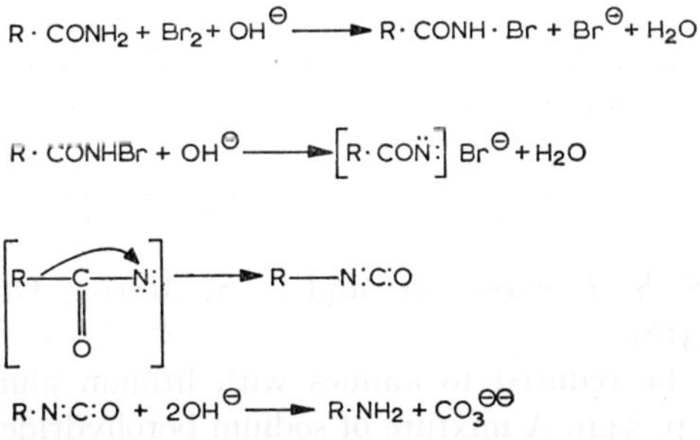

The *N*-bromoamide and the isocyanate can be isolated by proper choice of reaction conditions. If reaction is carried out in ethanolic solution a urethane is formed, which can be hydrolysed subsequently to amine:

$$R\cdot NCO + EtOH \longrightarrow R\cdot NH\cdot COOEt \xrightarrow{NaOH} R\cdot NH_2$$

Amides whose optical activity depends on the configuration of the group attached to the carbonyl group give optically active amines by Hofmann degradation. There is no racemization and, most probably, complete retention of configuration in the product:

$$(+)\ C_6H_5\cdot CH_2\cdot CH(CH_3)\cdot CONH_2 \longrightarrow (+)\ C_6H_5\cdot CH_2\cdot CH(CH_3)\cdot NH_2$$

(*E. S. Wallis* and *S. C. Nagel*, J. Amer. chem. Soc., 1931, **53**, 2787), showing that the transfer of the group from carbon to nitrogen proceeds with complete stereospecificity, and that the group does not separate as a free ion during the course of reaction.

(m) Curtius degradation (cf. *P. A. S. Smith*, Org. Reactions, 1946, 3, 337). *Acid azides* decompose on being heated to yield an isocyanate; if the decomposition is carried out in an inert solvent (benzene, chloroform) the isocyanate can be isolated and hydrolysed to amine. In ethanol a urethane results, and this can also be hydrolysed to amine. The Curtius degradation is similar in mechanism to the Hofmann degradation:

$$R\cdot CON_3 \longrightarrow N_2 + [R\cdot CO\ddot{N}:]$$

$$[R\cdot CO\ddot{N}:] \longrightarrow R\cdot NCO \xrightarrow{\text{Solvent}} \text{Products}$$

In aqueous solution the usual product is the urea, $(R\cdot NH)_2CO$, and this gives the amine on hydrolysis. Curtius degradation of an optically active azide yields an optically active amine (*L. W. Jones* and *Wallis*, J. Amer. chem. Soc., 1926, **48**, 169).

(n) Schmidt degradation (cf. *H. Wolff*, Org. Reactions, 1946, 3, 307). Treatment of a *carboxylic acid* with *hydrazoic acid* in presence of a strong mineral acid gives an amine:

$$R\cdot CO_2H + HN_3 \xrightarrow{H_2SO_4} R\cdot NH_2 + N_2 + CO_2$$

The mechanism of reaction involves various protonated intermediates:

$$R\cdot CO_2H + H^{\oplus} \rightleftharpoons [R\cdot CO_2H_2]^{\oplus} \rightleftharpoons R\overset{\oplus}{C}O + H_2O$$

$$R\cdot\overset{\oplus}{C}O + HN_3 \longrightarrow [R\cdot CO\cdot NH\cdot N\vdots N]^{\oplus} \longrightarrow R\cdot\overset{\oplus}{N}H{:}CO + N_2$$

$$R\cdot\overset{\oplus}{N}H{:}CO + H_2O \longrightarrow R\cdot\overset{\oplus}{N}H_3 + CO_2$$

(*M. S. Newman* and *H. N. Gildenhorn*, J. Amer. chem. Soc., 1948, **70**, 317).

The Schmidt reaction can be applied to ketones to give a similar result to that obtained by Beckmann rearrangement [see (p) below]:

$$R_2C{:}O + HN_3 \xrightarrow{H_2SO_4} R{\cdot}NH{\cdot}COR + N_2$$

$$R{\cdot}NH{\cdot}COR \longrightarrow R{\cdot}NH_2 + R{\cdot}CO_2H$$

(o) Lossen degradation (cf. *H. L. Yale*, Chem. Reviews, 1943, 33, 242). Heating a solution of a salt of an *acyl* or *aroyl hydroxamic* acid gives an amine:

$$[RCO\overset{\ominus}{N}{\cdot}O{\cdot}COR']\overset{\oplus}{Na} + H_2O \longrightarrow R{\cdot}NH_2 + CO_2 + R'CO_2Na$$

This reaction is much less useful preparatively than (l), (m), and (n), but it has a closely related mechanism:

$$[R{\cdot}CO\overset{\ominus}{N}{\cdot}O{\cdot}COR']Na^{\oplus} \longrightarrow [R{\cdot}CO{\cdot}\ddot{N}{:}] + R'{\cdot}COO^{\ominus}Na^{\oplus}$$

$$[R{\cdot}CO{\cdot}\ddot{N}{:}] \longrightarrow R{\cdot}NCO \longrightarrow R{\cdot}NH_2$$

(p) Beckmann rearrangement. Treatment of an oxime, particularly a *ketoxime*, with certain *acidic reagents*, *e.g.* concentrated sulphuric acid, or phosphorus pentachloride in ether, or hydrogen chloride in a mixture of acetic acid and acetic anhydride, gives an amide, which can be hydrolysed to an amine:

$$R_2C{:}N{\cdot}OH \longrightarrow R{\cdot}CONH{\cdot}R \longrightarrow RCO_2H + R{\cdot}NH_2$$

(*E. Beckmann*, Ber., 1886, 19, 988). With oximes of mixed ketones, R·C(:N·OH)R′, two amines are possible products. In general the product formed by Beckmann rearrangement is that from the *anti*-transposition of groups, if the oxime does not isomerise:

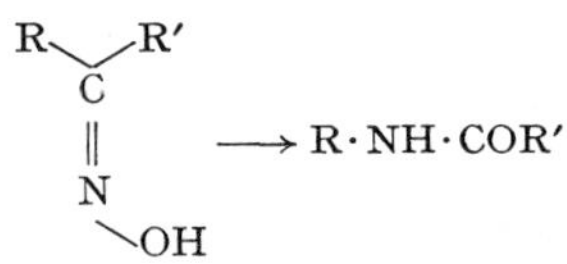

The Beckmann rearrangement has wide scope, both in synthesis and in degradation, and has been extensively studied (see *L. G. Donaruma* and *W. Z. Heldt*, Org. Reactions, 1960, 11, 1).

(q) *Alkaline cleavage* of *p-nitroso-N,N-dialkylarylamines* gives secondary alkylamines; this reaction is sometimes useful for preparing lower secondary alkylamines in a pure state:

$$p\text{-}(CH_3)_2N\cdot C_6H_4\cdot NO + NaOH \longrightarrow (CH_3)_2NH + p\text{-}NaO\cdot C_6H_4\cdot NO$$

(*A. Baeyer* and *H. Caro*, Ber., 1874, **7**, 963; *R. Munch, G. T. Thannhauser*, and *D. L. Cottle*, J. Amer. chem. Soc., 1946, **68**, 1297).

(r) *α-Amino-acids* can be decarboxylated to amines either by heating them:

$$R\cdot CH(NH_2)\cdot CO_2H \longrightarrow R\cdot CH_2\cdot NH_2 + CO_2$$

(*T. B. Johnson* and *F. G. Daschavsky*, J. biol. Chem., 1925, **62**, 725) or enzymatically. The latter process is the source of many amines found in nature (*E. Werle*, Angew. Chem., 1951, **63**, 550).

(4) *Miscellaneous methods*. (s) From *organo-metallic compounds*. Chloramine reacts with Grignard reagents and other organo-metallic compounds in a direct synthesis of primary amines:

$$n\text{-}C_4H_9\cdot MgCl + NH_2Cl \longrightarrow n\text{-}C_4H_9\cdot NH_2 + MgCl_2$$

(*G. H. Coleman* and *R. F. Blomquist*, J. Amer. chem. Soc., 1941, **63**, 1692, and earlier papers).

Grignard reagents also yield primary amines with *O*-methylhydroxylamine:

$$(CH_3)_2CH\cdot CH_2\cdot CH_2\cdot MgCl + CH_3ONH_2 \longrightarrow (CH_3)_2CH\cdot CH_2\cdot CH_2\cdot NH_2 + CH_3O\cdot MgCl$$

(*N. I. Sheverdina* and *K. A. Kocheshkov*, J. gen. Chem. U.S.S.R., 1938, **8**, 1825).

The addition products formed from Grignard reagents and nitriles are reduced to amines by lithium aluminium hydride:

$$C_2H_5\cdot MgBr + C_2H_5\cdot CN \longrightarrow (C_2H_5)_2\cdot C{:}N\cdot MgBr \xrightarrow{LiAlH_4} (C_2H_5)_2\cdot CH\cdot NH_2$$

(*A. Pohland* and *H. R. Sullivan*, J. Amer. chem. Soc., 1953, **75**, 5898).

(t) Reaction of *ammonia* (or a primary or secondary amine) with an *olefin, carbon monoxide, water*, and iron carbonyl hydride as catalyst, gives a mixture of amines:

$$NH_3 + CH_2{:}CH_2 + 3CO + H_2O \longrightarrow CH_3\cdot CH_2\cdot CH_2\cdot NH_2 + 2CO_2$$

(*W. Reppe* and *H. Vetter*, Ann., 1953, **582**, 133).

(u) Reaction between an *olefin, sulphuric acid*, and *hydrogen cyanide* or an *organic cyanide*, gives amides which can be hydrolysed to amines.

This reaction presumably proceeds by addition of a carbonium ion to the cyano group, and is a useful way of making primary amines carrying a tertiary alkyl group (*J. J. Ritter et al.*, J. Amer. chem. Soc., 1948, **70**, 4045, 4048; 1949, **71**, 4128):

$$(CH_3)_2C{:}CH_2 \xrightarrow{H^{\oplus}} (CH_3)_3\overset{\oplus}{C} \xrightarrow{RCN} (CH_3)_3C\cdot N{:}\overset{\oplus}{C}R$$

$$\xrightarrow{H_2O} (CH_3)_3C\cdot NHCOR \longrightarrow (CH_3)_3C\cdot NH_2$$

Technical production of amines

Many lower alkylamines are important commercial products and are produced on a large scale, mainly for use as intermediates for other chemicals. In general, they are prepared by ammonolysis of alcohols or by vapour phase alkylation of ammonia with alkyl chlorides. These methods produce mixtures of primary, secondary and tertiary amines which are separated by fractional distillation.

Methylamines are prepared by ammonolysis of methanol at 300–500° over catalysts (*C. A. Andrews*, U.S.P., 2,073,671/1936). Separation of the three methylamines is difficult because of the closeness of their boiling-points, but it is achieved, *e.g.*, by fractional distillation at two different pressures within the range 100–300 p.s.i.

(ii) Properties and reactions of amines

The lower alkylamines are gases or volatile liquids with strong smells (ammoniacal and "fishy"); they are soluble in water. As usual, in a homologous series, the higher members tend to resemble hydrocarbons and are insoluble in water and almost odourless.

Some physical constants of the lower members of the series (C_1–C_5) are given in Table 2.

The outstanding chemical properties of amines are due to the unshared pair of electrons on the nitrogen atoms. This makes amines both bases and nucleophiles. Aqueous solutions of amines are dissociated according to the equilibrium:

$$RR'R''N + H_2O \rightleftharpoons RR'R''\overset{\oplus}{N}H + OH^{\ominus}$$

Amines are weak bases, though the introduction of alkyl groups does, as would be expected from the electron-releasing tendency of alkyl groups, make them stronger bases than ammonia:

Base	NH_3	$CH_3\cdot NH_2$	$(CH_3)_2NH$	$(CH_3)_3N$
pK_b	4·7	3·36	3·29	4·28

Introduction of one and two methyl groups into ammonia steadily increases the base strength of the amine; introduction of a third methyl group,

TABLE 2

PHYSICAL PROPERTIES OF AMINES

R	RNH_2				R_2NH				R_3N			
	m.p.°	b.p.°	$d^{t°}$	$n_D^{t°}$	m.p.°	b.p.°	$d^{t°}$	$n_D^{t°}$	m.p.°	b.p.°	$d^{t°}$	$n_D^{t°}$
CH_3	—93	—6·3	0·662^{20}		—92	6·9	0·654^{20}		—117	2·9	0·632^{20}	
C_2H_5	—81	16·5	0·706^{0}		—50	55·4	0·711^{20}	1·387^{18}	—115	89·5	0·728^{20}	1·400^{20}
n-C_3H_7	—83	49	0·719^{20}	1·389^{20}	—40	111	0·738^{20}	1·405^{20}	—94	156	0·757^{20}	1·417^{20}
iso-C_3H_7	—101	32	0·694^{20}	1·377^{15}	—60	84	0·719^{20}			139	0·765^{20}	1·414^{20}
n-C_4H_9	—50·5	78	0·733^{26}	1·401^{20}	<—50	159	0·767^{20}			216	0·778^{20}	
iso-C_4H_9	—85·5	68	0·724^{25}	1·399^{17}	—74	138	0·742^{25}	1·409^{20}	—22	192	0·771^{17}	1·425^{17}
sec-C_4H_9 (*dl*)	—104	63	0·727^{15}	1·395^{17}		132	0·783^{0}					
tert-C_4H_9	—68	44	0·698^{18}	1·379^{18}		114/715 mm.						
n-C_5H_{11}	—55	104	0·766^{19}			91–3/14 mm.				240–5		
iso-C_5H_{11}		96	0·750^{20}			185–7	0·767^{21}			237	0·788^{13}	
tert-C_5H_{11}		77	0·748^{16}									

however, gives a sharp reduction in the base strength of trimethylamine. This variation of base strength has been interpreted in terms of steric strain in crowding three methyl groups around the nitrogen atom (*H. C. Brown*, J. chem. Soc., 1956, 1248). Amines combine not only with protons but with many other electron-deficient molecules to form stable complexes (*e.g.* $Me_3N \rightarrow BF_3$; $Et_3N \rightarrow SO_3$).

The presence of a free hydrogen atom on the nitrogen atom of primary and secondary amines influences most of the reactions of these compounds; consequently, reactions of tertiary amines are often qualitatively quite different from those of the primary and secondary compounds. The very wide variety of reactions which amines undergo can be classified conveniently in four groups: (1) where amines act as bases and nucleophiles; (2) degradative reactions; (3) addition reactions; (4) miscellaneous reactions.

(1) *Amines acting as bases and nucleophiles.* (a) *Salt formation.* Although alkylamines are weak bases they form well-defined salts with many strong acids. *Picrates* have been used to characterize amines and *chloroplatinates*, which decompose on combustion leaving a residue of pure platinum, can be used to estimate their equivalent weights. Salts of amines and carboxylic acids are dehydrated to *amides* on being heated alone or, better, with phosphorus pentoxide:

$$R\cdot\overset{\oplus}{N}H_3\ \overset{\ominus}{O}OCR' \longrightarrow R\cdot NH\cdot COR' + H_2O$$

(b) The basic properties of amines and their solubility in common organic solvents make them useful catalysts in certain reactions (*e.g.* Knoevenagel reaction) and efficient dehydrohalogenating agents.

(c) Amines act as nucleophiles towards alkyl halides and form *N*-alkylated derivatives [see Method (a) under Preparation of Amines, p. 113, and also Quaternary Ammonium Compounds, p. 133].

(d) *Acylation.* Primary and secondary amines react with acid halides, acid anhydrides, or esters to give the acylated amine (a *substituted amide*):

$$R\cdot NH_2 + R'CO\cdot X \longrightarrow R\cdot NH\cdot COR' + HX\ (X = \text{halogen},\ R''\cdot CO_2—,\ \text{or}\ R''O—)$$

The amides produced are almost entirely non-basic, and ordinarily a second acyl group is not introduced into amides derived from primary amines.

The acylation reaction is frequently used to prepare derivatives of amines for characterization or for protection of the amino group during subsequent reactions. Among the most frequently prepared derivatives are acetyl derivatives (from reaction between amines and acetyl chloride or acetic anhydride) and benzoyl or ring-substituted benzoyl derivatives (from reaction between amines and the appropriate acid chloride). Formyl derivatives of amines are prepared by the action of formic acid on the amine; alternatively formyl fluoride (formyl chloride is unknown) may be used (*C. Olah* and *I. Kuhn*, Ber., 1956, **89**, 2211). Acylated amines regenerate the parent amine on hydrolysis with either acids or bases.

Reaction of primary amines with phosgene is, formally, an acylation. The *alkylcarbamoyl chloride* formed initially loses hydrogen chloride when it is heated, giving an *alkyl isocyanate*:

$$R \cdot NH_2 + COCl_2 \longrightarrow HCl + R \cdot NH \cdot COCl \longrightarrow R \cdot NCO + HCl$$

This reaction can be carried out in a solvent, or in the vapour phase, and the intermediate carbamyl chloride can be isolated (*R. J. Slocombe et al.*, J. Amer. chem. Soc., 1950, **72**, 1888).

(e) *Sulphonylation*. The counterpart of acylation with the halides of sulphonic acids, particularly benzene-, *p*-bromobenzene-, and *p*-toluenesulphonyl chlorides, is widely used for preparing derivatives of primary and secondary amines, and for separating mixtures of all three classes of amines. Tertiary amines are unaffected by the sulphonyl chloride and may be extracted by acid from the reaction mixture. Secondary amines give *sulphonamides*, $Ar \cdot SO_2 \cdot NR_2$, carrying no hydrogen atom on the nitrogen atom, which are neutral and insoluble in alkali. Sulphonamides derived from primary amines, $Ar \cdot SO_2 \cdot NHR'$, are weakly acidic (the sulphonyl group activates the hydrogen atom on the adjacent nitrogen atom) and are alkali-soluble (*O. Hinsberg*, Ber., 1890, **23**, 2963; *Hinsberg* and *J. Kessler*, *ibid.*, 1905, **38**, 906). In the reaction with primary amines small amounts of alkali-insoluble disulphonyl compounds, $(Ar \cdot SO_2)_2N \cdot R$, may be obtained; they are readily hydrolysed by alkali to the normal monosulphonyl derivatives. Sulphonamides can be hydrolysed by acids or bases, or cleaved reductively by sodium in butanol, to give back the parent amine.

(f) Primary and secondary amines react with *halides of non-metals and metalloids*:

$$E{-}X + 2R_2N \cdot H \longrightarrow E{-}NR_2 + R_2\overset{\oplus}{N}H_2\overset{\ominus}{X}$$

E can be As, B, Bi, P, S, Si, etc., and may carry other functional groups. The reactions of primary amines are sometimes more complex than the above equation would suggest (see pp. 145 to 147).

(2) *Degradative reactions*. (g) *Oxidation*. All classes of amines are easily oxidised. With vigorous oxidising agents primary and secondary amines give breakdown products, particularly carboxylic acids, in which carbon-nitrogen bond cleavage has occurred. However, in a few special cases the carbon–nitrogen bond is retained intact and useful products are obtained.

Primary amines carrying a tertiary alkyl group attached to the amino group give N-alkylhydroxylamines on being oxidised with permonosulphuric acid (*E. Bamberger* and *R. Seligman*, Ber., 1903, **36**, 685) and nitroalkanes with potassium permanganate (*N. Kornblum* and *R. J. Cutter*, J. Amer. chem. Soc., 1954, **76**, 4494):

$$(CH_3)_3C\cdot NH_2 \longrightarrow (CH_3)_3C\cdot NHOH \longrightarrow (CH_3)_3C\cdot NO \longrightarrow (CH_3)_3C\cdot NO_2$$

Some secondary amines are oxidised to benzoyl esters of N,N-dialkylhydroxylamines by reaction with benzoyl peroxide:

$$(C_2H_5)_2NH + (C_6H_5CO\cdot O)_2 \longrightarrow (C_2H_5)_2N\cdot O\cdot COC_6H_5 + C_6H_5\cdot CO_2H$$

(*S. Gambarjan*, Ber., 1925, **58**, 1775). Tertiary amines give amine oxides (p. 142) when they are oxidised.

(h) *Action of nitrous acid.* The different classes of amines may be distinguished by their different reactions with this reagent. Primary amines react with nitrous acid (sodium nitrite and a mineral acid) and yield nitrogen quantitatively. This is the basis of a widely used method of estimating amino-acids (*Van Slyke*). Methylamine and ethylamine give, in addition, fair yields of the corresponding alcohol:

$$CH_3\cdot NH_2 + HONO \longrightarrow CH_3\cdot OH + H_2O + N_2$$

but this is not a general reaction. Higher amines give complex mixtures of products; *n*-butylamine, when treated with sodium nitrite and hydrochloric acid, give *n*-butanol (25%), *sec.*-butanol (13%), *n*-butenes (36%), *n*-butyl chloride (5%), *sec*-butyl chloride (3%), and higher-boiling products (*F. C. Whitmore* and *D. P. Langlois*, J. Amer. chem. Soc., 1932, **54**, 3441). The products obtained from primary amines are qualitatively in accord with those expected if an intermediate carbonium ion is formed in the reaction (see *L. G. Cannell* and *R. W. Taft*, *ibid.*, 1956, **78**, 5812; *J. H. Ridd*, Chem. Reviews, 1961, **15**, 418, for a discussion).

Secondary amines are converted into nitrosamines, $R_2N\cdot NO$, (p. 150), and tertiary amines are unaffected by nitrous acid in fairly concentrated acid solution ($pH < 3$). At pH 3–6 tertiary amines may undergo nitrosative cleavage to give N-nitroso derivatives of secondary amines (*G. E. Hein*, J. chem. Educ., 1963, **40**, 181).

(i) *Reaction with cyanogen bromide* (von Braun reaction). Tertiary amines react with, and are cleaved by, cyanogen bromide:

$$R_3N + CNBr \longrightarrow R\cdot Br + R_2N\cdot CN$$

This reaction has been widely used for degradation of natural products in structure determinations (*H. A. Hageman*, Org. Reactions, 1953, **7**, 198).

(j) *Dealkylation.* Amine salts, on being heated, are dealkylated:

$$(CH_3)_3\overset{\oplus}{N}H\overset{\ominus}{Cl} \longrightarrow (CH_3)_2NH + CH_3Cl$$

(*A. Hofmann,* Proc. roy. Soc., 1860, **10,** 595). A useful variant of this reaction is cleavage of lower alkyl groups, particularly methyl groups, from amines by hydrogen iodide. The conditions for cleavage (300–360°) are much more vigorous than for cleavage of alkoxyl groups:

$$>N\cdot CH_3 + HI \longrightarrow [>NH\cdot CH_3]^{\oplus}I^{\ominus} \longrightarrow >N\cdot H + CH_3I$$

(*J. Herzig* and *H. Meyer,* Ber., 1894, **27,** 319; Monatsh., 1897, **18,** 379). This reaction has been used for quantitative estimation of *N*-methyl groups in natural products.

(3) *Addition reactions.* (k) Primary and secondary amines react with *aldehydes* and, much less readily, with *ketones* (see also Methods (j) and (k) under Preparation of Amines, p. 117).

(1) With *formaldehyde* amines give hydroxymethyl derivatives which may react with a second molecule of amine:

$$CH_3\cdot NH_2 + HCHO \longrightarrow CH_3\cdot NH\cdot CH_2OH$$

$$CH_3\cdot NH\cdot CH_2OH + CH_3\cdot NH_2 \longrightarrow (CH_3\cdot NH)_2CH_2 + H_2O$$

(2) With higher aliphatic amines the hydroxyamine first formed is not isolable, but spontaneously loses water to give an azomethine, or Schiff's base:

$$CH_3\cdot NH_2 + CH_3\cdot CHO \longrightarrow [CH_3\cdot NH\cdot CH(OH)CH_3] \longrightarrow CH_3\cdot N{:}CH\cdot CH_3 + H_2O$$

Schiff's bases of this kind are often unstable and polymerize easily to polymers of uncertain structure.

(3) With *aromatic aldehydes* stable Schiff's bases are the usual product:

$$CH_3\cdot NH_2 + C_6H_5\cdot CHO \longrightarrow CH_3\cdot N{:}CH\cdot C_6H_5 + H_2O$$

(l) *Reaction with carbon dioxide and carbon disulphide.* Primary and secondary amines react with carbon dioxide (and will readily absorb it from air if they are exposed to it) to give carbamates:

$$2R_2NH + CO_2 \longrightarrow [R_2N\cdot CO_2]^{\ominus}H_2\overset{\oplus}{N}R_2$$

In a similar way they react with carbon disulphide to give dithiocarbamates:

$$2C_2H_5\cdot NH_2 + CS_2 \longrightarrow [C_2H_5\cdot NH\cdot CS_2]^{\ominus}H_3\overset{\oplus}{N}\cdot C_2H_5$$

Primary alkyl dithiocarbamates react with mercuric chloride to give alkyl isothiocyanates:

$$[C_2H_5\cdot NH\cdot CS_2]^{\ominus}H_3\overset{\oplus}{N}\cdot C_2H_5 + HgCl_2 \longrightarrow C_2H_5\cdot NCS + C_2H_5\overset{\oplus}{N}H_3\overset{\ominus}{Cl} + HgS + HCl$$

(m) *Reaction with unsaturated compounds.* Primary and secondary amines add to activated unsaturated compounds to give more highly alkylated amines (see Methods (e) and (t), pp. 115 and 120, under Preparation of Amines).

(n) *Inorganic cyanates* or *organic isocyanates* react with primary and secondary amines to give substituted ureas; inorganic thiocyanates or organic isothiocyanates give substituted thioureas:

$$R\cdot \overset{\oplus}{N}H_3CN\overset{\ominus}{O} \longrightarrow R\cdot NH\cdot CO\cdot NH_2$$

$$R\cdot NH_2 + R'\cdot NCO \longrightarrow R\cdot NH\cdot CO\cdot NH\cdot R'$$

$$R\cdot \overset{\oplus}{N}H_3CN\overset{\ominus}{S} \longrightarrow R\cdot NH\cdot CS\cdot NH_2$$

$$R\cdot NH_2 + R'\cdot NCS \longrightarrow R\cdot NH\cdot CS\cdot NH\cdot R'$$

Reactions with phenyl and 1-naphthyl isocyanates are often used to prepare crystalline derivatives for characterizing amines.

(o) Compounds containing *cyano* groups react with primary and secondary amines. Nitriles give amidines:

$$R\cdot NH_2 + R'\cdot CN \longrightarrow RNH\cdot C(:NH)R'$$

Substituted cyanamides give guanidine derivatives:

$$R\cdot NH_2 + R_2'N\cdot CN \longrightarrow RNH\cdot C(:NH)\cdot NR'_2$$

(p) Primary and secondary amines react readily with *epoxides* to give β-hydroxyalkylamines (alkanolamines):

$$R_2NH + \underbrace{CH_2{-}CH_2}_{\diagdown O \diagup} \longrightarrow R_2N\cdot CH_2\cdot CH_2OH$$

With homologues of ethylene oxide the product usually has the amino group attached to the carbon atom carrying the greater number of hydrogen atoms:

$$C_2H_5\cdot NH_2 + CH_3\cdot \underbrace{CH{-}CH_2}_{\diagdown O \diagup} \longrightarrow C_2H_5\cdot NH\cdot CH_2\cdot CH(OH)CH_3$$

(*R. E. Parker* and *N. S. Isaacs*, Chem. Reviews, 1959, 59, 737).

(q) Primary and secondary amines, usually in the form of their salts, react with a mixture of *formaldehyde* and a *compound containing an "active" hydrogen atom* (*e.g.* aldehydes, ketones, β-keto-esters, 1-alkynes, nitroalkanes) to give a so-called Mannich base, *e.g.*

$$C_6H_5\cdot CO\cdot CH_3 + HCHO + (C_2H_5)_2\overset{\oplus}{N}H_2Cl^{\ominus} \longrightarrow$$

$$C_6H_5\cdot CO\cdot CH_2\cdot CH_2\cdot \overset{\oplus}{N}H(C_2H_5)_2\overset{\ominus}{Cl} + H_2O$$

(For reviews of this important synthetic Mannich reaction see *F. F. Blicke*, Org. Reactions, 1942, 1, 303; *R. Jacquier, M. Mousseron* and *S. Boyer*, Bull. Soc. chim. Fr., 1956, 1653).

(4) *Miscellaneous reactions.* (r) Primary amines give carbylamines (isocyanides) on being warmed with *chloroform* and *alkali* (carbylamine reaction):

$$R\cdot NH_2 + CHCl_3 + 3KOH \longrightarrow R\cdot NC + 3KCl + 3H_2O$$

The toxic isocyanides are easily recognized by their intense nauseating smells; the reaction is sometimes used as a qualitative test for primary amines.

(s) Primary and secondary amines are attacked by *halogens*, but the reaction does not give useful products. When tertiary amines are chlorinated in aqueous solution, good yields of *N*-chloroamines are obtained:

$$(CH_3\cdot CH_2)_3N \xrightarrow{Cl_2,\ H_2O} (CH_3\cdot CH_2)_2N\cdot Cl + CH_3CHO + HCl$$

(*J. Meisenheimer*, Ber., 1913, 46, 1148).

(t) Treatment of primary and secondary amines with *nitric acid* or, better, acetone cyanhydrin nitrate gives nitramines (p. 148).

(u) Both primary and secondary amines react with *Grignard reagents* to form a magnesium derivative of the amine, and a hydrocarbon.

$$R_2NH + R'\cdot MgX \longrightarrow R_2N\cdot MgX + R'H$$

Methylmagnesium halides can, consequently, be used to estimate primary and secondary amines quantitatively; both classes give one mole of methane per mole of amine.

(v) Amines form *metal derivatives* (thus exhibiting acidic properties) with alkali and alkaline-earth metals by direct reaction with the metals:

$$2CH_3\cdot NH_2 + 2Li \longrightarrow 2CH_3\cdot NHLi + H_2$$

or by reaction with metal alkyls:

$$(CH_3)_2NH + CH_3Li \longrightarrow (CH_3)_2N\cdot Li + CH_4$$

These substituted metal amides have synthetic uses, *e.g.*, in introducing substituted amino groups into aromatic rings:

$$C_6H_5Cl + (CH_3)_2N\cdot Li \longrightarrow C_6H_5\cdot N(CH_3)_2 + LiCl$$

(see *H. Gilman* and *R. H. Kyle,* J. Amer. chem. Soc., 1952, 74, 3027).

Under somewhat different conditions certain alkali metals will dissolve unchanged in lower primary alkylamines (cf. ammonia). The solutions so formed are very powerful reducing agents and will, *e.g.*, reduce benzene to a mixture of cyclohexene and cyclohexane (see *A. J. Birch* and *H. Smith,* Quart. Reviews, 1958, **12,** 17).

*(iii) Individual amines**

Methylamine is used as an intermediate in the manufacture of drugs, photographic chemicals, dyes, and explosives.

Dimethylamine, the most valuable of the three methylamines at present, is used in the dehairing of skins and hides, and in the manufacture of vulcanization accelerators for rubber.

Trimethylamine is used principally to prepare various quaternary ammonium salts (*q.v.*) which are bactericides and antiseptics.

Ethylamines are prepared by ammonolysis either of ethanol over a metal catalyst at 150–230° and 200–300 p.s.i. (*J. F. Olin* and *J. F. McKenna,* U.S.P., 2,365,721/1944), or of diethyl ether over various types of catalyst at 300–450° and 200–400 p.s.i. (*R. H. Goshorn,* U.S.P., 2,389,500/1945), and are used as chemical intermediates for corrosion inhibitors, emulsifiers, vulcanization accelerators, etc.

Propylamines and **butylamines** are prepared from alcohols and ammonia in catalytic, vapour-phase, reactions, and find similar uses to ethylamines.

Amylamines are prepared by reaction between ethanolic ammonia and amyl chlorides (*C. K. Hunt,* Ind. Eng. Chem., 1943, **35,** 1048). They are used as dyestuffs intermediates, and as components of wetting agents for froth flotation of mineral ores.

(For fuller information on methods of preparation on an industrial scale, and industrial uses, of alkylamines see articles under names of individual amines etc. in the following: *R. A. Kirk* and *D. F. Othmer,* Editors, "Encyclopedia of Chemical Technology", The Interscience Encyclopedia Inc., New York, 1947–1954; *J. F. Thorpe* and *M. A. Whiteley,* Eds., "Thorpe's Dictionary of Applied Chemistry", 4th Edn., Longmans, Green and Co., London, 1937–1954; *W. Foerst,* Ed., "Ullmann Encyklopädie der technischen Chemie", 3rd Edn., Urban and Schwarzenberg, München-Berlin, 1951–1961).

* For physical properties see Table 2, p. 122.

(b) Halogen-substituted alkylamines

Since alkyl halides react with all classes of amines it is not surprising that the halogen-substituted alkylamines are unstable compounds, which readily undergo self-condensation either intramolecularly, to give cyclic compounds, or intermolecularly, to give either cyclic compounds or polymers; *e.g.* 1-amino-5-chloropentane gives piperidine hydrochloride very rapidly:

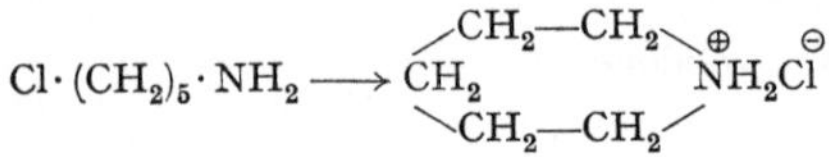

Piperidine hydrochloride

2-Chloro-*N*-methylethylamine, however, gives both *N*-methylethyleneimine and *N*,*N'*-dimethylpiperazine hydrochlorides:

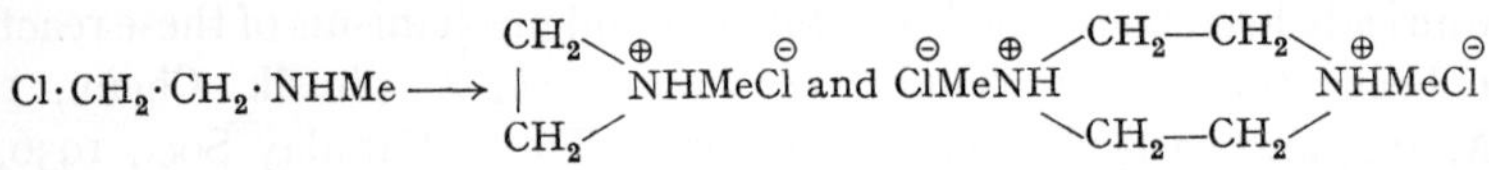

The β-chloroalkylamines have been the most fully studied class of halogen-substituted alkylamines. The vesicant activity of tris-(2-chloroethyl)amine was observed by *K. Ward* (J. Amer. chem. Soc., 1935, 57, 914) and the structural similarity between this compound, $(ClCH_2CH_2)_3N$, and "mustard gas", $(ClCH_2CH_2)_2S$, is apparent. Analogous amines were studied, particularly di-(2-chloroethyl)methylamine (*O. Eisleb*, Ber., 1941, 74, 1433), and large quantities of a number of these di- and tri-(β-chloroalkyl)amines, the so-called "nitrogen mustards", were prepared during World War II for possible use as chemical warfare agents.

More recently, these compounds have been used as chemotherapeutic agents for certain types of cancer, and their uses in this field are being actively studied. (For reviews see *A. Gellhorn*, Cancer Res., 1953, 13, 205; *W. Davis*, Mfg. Chemist, 1960, 31, 233).

Preparation. Generally the halogen-substituted alkylamines are made by a combination of reactions which introduce the amino group and the halogen atom separately; *e.g.* (1) by treating an alkanolamine with thionyl chloride or a halogen acid (*W. E. Hanby* and *H. N. Rydon*, J. chem. Soc., 1947, 513), or (2) alternatively by application of the Gabriel synthesis to a dihalide:

$$C_6H_4(CO)_2N\cdot K + Br\cdot(CH_2)_n\cdot Br \longrightarrow C_6H_4(CO)_2N\cdot(CH_2)_n\cdot Br \longrightarrow$$

$$H_2N\cdot(CH_2)_n\cdot F + C_6H_4(CO_2H)_2$$

or (3) by the addition of hydrogen halides to unsaturated amines (*C. Paal* and *C. Hermann*, Ber., 1889, **22,** 3076).

(4) An interesting new method is the cleavage of an oxazolidone with a halogen acid (*T. F. Wood*, U.S.P., 2,617,825):

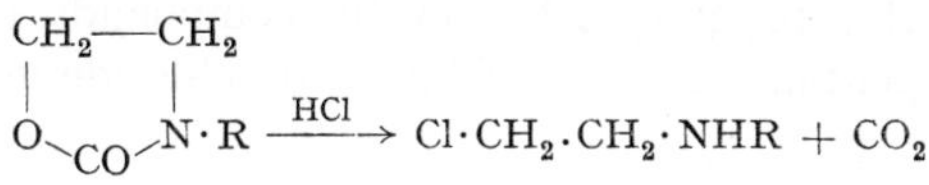

Oxazolidone

(5) 2-Fluoroethylamine has been prepared by lithium aluminium hydride reduction of fluoroacetamide, and this method may have wider applicability.

Properties and reactions. Since the free halogenoalkylamines are relatively unstable they are usually stored as salts. In addition to the usual reactions of alkylamines, the halogenoalkylamines undergo cyclization reactions which have already been mentioned. The kinetics and mechanisms of these reactions have been studied extensively (*H. Freundlich,* Z. physik. Chem., 1933, **166A,** 161, and earlier papers; *G. Salomon,* Trans. Faraday Soc., 1936, **32,** 153; *B. Cohen, E. R. van Artsdalen,* and *J. Harris,* J. Amer. chem. Soc., 1948, **70,** 281). The rates of solvolysis of 2-halogenoethylamines are considerably more rapid than those of ethyl halides and it is clear that an ethylene-imonium ion is involved as an intermediate (*P. D. Bartlett et al,* J. Amer. chem. Soc., 1947, **69,** 2971). This ion can form products by reacting with any nucleophilic species present, thus:

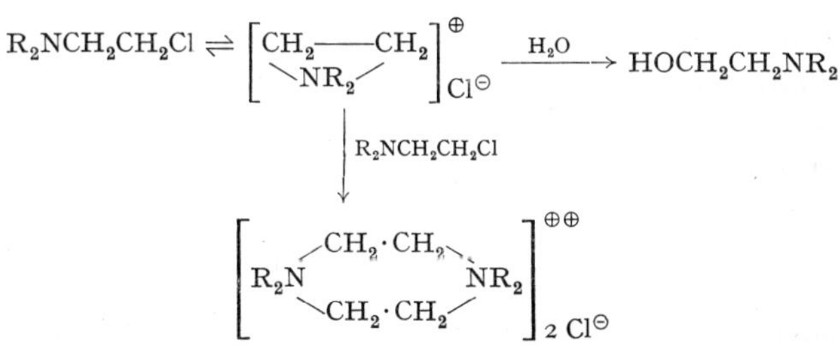

Instead of dimeric (piperazine) derivatives (see C.C.C., 1st Edn. 1959, Vol. IVB, p. 1338), linear polymers may be formed (*C. S. Marvel et al.*, J. Amer. chem. Soc., 1933, **55,** 753; 1934, **56,** 725; 1935, **57,** 117). For a discussion of the mechanism of solvolytic displacement reactions in these systems, see *A. Streitwieser,* Chem. Reviews, 1956, **56,** 677.

The reactions of the nitrogen mustards have been examined very fully. The halogen atom is quite reactive and these compounds react with α-amino groups of amino-acids, with the —SMe group of methionine, with the imidazole group of histidine, and with many vitamins (*M. Bergmann et al.*, J. org. Chem., 1946, **11**, 518 *et seq.*). Treatment of nitrogen mustards with per-acids yields N-*oxides* (*M. A. Stahmann* and *M. Bergmann*, J. org. Chem., 1946, **11**, 586); these substances have also been investigated as potential chemotherapeutic agents for cancer (*M. Ishidate et al.*, Proc. Japan Acad., 1951, **27**, 493).

2-**Fluoroethylamine,** b.p. 39–42°; 2-*chloroethylamine hydrochloride*, m.p. 144°; 2-*bromoethylamine hydrobromide*, m.p. 174–175°; 2-*iodoethylamine hydriodide*, m.p. 192–194°. 2-*Chloroethyldimethylamine*, $(CH_3)_2$-$NCH_2 \cdot CH_2Cl$, b.p. 110°; 2-*chloroethyldiethylamine*, b.p. 51–52°/16 mm; 2-*bromoethyldiethylamine* b.p. 63°/8 mm. 2-*Chloropropylamine hydrochloride*, m.p. 179°; 3-*chloropropyldimethylamine*, b.p. 135°; 3-*bromopropyldiethylamine*, b.p. 60–64°/7 mm. 4-*Bromobutyldiethylamine*, b.p. 68–70°/6 mm. *Bis*-(2-*chloroethyl*)*amine hydrochloride*, m.p. 216°; *bis*-(2-*chloroethyl*)*methylamine*, $(ClCH_2CH_2)_2NMe$, b.p. 87°/18 mm.; *bis*-(2-*chloroethyl*)*ethylamine*, b.p. 85°/10 mm.; *bis*-(2-*chloroethyl*)-n-*propylamine*, b.p. 88–89°/5·5 mm.; *tris*-(2-*chloroethyl*)*amine*, b.p. 144°/15 mm., *hydrochloride*, m.p. 130–131°.

(*c*) *Nitroalkylamines*

Preparation. (1) Addition of ammonia or a primary or secondary amine to a nitro-olefin:

$$CH_3 \cdot CH{:}CH \cdot NO_2 + NH_3 \longrightarrow CH_3 \cdot CH(NH_2) \cdot CH_2NO_2$$

(*R. L. Heath* and *J. D. Rose*, J. chem. Soc., 1947, 1486).

(2) The Mannich reaction using a nitroalkane with an amine and formaldehyde. Monoaminonitroalkanes are obtained from primary or secondary amines and secondary nitroparaffins, diaminonitroalkanes result from primary nitroparaffins, thus:

$$RNH_2 \xrightarrow{CH_2O} RNHCH_2OH \xrightarrow{R'_2CHNO_2} RNHCH_2C(R')_2NO_2$$

$$RCH_2NO_2 \xrightarrow{CH_2O} (HOCH_2)_2CRNO_2 \xrightarrow{R'NH_2} (R'NHCH_2)_2CRNO_2$$

$$R_2NH \xrightarrow{CH_2O} R_2NCH_2OH \xrightarrow{R'CH_2NO_2} (R_2NCH_2)_2CR'NO_2$$

(see p. 99 and *M. Senkus*, J. Amer. chem. Soc., 1950, **72**, 2069; *H. G. Johnson*, *ibid.*, 1946, **68**, 12).

Nitroalkylamines carrying a primary amino group are unstable and decompose to tarry products if they are stored at ordinary temperatures. However, they can be reduced to diamines. The Mannich bases produced in reaction (2) above can be used, like other Mannich bases, in alkylation reactions (*G. L. Shoemaker* and *R. W. Keown*, *ibid.*, 1954, 76, 6374).

2-**Nitroethylamine,** 1-*amino-2-nitroethane*, $NO_2\cdot CH_2\cdot CH_2\cdot NH_2$, is too unstable to be distilled; 2-*amino-1-nitropropane*, $CH_3CH(NH_2)\cdot CH_2NO_2$, b.p. 50–55°/10 mm., *hydrochloride*, m.p. 114°; 2-*amino-3-nitrobutane*, b.p. 75–78°/20 mm., *hydrochloride*, m.p. 115°; 2-*amino-2-methyl-1-nitro-propane*, b.p. 62·5°/11 mm., *hydrochloride*, m.p. 182°; 2-*methyl-1-methylamino-2-nitropropane*, b.p. 60–62°/6 mm; 1,3-*bis*(*diethylamino*)-2-*nitropropane*, b.p. 110°/4 mm.

(*d*) *Quaternary ammonium compounds*

Quaternary ammonium compounds are derivatives of ammonium ion, $NH_4^{\oplus}$, in which all four hydrogen atoms are replaced by alkyl groups. They are usually named as derivatives of ammonium, and where more than one kind of alkyl group is attached to the nitrogen atom alkyl radicals are given in alphabetical order:

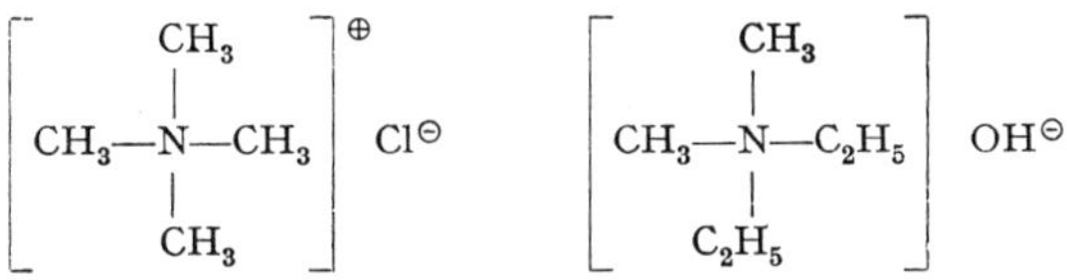

Tetramethylammonium chloride   Diethyldimethylammonium hydroxide

Quaternary ammonium compounds occur in nature, frequently in biologically vital roles, *e.g.*, acetylcholine, $(CH_3)_3\overset{\oplus}{N}\cdot CH_2\cdot CH_2\cdot O\cdot CO\cdot CH_3\overset{\ominus}{Cl}$, is involved in the transmission of mammalian nerve impulses.

Preparation. Quaternary ammonium salts are almost invariably prepared by direct nucleophilic attack of a tertiary amine on an alkyl halide or sulphate:

In these compounds all four nitrogen-carbon bonds are equivalent; reaction of trimethylamine with ethyl iodide gives the same compound as does reaction of ethyldimethylamine with methyl iodide:

$$(CH_3)_3N + C_2H_5I \longrightarrow (CH_3)_3(C_2H_5)\overset{\oplus\ominus}{NI} \longleftarrow (CH_3)_2(C_2H_5)N + CH_3I$$

The tetrahedral arrangement of the four carbon-nitrogen bonds was first demonstrated by resolution of allylbenzylmethylphenylammonium iodide into optical isomers (*W. J. Pope* and *S. J. Peachey*, J. chem. Soc., 1899, 75, 1127).

Quaternary ammonium halides or sulphates, produced by direct reaction between tertiary amine and alkyl halide or sulphate, can be converted into corresponding hydroxides, which are strong bases, by treatment with silver oxide, alcoholic potassium hydroxide, or the hydroxyl form of an anion-exchange resin. By neutralisation of a quaternary hydroxide with the appropriate acid, many salts which would not be available by direct synthesis may be prepared:

$$(C_2H_5)_3N + CH_3I \longrightarrow (C_2H_5)_3\overset{\oplus}{N}CH_3\overset{\ominus}{I} \xrightarrow[H_2O]{Ag_2O}$$

$$(C_2H_5)_3\overset{\oplus}{N}CH_3\overset{\ominus}{OH} \xrightarrow{HNO_3} (C_2H_5)_3\overset{\oplus}{N}CH_3\overset{\ominus}{NO_3}$$

Properties and reactions. Tetra-alkylammonium salts are usually crystalline solids which are frequently soluble in water or other polar solvents, and which show the reactions of the anion in solution. The salts frequently decompose on being heated, and so cannot be characterized by their melting points. Quaternary ammonium hydroxides are deliquescent solids or liquids which form stable hydrates; it is difficult to prepare them in a completely anhydrous form. The corresponding alkoxides are formed when tetra-alkylammonium salts are reacted with metal alkoxides in alcoholic solution. Quaternary ammonium compounds containing long-chain alkyl groups are surface-active agents and have the properties of soaps.

Quaternary ammonium compounds are thermally unstable. Halides generally decompose on being heated with loss of alkyl halide; if the compound carries a methyl group attached to nitrogen, methyl halide is preferentially eliminated:

$$(C_2H_5)_3\overset{\oplus}{N}CH_3Cl^{\ominus} \longrightarrow (C_2H_5)_3N + CH_3Cl$$

Quaternary hydroxides give olefins if an elimination reaction is possible; otherwise alcohols result:

$$(CH_3)_3\overset{\oplus}{N}\cdot CH_2\cdot CH_3\overset{\ominus}{OH} \longrightarrow (CH_3)_3N + CH_2{:}CH_2 + H_2O$$

$$(CH_3)_4\overset{\oplus}{N}\overset{\ominus}{OH} \longrightarrow (CH_3)_3N + CH_3OH$$

The elimination reaction is the basis of Hofmann's "exhaustive methylation" procedure for determining the structures of naturally occurring amines. (For the mechanism of thermal decomposition of the hydroxide see Vol. IA, p. 287 and *C. K. Ingold*, "Structure and Mechanism in Organic Chemistry", Cornell University Press, Ithaca, N.Y., 1953, p. 149.)

Quaternary ammonium compounds can act as alkylating agents and have been widely used in organic synthesis for this purpose. A simple example of the reaction using an alkoxide is as follows:

$$(CH_3)_4\overset{\oplus\ominus}{N}OC_2H_5 + CH_2(CO_2C_2H_5)_2 \longrightarrow CH_3{\cdot}CH(CO_2C_2H_5)_2 + (CH_3)_3N + C_2H_5OH$$

(*G. Wittig, M. Heintzeler* and *M. H. Wetterling*, Ann., 1945, 557, 201. For full reviews see *J. H. Brewster* and *E. L. Eliel*, Org. Reactions, 1953, 7, 99; *H. Hellman*, Angew. Chem., 1953, 65, 473).

A number of attempts to prepare derivatives of 5-covalent nitrogen from quaternary ammonium compounds have been unsuccessful. Alkylation of quaternary halides with metal derivatives of hydrocarbons does give compounds containing nitrogen and five hydrocarbon radicals; the properties of these compounds, however, show them to be quaternary ammonium salts of carbanions, *e.g.*:

$$(CH_3)_4\overset{\oplus\ \ominus}{NCl} + (C_6H_5)_3\overset{\ominus\ \oplus}{CNa} \longrightarrow (CH_3)_4\overset{\oplus\ominus}{N}C(C_6H_5)_3 + NaCl$$

(*W. Schlenk* and *J. Holtz*, Ber., 1916, 49, 604).

Electrolysis of quaternary ammonium salts with a mercury cathode gives tetra-alkylammonium amalgams, presumably containing $R_4N{\cdot}$ radicals (*G. B. Porter*, J. chem. Soc., 1954, 760). Tetramethylammonium radicals produced in this way decompose to give some methyl radicals:

$$(CH_3)_4N{\cdot} \longrightarrow CH_3{\cdot} + (CH_3)_3N$$

Technically important quaternary ammonium salts. Many quaternary ammonium salts have been tested as surface-active agents and bactericides. Long-chain alkyl compounds are particularly active, and in compounds of the type $R\overset{\oplus}{N}(CH_3)_3\overset{\ominus}{X}$ the highest germicidal activity is found when R is a straight-chain alkyl group containing 16–18 carbon atoms; the nature of the anion, $X^{\ominus}$, is relatively unimportant.

Tetramethylammonium iodide, $(CH_3)_4\overset{\oplus\ominus}{NI}$, m.p. 230° (dec.), has been used as a disinfectant.

Tetraethylammonium chloride, [*tetrahydrate,* $(C_2H_5)_4\overset{\oplus\ \ominus}{NCl}{\cdot}4H_2O$, m.p. 37·5°] and *bromide,* are deliquescent solids; they have been used in medicine for their pronounced effects on the nervous system.

Cetyltrimethylammonium bromide ("Cetrimide", "Cetavlon") is a mixture of alkylammonium bromides, principally $C_{16}H_{33}\overset{\oplus}{N}(CH_3)_3\overset{\ominus}{Br}$, prepared from cetyl bromide and trimethylamine. Its solutions are widely used as antiseptics.

(e) Unsaturated amines and ammonium compounds

In this class of compounds only derivatives of vinylamine, $CH_2{:}CH{\cdot}NH_2$, will be discussed, since other unsaturated amines, *e.g. allylamine,* $CH_2{:}CH{\cdot}$

CH_2NH_2, b.p. 58°, can be prepared by extensions of the methods already given for saturated amines, and their properties are not exceptional (cf. Org. Synth. Coll., Vol. II, p. 24).

Vinylamine itself is not known, though its polymers

$$CH_3\cdot\overset{\overset{\displaystyle NH_2}{|}}{C}H\cdot[CH_2\cdot CH(NH_2)]_n\cdot CH{:}CHNH_2$$

can be prepared indirectly (*D. D. Reynolds* and *W. W. Kenyon*, J. Amer. chem. Soc., 1947, **69**, 911). Many substituted vinylamines are known. Primary vinylamines are stable if an electron-attracting group is attached to the double bond:

$$n\text{-}C_3H_7\cdot CO\cdot CH{:}CHCl + NH_3 \longrightarrow n\text{-}C_3H_7\cdot CO\cdot CH{:}CH\cdot NH_2$$

(*A. N. Nesmeyanov, N. K. Kochetkov* and *Y. V. Dombrovskii*, Izvest. Akad. Nauk S.S.S.R., Otd. Khim. Nauk, 1955, 179):

$$p\text{-}O_2N\cdot C_6H_4\cdot SO_2CH{:}CHCl + NH_3 \longrightarrow p\text{-}O_2N\cdot C_6H_4\cdot SO_2\cdot CH{:}CH\cdot NH_2$$

(*F. Montanani*, Gazz., 1956, **86**, 406). Other compounds of this type have been used as reactive intermediates in synthesis, *e.g.*:

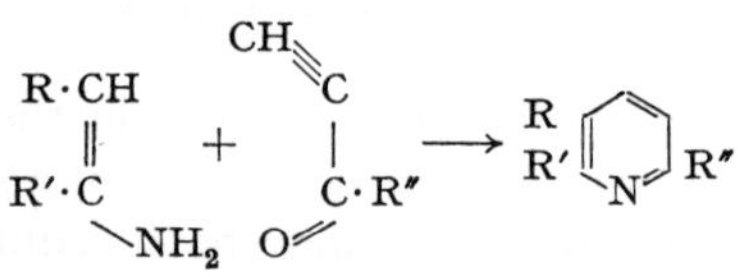

(*F. Bohlmann* and *D. Rahtz*, Ber., 1957, **90**, 2265).

Tertiary vinylamines (**enamines**) have been prepared by reaction between aldehydes and secondary amines:

$$C_6H_{13}\cdot CHO + 2(C_2H_5)_2NH \longrightarrow C_6H_{13}CH[N(C_2H_5)_2]_2 \longrightarrow$$
$$C_5H_{11}\cdot CH{:}CH\cdot N(C_2H_5)_2 + (C_2H_5)_2NH$$

(*C. Mannich et al., ibid.*, 1936, **69**, 2106, 2112).

They can be reduced to saturated amines by reaction with formic acid, and they may be intermediates in the Leuckart alkylation of amines (p. 117):

$$C_7H_{15}\cdot CH{:}CH\cdot N(CH_3)_2 + HCO_2H \longrightarrow C_7H_{15}\cdot CH_2\cdot CH_2\cdot N(CH_3)_2 + CO_2$$

(*P. L. De Benneville* and *J. H. Macartney*, J. Amer. chem. Soc., 1950, **72**, 3073). Enamines are generally unstable liquids which can be distilled without decomposition only under reduced pressure. (For a review of their chemistry see *J. Kloubek*, Chem. Listy, 1959, **53**, 821; *J. Smuszkovicz*, Adv. in org. Chem., 1963, **4**, 1). Enamines are useful in the alkylation of carbonyl compounds.

Reaction of acetylene with primary or secondary amines probably gives vinylamines as transient intermediates which react rapidly with a further molecule of acetylene:

$$(C_2H_5)_2NH + C_2H_2 \longrightarrow [(C_2H_5)_2N\cdot CH{:}CH_2] \xrightarrow{C_2H_2} (C_2H_5)_2N\cdot CH(CH_3)\cdot C{\vdots}CH$$

(*C. Gardner et al.*, J. chem. Soc., 1949, 780; *W. Reppe et al.*, Ann., 1955, **596**, 14).

Trimethylvinylammonium hydroxide (*neurin*), $[(CH_3)_3\overset{\oplus}{N}CH{:}CH_2]\overset{\ominus}{O}H$, is found in the brain, in egg yolk, and also in putrefying animal matter, where it is probably formed by dehydration of choline (cf. *R. R. Renshaw* and *J. C. Ware*, J. Amer. chem. Soc., 1925, **47**, 2992). It can be prepared by reaction between acetylene and aqueous trimethylamine below 50°:

$$(CH_3)_3N + CH{\vdots}CH + H_2O \longrightarrow [(CH_3)_3\overset{\oplus}{N}\cdot CH{:}CH_2]\overset{\ominus}{O}H$$

Thermal decomposition of neurin chloride, $[(CH_3)_3\overset{\oplus}{N}CH{:}CH_2]\overset{\ominus}{Cl}$ gives a very unstable liquid which polymerises rapidly, and is probably N,N-dimethylvinylamine, $(CH_3)_2N\cdot CH{=}CH_2$ (*Gardner et al., loc. cit.*).

3. *N*-Substituted derivatives of the alkylamines

(*a*) N-*Halogenoalkylamines*

Either one or both of the hydrogen atoms on the nitrogen atom of a primary amine, or the one on the nitrogen atom of a secondary amine, can be replaced by halogen to give an N-halogenoalkylamine:

$$R\cdot NH_2 \longrightarrow R\cdot NHX \longrightarrow R\cdot NX_2$$

$$R_2NH \longrightarrow R_2NX$$

Most of the available information in these series concerns the chloro-compounds, which are more stable than the bromo- or iodo-compounds; there is a little, very recent, work on fluoro-compounds. In the primary series the compounds, $R\cdot NCl_2$, are more stable than the corresponding compounds, $R\cdot NHCl$.

(*i*) *Preparation*

N,N-Difluoroamines have been prepared only recently by a reaction which has no counterpart for the other N-halogenoamines. Tetrafluorohydrazine, N_2F_4, dissociates in the gas phase into free difluoroamino radicals, $\cdot NF_2$, which can combine with alkyl free-radicals, generated, *e.g.*, by photolysis of alkyl iodides:

$$2CH_3\cdot I + N_2F_4 \xrightarrow{h\nu} 2CH_3\cdot NF_2 + I_2$$

The three known *N,N*-difluoroamines (p. 139) are all unstable liquids, sensitive to shock (*J. W. Frazer*, J. inorg. nucl. Chem., 1960, **16**, 63; *R. C. Petry* and *J. P. Freeman*, J. Amer. chem. Soc., 1961, **83**, 3912).

For the other halogens conversion of the amine into *N*-halogenoamine can be effected by elemental halogen or alkali-metal or alkaline-earth hypohalite solutions:

$$n\text{-}C_4H_9\cdot NH_2 \xrightarrow[NaHCO_3]{Cl_2} n\cdot C_4H_9\cdot NCl_2$$

$$(C_2H_5)_2NH \xrightarrow{NaOCl} (C_2H_5)_2NCl$$

(*J. Tcherniak*, Bull. Soc. chim. Fr., 1875, **24**, 451; 1876, **25**, 160; *L. K. Jackson, G. N. R. Smart*, and *G. F. Wright*, J. Amer. chem. Soc., 1947, **69**, 1539). Tertiary butyl hypochlorite has also been used to prepare *N,N*-dichloroamines:

$$(CH_3)_3C\cdot NH_2 + 2(CH_3)_3C\cdot OCl \longrightarrow (CH_3)_3C\cdot NCl_2 + 2(CH_3)_3C\cdot OH$$

(*H. Zimmer* and *L. F. Audrieth*, J. Amer. chem. Soc., 1954, **76**, 4561). Reaction between tertiary amines and sodium hypochlorite gives the corresponding *N*-chloro secondary amine:

$$(CH_3)_3N \xrightarrow{HOCl} (CH_3)_2NCl + HCHO$$

(*J. Meisenheimer*, Ber., 1913, **46**, 1148; *A. J. Ellis* and *F. G. Soper*, J. chem. Soc., 1954, 1750).

(ii) Properties and reactions

N-Halogenoalkylamines are unstable liquids, but they show less tendency to explode than do the nitrogen trihalides; they can be distilled at reduced pressure. They are easily hydrolysed by acids to amines and hypochlorous acid:

$$R_2NCl + H_2O \longrightarrow R_2NH + HOCl$$

The halogen atoms in *N*-halogenoamines are rather like those in a hypohalite, and the compounds can be used as oxidising or halogenating agents (*A. Pierson* and *K. Heumann*, Ber., 1883, **16**, 1047).

N-Halogenoalkylamines can undergo cyclization reactions when they are reacted with acids, or irradiated with u.v. light (Hoffman-Loffler reaction, see *M. E. Wolff*, Chem. Reviews, 1963, **63**, 55). These reactions are useful in the synthesis of certain heterocyclic compounds, *e.g.*:

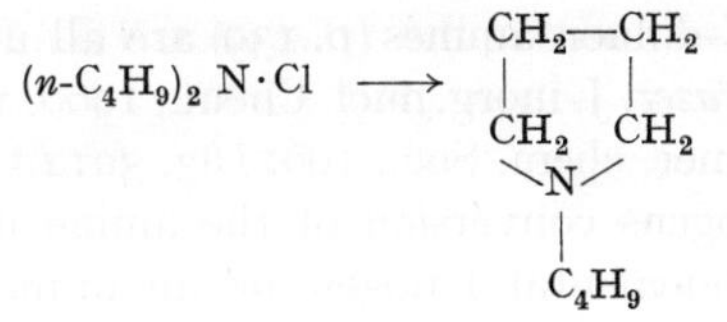

N,*N*-Dihalogenoamines carrying a primary alkyl group give nitriles when they are treated with bases:

$$R\cdot CH_2\cdot NX_2 + 2NaOCH_3 \longrightarrow R\cdot CN + 2NaX + 2CH_3OH$$

Similar compounds carrying a secondary alkyl group give rearranged products:

$$CH_3\cdot CH(NCl_2)CH_3 \xrightarrow[2.\ HCl]{1.\ NaOCH_3} CH_3\cdot CO\cdot CH_2\cdot \overset{\oplus}{N}H_3\overset{\ominus}{Cl}$$

(*H. E. Baumgarten* and *F. A. Bower*, J. Amer. chem. Soc., 1954, **76**, 4561). Secondary *N*-chloroamines undergo an elimination reaction when they are treated with alcoholic alkali, and yield aldehyde alkylimines, which can be hydrolysed by acids to aldehydes and primary amines:

$$CH_3\cdot CH_2\cdot NCl\cdot CH_2\cdot CH_2\cdot CH(CH_3)_2 \xrightarrow{KOH} CH_3CH{:}N\cdot CH_2\cdot CH_2\cdot CH(CH_3)_2$$
$$+ CH_3\cdot CH_2\cdot N{:}CH\cdot CH_2\cdot CH(CH_3)_2$$

(*A. Berg*, Bull. Soc. chim. Fr., 1897, [iii], **17**, 297).

(iii) Individual compounds

N,*N*-**Difluoromethylamine,** $CH_3\cdot NF_2$, b.p. —16°; *N,N-difluoroethylamine,* b.p. 15°; *N,N-difluoro*-tert-*butylamine,* b.p. 54–55°.

N-**Chloromethylamine,** $CH_3\cdot NHCl$, has been prepared in ethereal solution, but is too unstable to be isolated.

N,*N*-**Dichloromethylamine,** $CH_3\cdot NCl_2$, b.p. 58–60°; *N,N-dichloroethylamine,* b.p. 88°; *N,N-dichloro-1-propylamine,* b.p. 117°; *N,N-dichloro-2-propylamine,* b.p. 41–43°/15 mm.; *N,N-dichloro-1-butylamine,* b.p. 39–40°/17 mm.; *N,N-dichloro-2-methyl-2-propylamine,* $(CH_3)_3C\cdot NCl_2$, b.p. 30°/16 mm.

N-Bromomethylamine, $CH_3\cdot NHBr$, b.p. 47–48°; *N,N-dibromomethylamine,* $CH_3\cdot NBr_2$, b.p. 66–68°.

(b) N-*Oxygen derivatives of the alkylamines*

(i) Alkylhydroxylamines

Hydroxylamine, $NH_2\cdot OH$, can give rise to five series of alkylated derivatives:

(1) *N*-alkylhydroxylamines, $RNH\cdot OH$

(2) alkoxyamines or *O*-alkylhydroxylamines $NH_2\cdot OR$
(3) *N,N*-dialkylhydroxylamines, $RR'N\cdot OH$
(4) alkoxyalkylamines or *N,O*-dialkylhydroxylamines, $RNH\cdot OR'$
(5) alkoxydialkylamines or trialkylhydroxylamines, $RR'N\cdot OR''$.

Representatives of all classes have been prepared, and a few of their reactions have been examined, but, in general, alkylhydroxylamines are not well-known compounds. Recently all five *methylhydroxylamines* have been prepared, and their physical properties have been compared. Their boiling points show interesting variations, largely attributable to the various possibilities which exist for hydrogen bonding:

$(CH_3)_2N\cdot OCH_3$, b.p. 30°; $CH_3NH\cdot OCH_3$, b.p. 42°; $NH_2\cdot OCH_3$, b.p. 48°; $(CH_3)_2N\cdot OH$, b.p. 101°; $CH_3NH\cdot OH$, b.p. 115° (*T. C. Bissot, R. W. Parry* and *D. H. Campbell*, J. Amer. chem. Soc., 1957, **79**, 796). Their spectra have also been determined and interpreted (*M. Davies* and *N. A. Spiers*, J. chem. Soc., 1959, 3971).

(1) N-*Alkylhydroxylamines*, $RNH\cdot OH$. These compounds have been prepared by reduction of nitroalkanes with zinc dust and ammonium chloride solution (*E. Bamberger*, Ber., 1894, **27**, 1347), or with hydrogen, catalytically (*W. Traube* and *A. P. Schultz*, *ibid.*, 1923, **56**, 1856), or electrolytically (*A. I. Ryer* and *G. B. L. Smith*, J. Amer. chem. Soc., 1951, **73**, 5675). A new method of preparation, which is especially useful for tertiary alkyl compounds, is acid hydrolysis of an oxazirane, prepared by oxidation of a Schiff's base:

$$(CH_3)_3C\cdot NH_2 + C_6H_5\cdot CHO \longrightarrow (CH_3)_3C\cdot N{:}CH\cdot C_6H_5 \xrightarrow{CH_3CO_3H}$$

$$(CH_3)_3C\cdot \underset{\diagdown O \diagup}{N\text{——}CH}\cdot C_6H_5 \xrightarrow{H^{\oplus},\,H_2O} (CH_3)_3C\cdot NH\cdot OH$$

(*W. D. Emmons*, J. Amer. chem. Soc., 1957, **79**, 5739).

N-Alkylhydroxylamines are low-melting solids, soluble in water as mono-acid bases; they are reducing agents. They can be reduced by hydrogen iodide to primary amines, and they condense with aldehydes to give *N*-alkyloximes:

$$RNH\cdot OH + R'\cdot CHO \longrightarrow R'CH{:}\underset{\substack{\downarrow\\ O}}{N}R$$

***N*-Methylhydroxylamine,** $CH_3NH\cdot OH$, m.p. 38°; N-*ethylhydroxylamine*, m.p. 55–56°; N-*propylhydroxylamine*, m.p. 46°; N-*isopropylhydroxylamine*, m.p. 86–87°; N-tert-*butylhydroxylamine*, m.p. 64–65°.

(2) *Alkoxyamines, O-alkylhydroxylamines*, NH_2OR, are best prepared by alkylation of hydroxyurethane and hydrolysis of the resulting ester:

$$C_2H_5OOC{\cdot}NH{\cdot}OH \xrightarrow{RI} C_2H_5OOC{\cdot}NH{\cdot}OR \longrightarrow NH_2OR$$

(see, *e.g.*, *G. Palazzo, E. F. Rogers* and *G. B. Marini-Bettolo*, Gazz., 1954, 84, 915). They are water-soluble liquids and monoacid bases which are cleaved by hydrogen chloride yielding an alkyl chloride and hydroxylamine. *O*-Alkylhydroxylamines are easily alkylated further to trialkylhydroxylamines.

Methoxyamine, *O-methylhydroxylamine*, $NH_2{\cdot}OCH_3$, b.p. 48°; *ethoxyamine*, b.p. 75°; *butoxyamine*, b.p. 118°; *hexoxyamine*, b.p. 160°; *dodecoxyamine*, b.p. 120°/15 mm.

(3) N,N-*Dialkylhydroxylamines*, $R_2N{\cdot}OH$. Two general methods are available for the preparation of these compounds. They result from reaction between a Grignard reagent and an alkyl nitrate, followed by hydrolysis of the resulting complex:

$$2RMgX \xrightarrow[2.\ H_2O]{1.\ R'{\cdot}ONO_2} R_2N{\cdot}OH$$

(*H. Hepworth*, J. chem. Soc., 1921, 99, 251). A widely applicable method is thermal decomposition of tertiary amine oxides, which gives the hydroxylamine plus an olefin:

$$(CH_3{\cdot}CH_2{\cdot}CH_2)_3\overset{\oplus}{N}{-}\overset{\ominus}{O} \longrightarrow (CH_3{\cdot}CH_2{\cdot}CH_2)_2N{\cdot}OH + CH_3{\cdot}CH{:}CH_2$$

(*M. A. T. Rogers, ibid.*, 1955, 769; *A. C. Cope* and *E. R. Trumbull*, Org. Reactions, 1960, 11, 317). Direct alkylation of hydroxylamine would appear to give, as intermediate isolable products, *N,N*-dialkylhydroxylamines (*W. R. Dunstan* and *E. Goulding*, J. chem. Soc., 1899, 75, 802) but this method has not been widely used.

N,N-Dialkylhydroxylamines are usually liquids and strong monoacid bases. They can be reduced to secondary amines but are also very easily oxidised, even by air (*D. H. Johnson, M. A. T. Rogers* and *G. Trappe, ibid.*, 1956, 1093).

N,N-**Dimethylhydroxylamine,** $(CH_3)_2N{\cdot}OH$, *hydrochloride*, m.p. 107–109°; N,N-*diethylhydroxylamine*, b.p. 50°/17 mm.; N,N-*di*-n-*propylhydroxylamine*, b.p. 70°/17 mm.; N-*ethyl*-N-*methylhydroxylamine*, b.p. 30°/10 mm.; N-*methyl*-N-*propylhydroxylamine*, b.p. 99°.

(4) *Alkoxyalkylamines* or N,O-*Dialkylhydroxylamines*, $RNH{\cdot}OR'$ can be prepared by dialkylation of hydroxyurethane. It should be possible to

prepare mixed dialkyl compounds in this way, but so far only simple ones have been made:

$$C_2H_5OOC\cdot NH\cdot OH \xrightarrow{RI} C_2H_5OOC\cdot NH\cdot OR \xrightarrow{R'I} C_2H_5OOC\cdot NR'\cdot OR$$

$$\longrightarrow R'NH\cdot OR$$

(*R. T. Major* and *E. E. Fleck*, J. Amer. chem. Soc., 1928, **50**, 1479).

They are volatile liquids and reducing agents.

N-**Methoxymethylamine** (N,O-*dimethylhydroxylamine*), $CH_3NH\cdot OCH_3$, b.p. 42–43°; N-*ethoxyethylamine*, b.p. 83°.

(5) *Alkoxydialkylamines* or *trialkylhydroxylamines*, $R_2N\cdot OR'$. Further alkylation of alkoxyamines or *N*-alkoxyalkylamines gives the trialkyl compounds, which are monoacid bases but not reducing agents. They are hydrolysed by acids to give both *N,N*-dialkylhydroxylamines and secondary amines. They form crystalline methiodides, $[R_2(CH_3)N\cdot OR']^{\oplus}I^{\ominus}$, identical with those formed from the corresponding trialkylamine oxide, $R_2R'\overset{\oplus}{N}—\overset{\ominus}{O}$, and these salts have anticholinesterase activity (*G. Palazzo*, *E. F. Rogers* and *G. B. Marini-Bettolo*, Gazz., 1954, **84**, 915).

N-**Methoxydimethylamine,** *trimethylhydroxylamine*, $(CH_3)_2N\cdot OCH_3$, b.p. 30°; N-*butoxydimethylamine*, b.p. 113°; N-*amyloxydimethylamine*, b.p. 134°; N-*hexyloxydimethylamine*, b.p. 155°; N-*decyloxyldimethylamine*, b.p. 125°/15 mm.

(ii) Trialkylamine oxides

Trialkylamine oxides, $R_3N\rightarrow O$ or $R_3\overset{\oplus}{N}–\overset{\ominus}{O}$, are the primary products of oxidation of tertiary amines. They occur naturally, *e.g.* trimethylamine oxide in fish and other sources, where they are probably produced by bacterial oxidation of amines (see *C. C. J. Culvenor*, Reviews pure and appl. Chem. [Australia], 1953, **3**, 83; *Houben-Weyl*, "Methoden der Organischen Chemie", Vol. XI/2, Georg Thieme Verlag, Stuttgart, 1958, p. 190, for full reviews).

Preparation. The oxides are almost invariably produced by direct oxidation of tertiary amines.

Hydrogen peroxide is the favoured oxidising agent for the lower alkylamines (*W. R. Dunstan* and *E. Goulding*, J. chem. Soc., 1899, **75**, 792, 1004; cf. *A. C. Cope* and *P. H. Towle*, J. Amer. chem. Soc., 1949, **71**, 3426). For longer chain alkylamines, per-acids, *e.g.* 40% peroxyacetic acid, are preferred (*A. C. Cope* and *Hiok-Huang Lee*, J. Amer. chem. Soc., 1957, **79**, 964). Ozone is also a useful general oxidant for the preparation of amine oxides (*L. Horner*, *H. Schaefer* and *W. Ludwig*, Ber., 1958, **91**, 75).

Properties and reactions. Trialkylamine oxides are usually hygroscopic liquids or solids which form stable hydrates. The highly polar nature of the nitrogen–oxygen bond is shown by their large dipole moment (~ 5D for lower alkylamine oxides). Amine oxides bearing three different alkyl groups, *e.g.* allylethylmethylamine oxide, can be resolved into optical isomers, showing the tetrahedral form of the molecule (*J. Meisenheimer*, Ber., 1908, 44, 3966). The amine oxides are weak bases (pKa for $Me_3N \rightarrow O$ is 4.65) which quaternize readily with methyl iodide giving *N*-alkoxytrialkylammonium iodides:

$$(C_2H_5)_3N \rightarrow O + CH_3I \longrightarrow [(C_2H_5)_3N\text{—}OCH_3]^{\oplus}I^{\ominus}$$

The free bases corresponding to the quaternary salts, *N*-alkoxytrialkylammonium hydroxides, are much stronger bases than the amine oxides. The latter can be reduced to the parent amines by catalytic hydrogenation or by a wide variety of reducing agents.

In general, amine oxides are thermally unstable and decompose, by an intramolecular *cis*-elimination process, to give an olefin and a hydroxylamine:

$$\begin{array}{c} R_2N \rightarrow O \\ | \\ R'CH_2CH_2 \end{array} \longrightarrow R_2N{\cdot}OH + R'CH{=}CH_2$$

(*A. C. Cope* and *E. R. Trumbull*, Org. Reactions, 1960, 11, 361). If none of the attached alkyl groups can yield an olefin then a secondary amine and an aldehyde are produced:

$$(CH_3)_3N \rightarrow O \longrightarrow (CH_3)_2NH + H{\cdot}CHO$$

Treatment of methyl-substituted amine oxides with acetic anhydride brings about a similar degradation:

$$R_2CH_3N \rightarrow O \xrightarrow{(CH_3CO)_2O} R_2NH + H{\cdot}CHO$$

(see *J. C. Craig et al.*, J. Amer. chem. Soc., 1961, 83, 1871, for a discussion of this reaction).

Amine oxides containing an allyl or substituted allyl group undergo a rearrangement on being heated, and yield a substituted hydroxylamine:

$$\begin{array}{c} (C_2H_5)_2N \rightarrow O \\ | \\ CH_2{=}CH{\cdot}CH_2 \end{array} \longrightarrow (C_2H_5)_2N{\cdot}O{\cdot}CH_2{\cdot}CH{:}CH_2$$

(*Cope* and *Towle, loc. cit.*).

A similar reaction takes place with β-halogenoethylamine oxides:

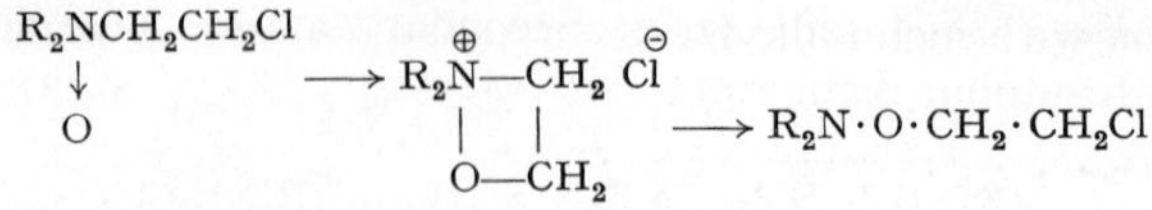

(*S. Owari*, Pharmaceutical Bulletin [Tokyo], 1953, **1**, 353; C.A., 1955, **49**, 11616).

Trimethylamine oxide, $Me_3N \rightarrow O$, m.p. 208° (sealed tube); *hydrate* ($2H_2O$), m.p. 96°, *picrate*, m.p. 200°; (because of their hygroscopic nature the free oxides have almost invariably been characterized as their picrates); *triethylamine oxide picrate*, m.p. 165°; *tri-*n*-propylamine oxide picrate*, m.p. 130°; *tri-*n*-butylamine oxide picrate*, m.p. 111°; *tri-*n*-hexylamine oxide picrate*, m.p. 51–52°; *allyldimethylamine oxide picrate*, m.p. 135·5–136·5°; *allyldi-*n*-propylamine oxide picrate*, m.p. 97°.

(c) N-*Sulphur derivatives of the alkylamines*

The action of the various sulphur halides, oxyhalides, and oxides on amines gives rise to many classes of compounds containing nitrogen–sulphur bonds. A brief survey of the types of compound formed by these reactions is given here classified under the starting material containing sulphur. (For a fuller review see *Houben-Weyl*, "Methoden der Organischen Chemie", Vol. XI/2, Georg Thieme Verlag, Stuttgart, 1958, p. 641).

(1) *Disulphur dichloride* (*sulphur monochloride*) reacts with primary amines under mild conditions, and especially conditions of high dilution, to give cyclic **tetrasulphur dialkyldiimides**:

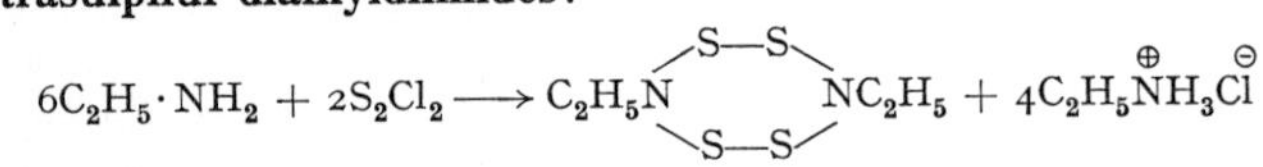

(*T. G. Levi*, Gazz., 1931, **61**, 294; *M. Becke-Goehring* and *H. Jenne*, Ber., 1959, **92**, 1149). With secondary amines disulphur dichloride gives **tetraalkyldiaminodisulphides**:

$$4\,(CH_3)_2NH + S_2Cl_2 \longrightarrow (CH_3)_2N\cdot S\cdot S\cdot N(CH_3)_2 + 2(CH_3)_2\overset{\oplus}{N}H_2\cdot\overset{\ominus}{Cl}$$

(*A. Michaelis* and *K. Luxembourg*, Ber., 1895, **28**, 165), which can be cleaved by halogens to give *N*-**dialkylaminosulphenyl halides**:

$$(CH_3)_2N\cdot S\cdot S\cdot N(CH_3)_2 + Cl_2 \longrightarrow 2(CH_3)_2N\cdot SCl$$

(*H. Freytag* and *F. Lober*, G.P., 965,968; B.P., 790,021/1958).

(2) *Sulphur dichloride*, in the one reaction reported with a primary amine, ethylamine, gave a tetrameric *compound*, $(C_2H_5\cdot N\cdot S)_4$ (*F. Lengfeld* and

J. Stieglitz, Ber., 1895, **28**, 2742). Secondary amines give **dialkylamino-sulphides** with sulphur dichloride:

$$4R_2NH + SCl_2 \longrightarrow R_2N{\cdot}S{\cdot}NR_2 + 2R_2\overset{\oplus}{N}H_2\overset{\ominus}{Cl}$$

(*E. S. Blake*, J. Amer. chem. Soc., 1943, **65**, 1267). These are stable liquids, insoluble in water, which are cleaved by hydrogen chloride, regenerating the amine and sulphur dichloride.

(3) *Sulphur dioxide* gives, with primary and secondary amines, mono- and di-**alkyl sulphinamic acids**, hygroscopic, rather unstable, solids:

$$RR'NH + SO_2 \longrightarrow RR'N{\cdot}SO_2H$$

(*Michaelis* and *O. Storbeck*, Ann., 1893, **274**, 192).

(4) *Thionyl chloride* gives **alkylsulphinylimines**, on reaction with primary amines:

$$3RNH_2 + SOCl_2 \longrightarrow R{\cdot}N{:}S{:}O + 2R\overset{\oplus}{N}H_3\overset{\ominus}{Cl}$$

Lower alkyl sulphinylimines are liquids (*ethyl thionylimine*, b.p. 70–75°, *1-propylsulphinylimine*, b.p. 104°) which are readily hydrolysed to sulphinamic acids and then to amines. They are decomposed by hydrogen chloride to thionyl chloride and the primary amine.

With secondary amines thionyl chloride gives **thionylamides**, or **sulphinyl-bis-dialkylamides**, $R_2N{\cdot}SO{\cdot}NR_2$ (*Michaelis*, Ber., 1895, **28**, 1015), which react further with thionyl chloride to give **dialkylsulphinamoyl chlorides**:

$$R_2N{\cdot}SO{\cdot}NR_2 + SOCl_2 \longrightarrow 2R_2N{\cdot}SOCl$$

The corresponding *dialkylsulphinamoyl fluorides* are made by reaction between thionyl fluoride and secondary amines at low temperature:

$$2(C_2H_5)_2NH + SOF_2 \longrightarrow (C_2H_5)_2N{\cdot}SOF + (C_2H_5)_2\overset{\oplus}{N}H_2\overset{\ominus}{F}$$

(*M. Goehring* and *G. Voigt*, Ber., 1956, **89**, 1050).

(5) *Sulphur trioxide* reacts with all classes of amine, in the vapour phase, to give **alkylsulphamic acids**:

$$C_2H_5NH_2 + SO_3 \longrightarrow C_2H_5NH{\cdot}SO_3H$$

$$(C_2H_5)_2NH + SO_3 \longrightarrow (C_2H_5)_2N{\cdot}SO_3H$$

$$(C_2H_5)_3N + SO_3 \longrightarrow (C_2H_5)_3\overset{\oplus}{N}{\cdot}\overset{\ominus}{S}O_3$$

Those derived from tertiary amines are zwitterionic (*F. Beilstein* and *E. Wiegand*, Ber., 1883, **16**, 1264; *W. Traube et al.*, *ibid.*, 1924, **57**, 1049).

They are remarkably stable compounds towards hydrolysing agents considering their composition and they are valuable sulphonating agents; "*anhydrotrimethylaminesulphamic acid*", $(CH_3)_3\overset{\oplus}{N}\cdot\overset{\ominus}{SO_3}$, m.p. 239°; "*anhydrotriethylaminesulphamic acid*", $(C_2H_5)_3\overset{\oplus}{N}\cdot\overset{\ominus}{SO_3}$, m.p. 91·5°. A convenient alternative synthesis for primary sulphamic acids is reaction between an isocyanate, or a carbamic ester, and concentrated sulphuric acid:

$$R\cdot N{:}C{:}O + H_2SO_4 \longrightarrow R\cdot NH\cdot SO_3H + CO_2$$

(*T. Bieber*, J. Amer. chem. Soc., 1953, **75**, 1405, 1409). Primary and secondary sulphamic acids can also be prepared by reaction between *N*-alkylhydroxylamines and sulphur dioxide:

$$RR'N\cdot OH + SO_2 \longrightarrow RR'N\cdot SO_3H$$

(*A. I. Ryer* and *G. B. L. Smith, ibid.*, 1951, **73**, 5675).

Tertiary sulphamic acids also result from reaction between sulphur dioxide and a tertiary amine oxide:

$$R_3\overset{\oplus}{N}\text{—}\overset{\ominus}{O} + SO_2 \longrightarrow R_3\overset{\oplus}{N}\cdot\overset{\ominus}{SO_3}$$

(*L. Mamlock* and *R. Wolffenstein*, Ber., 1901, **34**, 2499).

(6) *Sulphuryl chloride* reacts with secondary amines to give **dialkylsulphamoyl chlorides**:

$$2R_2NH + SO_2Cl_2 \longrightarrow R_2N\cdot SO_2Cl + R_2\overset{\oplus}{N}H_2\overset{\ominus}{Cl}$$

(*W. W. Binkley* and *E. F. Degering*, J. Amer. chem. Soc., 1939, **61**, 3250). The same compounds can also be prepared from *N*-chloroamines and sulphur dioxide:

$$(C_2H_5)_2N\cdot Cl + SO_2 \longrightarrow (C_2H_5)_2N\cdot SO_2Cl$$

(*K. Bodenbenner* and *R. Wegler*, G.P., 1,028,129; C.A., 1960, **54**, 14276). Sulphamoyl chlorides yield corresponding *fluorides* when they are treated with potassium fluoride or antimony trifluoride.

Sulphamoyl halides are lachrymatory liquids which can be distilled under reduced pressure. They give **sulphamoyl esters** on reaction with alkoxides, and **sulphuryl-** or **sulphonyl-bis-dialkylamides** by reaction with amines. These amides also result when sulphuryl chloride is treated with an excess of a secondary amine which is fairly resistant to oxidation and chlorination:

$$4R_2NH + SO_2Cl_2 \longrightarrow R_2N\cdot SO_2\cdot NR_2 + 2R_2\overset{\oplus}{N}H_2\overset{\ominus}{Cl}$$

(*R. Behrend,* Ann., 1884, **222,** 116). Sulphurylamides are hydrolysed by bases to sulphamic acids and amines.

(d) Boron, silicon, phosphorus, arsenic, and antimony derivatives of alkylamines

Hydrides, halides, oxyhalides etc. of the above-mentioned elements react with primary and secondary amines to give compounds containing bonds between nitrogen and the element. The preparation and chemistry of these compounds is far too large to be dealt with fully here. A brief outline of some of the more important classes of compounds is given below with references to recent reviews and monographs in which fuller accounts appear.

Boron. Various boron hydrides react with amines; tertiary amines usually give stable complexes, while secondary amines give unstable complexes which decompose into hydrogen and **aminoboranes (borazenes)**, $R_2N \cdot BH_2$, some of which are dimeric. Primary amines and boron hydrides often give **borazoles,** $R_3N_3B_3H_3$, planar, "aromatic", molecules. (Reactions between boron hydrides and amines are reviewed by *F. G. A. Stone* in "Advances in Inorganic Chemistry and Radiochemistry", Vol. 2, Academic Press Inc., New York, 1960, p. 279; borazoles are reviewed by *J. C. Sheldon* and *B. C. Smith,* Quart. Reviews, 1960, **14,** 200).

Boron halides react with secondary amines with replacement of halogen by the amino group:

$$BX_3 \xrightarrow{R_2NH} R_2N \cdot BX_2 \xrightarrow{R_2NH} (R_2N)_2BX \xrightarrow{R_2NH} (R_2N)_3B$$

Alkoxy and aryloxy boron halides react similarly. Boron trichloride reacts with primary amines, or their hydrochlorides, to give substituted borazoles. (*W. Gerrard,* "The Organic Chemistry of Boron", Academic Press, London, 1961, reviews these reactions).

Silicon. Alkyl- and alkoxy- silicon halides react with primary and secondary amines with replacement of halogen by the amino group. In general the intermediate products, containing both amino groups and halogens, are stable:

$$RSiCl_3 + R'R''NH \longrightarrow RSiCl_2NR'R'' \xrightarrow{R'R''NH} RSiCl(NR'R'')_2$$

$$\xrightarrow{R'R''NH} RSi(NR'R'')_3$$

Similar schemes apply for compounds of the types R_2SiCl_2 and R_3SiCl, and for the silicon tetrahalides. (For a full review see *R. Fessenden* and *J. S. Fessenden,* Chem. Reviews, 1961, **61,** 361).

Phosphorus. In general *phosphorus halides* and *oxyhalides,* and substituted derivatives of these compounds, react with secondary amines with step-wise replacement of halogen by amino groups:

$$PCl_3 + (CH_3)_2NH \longrightarrow (CH_3)_2 \cdot N \cdot PCl_2 \xrightarrow{(CH_3)_2NH} [(CH_3)_2N]_2PCl$$

$$\xrightarrow{(CH_3)_2NH} [(CH_3)_2N]_3P$$

$$POCl_3 + R_2NH \longrightarrow R_2N \cdot POCl_2$$

$$RPBr_2 + (CH_3)_2NH \longrightarrow R \cdot PBrN(CH_3)_2 \text{ etc.}$$

Reactions between the halides and primary amines are more complex and frequently polymeric products are obtained; few definitive experiments have been carried out in this area. (For reviews of the available literature see *G. M. Kosolapoff*, "Organophosphorus Compounds", John Wiley and Sons Inc., New York, 1950; *J. Van Wazer*, "Phosphorus and its Compounds", Vol. I, Interscience Publishers Inc., New York, 1958; *A. W. Frank*, Chem. Reviews, 1961, **61**, 389; *R. Burgada*, Bull. Soc. chim. Fr., 1963, 2335.)

Arsenic and **antimony.** The pattern of reactions of the arsenic and antimony halides resembles that of the phosphorus halides. Some relatively recent work on the reactions of arsenic compounds with amines has appeared (*G. O. Doak*, J. Amer. pharm. Assoc., 1935, **24**, 453; *G. Kamai* and *Z. L. Khisamova*, Doklady Akad. Nauk U.S.S.R., 1955, **105**, 489). For earlier work on arsenic and antimony compounds the following should be consulted: *E. Krause* and *A. von Grosse*, "Die Chemie der Metallorganischen Verbindungen", Borntraeger, Berlin, 1937; *W. G. Christiansen*, "Organic Derivatives of Antimony", Chemical Catalog Co., New York, 1923.

(e) Derivatives of alkylamines with a nitrogen–nitrogen linkage

(i) Nitramines (nitroamines)

Alkylnitramines, $R_2N \cdot NO_2$ and $RNH \cdot NO_2$, may be regarded as amides of nitric acid. They have attracted some interest as potential explosives and have been fully investigated for this purpose. (For a review see *A. H. Lamberton*, Quart. Reviews, 1951, **5**, 75).

Monoalkylnitramines, or primary nitramines, $R \cdot NH \cdot NO_2$.

Preparation. (1) Nitration of an acylated amine particularly a urethane, to an *acylnitramine* (*nitramide*) which gives the nitramine on hydrolysis:

$$CH_3 \cdot NH_2 \xrightarrow{Cl \cdot CO_2C_2H_5} CH_3 \cdot NH \cdot CO_2C_2H_5 \xrightarrow{HNO_3} CH_3 \cdot N(NO_2)CO_2C_2H_5$$

$$\xrightarrow[2.\ HCl]{1.\ NH_3} CH_3 \cdot NH \cdot NO_2$$

(*J. Thiele* and *A. Lachmann*, Ann., 1895, **288**, 267).

(2) Nitration of an *N,N*-dichloroamine and reduction of the intermediate chloronitramine, is an alternative route:

$$C_4H_9\cdot NCl_2 \xrightarrow[(CH_3CO)_2O]{HNO_3} C_4H_9\cdot NCl\cdot NO_2 \xrightarrow{Na_2S_2O_4} C_4H_9\cdot NH\cdot NO_2$$

(*G. N. R. Smart* and *G. F. Wright,* Canad. J. Res., 1948, **26B**, 284).

(3) Probably the best route to these compounds is direct nitration of the amine with acetonecyanohydrin nitrate in basic solution:

$$C_2H_5NH_2 + (CH_3)_2C(ONO_2)CN \xrightarrow{NaOH} C_2H_5\cdot NH\cdot NO_2 + (CH_3)_2C(OH)CN$$

(*W. D. Emmons* and *J. P. Freeman,* J. Amer. chem. Soc., 1955, 77, 4387).

Properties and reactions. Monoalkylnitramines are weak monobasic acids. They may decompose on heating but can usually be distilled under reduced pressure. They are rapidly decomposed by acids to give alcohols:

$$RNH\cdot NO_2 \longrightarrow ROH + N_2O$$

but, except for salt-formation, are unaffected by bases. Mild reduction, gives, among other products, poor yields of alkylhydrazines, $RNH\cdot NH_2$.

The anion of a monoalkylnitramine is ambident:

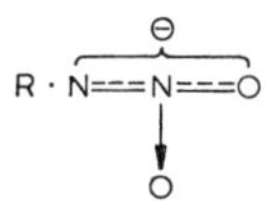

and can be alkylated on nitrogen or oxygen. Alkali metal salts are *N*-alkylated by alkyl halides, giving dialkyl nitramines:

$$(C_2H_5\cdot N\cdot NO_2)^{\ominus}Na^{\oplus} + CH_3I \longrightarrow C_2H_5\cdot N(CH_3)NO_2 + NaI$$

Silver salts are *O*-alkylated:

$$(C_2H_5\cdot N\cdot NO_2)^{\ominus}Ag^{\oplus} + CH_3I \longrightarrow C_2H_5N{:}NO\cdot OCH_3 + AgI$$

Methylnitramine, $CH_3NH\cdot NO_2$, m.p. 38°; *ethylnitramine,* m.p. 3°; n-*propylnitramine,* b.p. 128°/40 mm.; n-*butylnitramine,* b.p. 124°/20 mm.; n-*octylnitramine,* b.p. 140°/10 mm.

Dialkylnitramines, secondary nitramines, $RR'N\cdot NO_2$, have been more fully studied than monoalkylnitramines.

Preparation. (1) Alkylation of monoalkylnitramines (see above).

(2) Nitration of dialkylamides:

$$CH_3CONR_2 + HNO_3 \longrightarrow R_2N\cdot NO_2 + CH_3CO_2H$$

(3) The action of dinitrogen tetroxide on amines at —80°:

$$2(C_2H_5)_2NH + N_2O_4 \longrightarrow 2(C_2H_5)_2N \cdot NO_2$$

(*E. H. White* and *W. R. Feldman, ibid.*, 1957, **79**, 5832).

(4) Unbranched secondary amines can be converted into nitramines by action of acetonecyanohydrin nitrate:

$$(n\text{-}C_4H_9)_2NH + (CH_3)_2C(ONO_2)CN \longrightarrow (n\text{-}C_4H_9)_2N \cdot NO_2 + (CH_3)_2C(OH)CN$$

(*Emmons* and *Freeman, loc. cit.*). This reagent is less effective with branched-chain amines, but these react with dinitrogen pentoxide at —30° to give nitramines:

$$[(CH_3)_2CH]_2NH \xrightarrow{N_2O_5} [(CH_3)_2CH]_2N \cdot NO_2$$

(*Emmons, A. S. Pagano*, and *T. E. Stevens*, J. org. Chem., 1958, **23**, 311).

(5) Nitrosamines can be oxidised to nitramines by hydrogen peroxide in trifluoroacetic acid:

$$(C_3H_7)_2N \cdot NO \xrightarrow{H_2O_2/CF_3CO_2H} (C_3H_7)_2N \cdot NO_2$$

(*Emmons*, J. Amer. chem. Soc., 1954, **76**, 3468).

Properties and reactions. Dialkylnitramines are neutral liquids or crystalline solids, insoluble in water and showing no basic properties. They are stable to acids but are generally decomposed by bases, initially to aldimines which may be further hydrolysed:

$$RN(NO_2) \cdot CH_2R' + NaOH \longrightarrow R \cdot N{:}CHR' + NaNO_2 + H_2O$$

$$R \cdot N{:}CHR' \longrightarrow R \cdot NH_2 + R'CHO$$

Dialkyl nitramines resist oxidation; they can be reduced in good yield to *N,N*-dialkylhydrazines, $R_2N \cdot NH_2$.

Dimethylnitramine, $(CH_3)_2N \cdot NO_2$, m.p. 58°, b.p. 187°; *diethylnitramine,* b.p. 48°/0·5 mm.; *di-n-propylnitramine,* b.p. 77°/10 mm.; *di-isopropylnitramine,* m.p. 108°; *dioctylnitramine,* m.p. 20°.

(ii) Nitrosoamines (nitrosamines)

Preparation. Secondary amines can be converted into nitrosamines, $R_2N \cdot NO$, by a variety of methods (see p. 125). Reaction of a secondary amine with nitrous acid, generated in aqueous or acetic acid solution from an alkali metal nitrite, gives the corresponding nitrosamine. Other reagents which give the same product and can be used in non-aqueous neutral solution include nitrosyl chloride; dinitrogen tetroxide (*E. H. White* and *W. R.*

Feldman, J. Amer. chem. Soc., 1957, 79, 5832); and nitrosonium tetrafluoroborate, $NO^{\oplus}BF_4^{\ominus}$ (*G. Olah et al.*, Ber., 1956, 89, 2374).

Properties and reactions. The nitrosamines are usually yellow liquids, relatively insoluble in water. They may be distilled under reduced pressure without decomposition. Unlike the *C*-nitroso compounds they are monomeric in the liquid state:

Nitrosamines can be hydrolysed by heating them with acids; they give the parent amine and nitrous acid. In consequence they give a positive *Liebermann reaction* when they are heated with phenol and sulphuric acid. Oxidation of nitrosamines under certain conditions gives nitramines (*q.v.*). Reduction of nitrosamines with mild reducing agents, *e.g.* zinc and acetic acid, gives dialkylhydrazines; stronger reduction gives amines:

$$R_2N\cdot NO \xrightarrow{Zn/HOAc} R_2N\cdot NH_2 \longrightarrow R_2NH + NH_3$$

Dimethylnitrosamine, $(CH_3)_2N\cdot NO$, b.p. 151°/767 mm.; *diethylnitrosamine,* $(C_2H_5)_2N\cdot NO$, b.p. 174·5°/777 mm.; *di-n-propylnitrosamine,* b.p. 89°/13 mm.; *di-n-butylnitrosamine,* b.p. 125°/21 mm. For other physical properties of nitrosamines see *A. Vogel*, J. chem. Soc., 1948, 1833.

(iii) Alkylhydrazines

Hydrazine, $NH_2\cdot NH_2$, has four hydrogen atoms which can be replaced by alkyl groups; both simple and mixed di-, tri-, and tetra-alkylhydrazines have been prepared.

Monoalkylhydrazines

Preparation. Direct alkylation of hydrazine usually gives a trialkylhydrazinium salt, $R_3\overset{\oplus}{N}\cdot NH_2\overset{\ominus}{X}$, since the introduction of an alkyl group on to one nitrogen atom makes that nitrogen atom more basic, and more easily alkylated further (*C. Harries* and *T. Haga*, Ber., 1898, 31, 56). Nevertheless by using longer chain alkyl chlorides, and by carefully controlling the reaction conditions, it is possible to get useful yields of monoalkylhydrazines by direct alkylation of hydrazine (*O. Westphal, ibid.*, 1941, 74, 759). Methylhydrazine can be prepared by methylation of benzalazine and hydrolysis of the product:

$$C_6H_5CH{:}N\cdot N{:}CHC_6H_5 + (CH_3)_2SO_4 + 3H_2O \longrightarrow CH_3NH\cdot \overset{\oplus}{N}H_3HS\overset{\ominus}{O_4}$$
$$+ CH_3OH + 2C_6H_5\cdot CHO$$

(*H. H. Hatt*, Org. Synth. Collective, Vol. II, 1943, p. 395), and this method is probably applicable to higher homologues.

The best general methods for preparing monoalkylhydrazines, however, are the reaction between amines and hydroxylamine-*O*-sulphonic acid:

$$RNH_2 + NH_2 \cdot O \cdot SO_3H \longrightarrow RNH \cdot \overset{\oplus}{N}H_3 H\overset{\ominus}{SO_4}$$

(*G. Gever* and *K. Hayes*, J. org. Chem., 1949, **14**, 813), and the modified Raschig reaction between chloramine and a primary amine:

$$RNH_2 + NH_2Cl \longrightarrow RNH \cdot \overset{\oplus}{N}H_3 \overset{\ominus}{Cl}$$

(*L. F. Audrieth* and *L. H. Diamond*, J. Amer. chem. Soc., 1954, **76**, 4869; 1955, **77**, 3131; *H. H. Sisler et al.*, *ibid.*, 1956, **78**, 3874).

Properties and reactions. Monoalkylhydrazines resemble hydrazine. They are water-soluble liquids which fume in air, strong monoacid bases, and powerful reducing agents.

Dialkylhydrazines

These may be divided into two groups, symmetrical (*s*-), RNH·NHR, and asymmetrical (*as*-), $R_2N \cdot NH_2$.

s-Dialkylhydrazines are formed by alkylation of a diacylated hydrazine, and subsequent hydrolysis:

$$C_6H_5CONH \cdot NHCOC_6H_5 \xrightarrow{(CH_3)_2SO_4} C_6H_5CON(CH_3) \cdot N(CH_3)COC_6H_5$$
$$\longrightarrow CH_3NH \cdot NHCH_3 + C_6H_5CO_2H$$

(*Hatt, loc. cit.*, p. 208; *C. D. Harries*, Ber., 1894, **27**, 2276). Catalytic reduction of aldazines or ketazines also gives *s*-dialkylhydrazines:

$$(CH_3)_2C{:}N \cdot N{:}C(CH_3)_2 \longrightarrow (CH_3)_2CHNH \cdot NHCH(CH_3)_2$$

(*H. L. Lochte, J. R. Bailey* and *W. A. Noyes*, J. Amer. chem. Soc., 1921, **43**, 2597; *L. D. Barrick, G. W. Drake,* and *Lochte, ibid.*, 1936, **58**, 160). Azines are also reduced by lithium aluminium hydride to *s*-dialkylhydrazines (*R. Renaud* and *L. C. Leitch*, Canad. J. Chem., 1954, **32**, 545).

Properties and reactions. *s*-Dialkylhydrazines are strong monoacid bases and reducing agents. They are oxidised to alkylazo-alkanes (p. 155).

as-Dialkylhydrazines are most easily prepared by reduction of nitrosamines with zinc and acetic acid, sodium and ethanol, lithium aluminium hydride etc.:

$$R_2N \cdot NO \longrightarrow R_2N \cdot NH_2$$

(*Hatt, loc. cit.*, p. 211; *L. F. Audrieth et al.*, J. Amer. chem. Soc., 1955, **77**, 790). They can also be prepared by reaction between chloramine and a secondary amine:

$$(n\text{-}C_4H_9)_2NH + NH_2Cl \longrightarrow (n\text{-}C_4H_9)_2N \cdot \overset{\oplus}{N}H_3 \overset{\ominus}{Cl}$$

(*R. A. Rowe* and *Audrieth*, J. Amer. chem. Soc., 1956, **78**, 563; *Sisler et al.*, *loc. cit.*).

as-Dialkylhydrazines resemble the symmetrical compounds in being strong bases and good reducing agents. On careful oxidation they give tetra-alkyl 2-tetrazenes (p. 163).

Tri- and tetra-alkylhydrazines

These compounds are less well known than the mono- and di-alkylhydrazines. They can be prepared by vigorous alkylation of hydrazine (*O. Westphal*, Ber., 1941, **68**, 759) or by reduction of a formyl hydrazine:

$$(CH_3)_2N\cdot NH_2 \xrightarrow{H\cdot CO_2C_2H_5} (CH_3)_2N\cdot NH\cdot CHO \xrightarrow{LiAlH_4} (CH_3)_2N\cdot NHCH_3$$

$$\xrightarrow{H\cdot CO_2C_2H_5} (CH_3)_2N\cdot N(CH_3)CHO \xrightarrow{LiAlH_4} (CH_3)_2N\cdot N(CH_3)_2$$

(*R. T. Beltrami* and *E. R. Bissell*, J. Amer. chem. Soc., 1956, **78**, 2467).

Tri- and tetra-alkylhydrazines are weakly basic liquids and are not strong reducing agents.

General reactions of alkylhydrazines. Alkylhydrazines having free NH or NH_2 groups undergo acylation reactions with acyl halides. All classes of alkylhydrazines react with alkyl halides on the more basic (more highly alkylated) nitrogen atom to produce *hydrazinium salts* which are analogous in many ways to quaternary ammonium salts.

Alkylhydrazines are useful reagents in the synthesis of various heterocyclic systems containing two adjacent nitrogen atoms (see *G. D. Byrkit* and *G. A. Michalek*, Ind. Eng. Chem., 1950, **42**, 1862; *R. A. Reed*, Royal Institute of Chemistry Monographs, 1957, No. 5).

Methylhydrazine, $CH_3NH\cdot NH_2$, b.p. 87°; *ethylhydrazine*, b.p. 100°; n-*hexylhydrazine*, b.p. 80–81°/14 mm.; n-*octylhydrazine*, b.p. 112–114°/12 mm. (Many monoalkylhydrazines, being hygroscopic liquids which absorb carbon dioxide from air, have been characterized as hydrogen oxalate or hydrogen sulphate salts, rather than as pure liquids).

s-**Dimethylhydrazine,** $CH_3NH\cdot NHCH_3$, b.p. 81°; s-*diethylhydrazine*, b.p. 106°; s-*di*-n-*propylhydrazine*, b.p. 150°; s-*di-isopropylhydrazine*, b.p. 124°/740 mm.; s-*di*-sec-*butylhydrazine*, b.p. 168–170°; s-*diheptylhydrazine*, m.p. 47–48°, b.p. 118°/3 mm.; s-*methylisopropylhydrazine*, $CH_3NH\cdot NHCH(CH_3)_2$, b.p. 100°.

as-**Dimethylhydrazine,** $(CH_3)_2N\cdot NH_2$, b.p. 62°; as-*diethylhydrazine*, b.p. 96–99°; as-*di*-n-*propylhydrazine*, b.p. 43–45°/16 mm.; as-*di-isopropylhydrazine*, b.p. 36–38°/12 mm.; as-*di*-n-*butylhydrazine*, b.p. 87–90°/21 mm.; as-*di*-n-*amylhydrazine*, b.p. 108–110°/14 mm.

Trimethylhydrazine, $(CH_3)_2N\cdot NHCH_3$, b.p. 59–61°; *triethylhydrazine*, b.p.

43°/30 mm.; *tri-n-propylhydrazine*, b.p. 59–61°/11 mm.; *tri-n-octylhydrazine*, b.p. 186°/4 mm.

Tetramethylhydrazine, $(CH_3)_2N\cdot N(CH_3)_2$, b.p. 71–72°; *tetra-n-propylhydrazine*, b.p. 89°/11 mm.; *tetra-n-butylhydrazine*, b.p. 133°/12 mm.; *tetra-n-dodecylhydrazine*, m.p. 52·5°.

(iv) *Alkylazoxyalkanes*

Alkylazoxyalkanes, RN(O):NR′, have been very little studied, in comparison with their aromatic analogues, but the recent isolation of two natural products containing azoxyalkyl groups has stimulated a little interest in these compounds.

Nomenclature. In accordance with IUPAC rules R·N:N(O)— is the alkylazoxy group and as such is used as a normal prefix in naming azoxy compounds; thus $CH_3N{:}N(O)CH_3$ is methylazoxymethane (I); $CH_3\cdot CH_2N{:}N(O)CH(CH_3)_2$ is 2-(1′-ethylazoxy)propane. The shortened form, azoxyalkane, is frequently used for symmetrical compounds; thus I becomes azoxymethane but this simplified nomenclature can be ambiguous.

Macrozamin, a toxic compound isolated from an Australian plant, has the structure

$$\underset{\underset{O}{\downarrow}}{CH_3N}{:}NCH_2OR \quad (R = \text{a primeverose residue})$$

(*B. W. Langley, B. Lythgoe,* and *N. V. Riggs*, J. chem. Soc., 1951, 2309).

Elaiomycin, a tuberculostatic antibiotic, is an unsaturated azoxy compound:

$$n\text{-}C_6H_{13}\cdot CH{:}CH\underset{\underset{O}{\downarrow}}{N}{:}NCH(CH_2\cdot OCH_3)\cdot CH(OH)CH_3$$

(*C. L. Stevens et al.*, J. Amer. chem. Soc., 1958, **80**, 6088).

Preparation. Simple alkylazoxyalkanes are best prepared by perbenzoic acid oxidation of the corresponding alkylazoalkane:

$$C_3H_7N{:}NC_3H_7 + C_6H_5\cdot CO_3H \longrightarrow C_3H_7N{:}\underset{\underset{O}{\downarrow}}{N}C_3H_7 + C_6H_5\cdot CO_2H$$

(*Langley, Lythgoe* and *L. S. Rayner*, J. chem. Soc., 1952, 4191). Tertiary alkyl derivatives can be prepared by condensation reactions between nitrosoalkanes and *N*-alkylhydroxylamines:

$$CH_3NHOH + (CH_3)_2C(NO)\cdot CH_2\cdot CH_2\cdot CH(CH_3)_2$$

$$\longrightarrow CH_3N{:}\underset{\underset{O}{\downarrow}}{N}C(CH_3)_2CH_2\cdot CH_2\cdot CH(CH_3)_2$$

(*J. G. Aston* and *D. M. Jenkins*, Nature, 1951, 167, 863, and refs. therein).

Properties and reactions. Alkylazoxyalkanes are relatively high-boiling liquids which can be distilled without decomposition. They can be reduced catalytically to corresponding *s*-dialkylhydrazines. They are decomposed by both alkalis and acids.

Secondary alkylazoxy compounds react with hydrochloric acid to give hydrazine and a ketone:

$$(CH_3)_2CH\underset{\underset{O}{\downarrow}}{N}:NCH(CH_3)_2 \xrightarrow{HCl} 2(CH_3)_2C{=}O + NH_2\cdot NH_2$$

Primary alkylazoxy compounds, under similar conditions, give a carboxylic acid and a substituted hydrazine:

$$CH_3\cdot CH_2\cdot CH_2\underset{\underset{O}{\downarrow}}{N}:NCH_2\cdot CH_2\cdot CH_3 \xrightarrow{HCl} CH_3\cdot CH_2\cdot CO_2H + CH_3\cdot CH_2\cdot CH_2NH\cdot NH_2$$

Methylazoxymethane, $CH_3\overset{\overset{O}{\uparrow}}{N}:NCH_3$, b.p. 98°; 1-*propylazoxypropane*, b.p. 67°/20 mm.; 2-*propylazoxypropane*, b.p. 38°/14 mm.; 2-*azoxy-2,5-dimethyl-2-(2,5'-dimethylhexyl-2'-azoxy)hexane*, b.p. 110°/5 mm.; 2-(*methylazoxy*)-2,5-*dimethylhexane*, b.p. 184°/738 mm.

(*v*) *Alkylazoalkanes*

Alkylazoalkanes, RN:NR'* at one time chemical curiosities, have recently been receiving considerable attention as initiators of polymerisation. Some azoalkane derivatives, *e.g.* azo-bis-isobutyronitrile, $(CH_3)_2C(CN)N{:}N\cdot C(CN)(CH_3)_2$, are already in commercial use for this purpose. Simple azoalkanes are mainly of interest as sources of free-radicals in the gas phase. They are reviewed by *H. Zollinger*, "Azo and Diazo Chemistry", Interscience Publishers Ltd., London, 1961.

Preparation. The most widely applicable method for preparing azoalkanes is oxidation of *s*-dialkylhydrazines. Mercuric oxide is the reagent of choice:

$$C_2H_5NH\cdot NHC_2H_5 + HgO \longrightarrow C_2H_5N{:}NC_2H_5 + Hg + H_2O$$

* The nomenclature is similar to that of the azoxy-compounds (see p. 154), RN:N- being the alkylazo group. Again, sometimes symmetrical compounds are described by the trivial name, azoalkane, methylazomethane becoming azomethane.

(*R. Renaud* and *L. C. Leitch*, Canad. J. Chem., 1954, **32**, 545), but cupric chloride, cupric oxide, and hydrogen peroxide have also been used. Chlorine reacts with ketazines to produce chloroazoalkanes:

$$(CH_3)_2C{:}N{\cdot}N{:}C(CH_3)_2 + Cl_2 \longrightarrow (CH_3)_2CCl{\cdot}N{:}N{\cdot}CCl(CH_3)_2$$

(*St. Goldschmidt* and *B. Acksteiner*, Ann., 1958, **618**, 173).

Most known alkylazoalkanes are symmetrical compounds, but a few unsymmetrical ones, *e.g.* $CH_3N{:}NCH(CH_3)_2$, have been prepared (see *H. C. Ramsperger*, J. Amer. chem. Soc., 1929, **51**, 918; *C. G. Overberger* and *A. V. Di Giulio*, *ibid.*, 1958, **80**, 6562).

Reactions. The most noteworthy, and most fully studied, reaction of alkylazoalkanes is their homolytic fission in the gas phase, initiated either photochemically or thermally, into nitrogen and free alkyl radicals:

$$RN{:}NR' \longrightarrow R\cdot + N_2 + R'\cdot$$

In this homolysis both carbon–nitrogen bonds break simultaneously. The kinetics of decomposition, and the reactions of the free-radicals produced by it, have been examined in detail. (See *E. W. R. Steacie*, "Atomic and Free Radical Reactions", 2nd. Edn., Reinhold, New York, 1954, for a summary).

Apart from this aspect of their chemistry azoalkanes have been little studied. They are liquids which usually decompose smoothly on being heated (the half-life of ethylazoethane is about one hour at 250°) but may sometimes explode. Alkylazoalkanes are easily reduced to *s*-dialkylhydrazines. Alkylazoalkanes could exist in *cis*- and *trans*-modifications:

cis *trans*

but in no simple case have both isomers been obtained.

Methylazomethane, *azomethane*, $CH_3N{:}NCH_3$, b.p. 1·5°, has been shown to have the *trans* configuration; (*H. Boersch*, Monatsh., 1935, **65**, 327).

Ethylazoethane, b.p. 58°; 1,1′-*propylazopropane*, b.p. 113·5°; 2,2′-*propylazopropane*, b.p. 88·5°; 2,2′-*butylazobutane*, b.p. 141°; tert-*butylazo*-tert-*butane*, $(CH_3)_3CN{:}NC(CH_3)_3$, b.p. 109°; 1,1′-*heptylazoheptane*, b.p. 113°/3 mm.; 2-(*methylazo*)*propane*, $CH_3{\cdot}N{:}N{\cdot}CH(CH_3)_2$, b.p. 46°.

(vi) Diazoalkanes

Diazoalkanes, $RR'C{\cdot}N_2$, are formally derived from carbonyl compounds and not from alcohols. However they are discussed in this chapter since they are closely related to amines and other nitrogen derivatives of aliphatic hydrocarbons. The lowest member of the series, diazomethane, CH_2N_2, was the first diazoalkane prepared (*H. von Pechmann*, Ber., 1894, **27**, 1888) and its chemistry has been much more fully investigated than that of any of the higher homologues.

The structure of diazomethane was a matter of dispute for some time; it is now generally agreed that the molecule cannot be represented by any single valence-bond formula but must be regarded as a resonance hybrid, the most important contributing structures being:

$$\overset{\ominus}{C}H_2{-}\overset{\oplus}{N}{\equiv}N \longleftrightarrow CH_2{=}\overset{\oplus}{N}{=}\overset{\ominus}{N} \longleftrightarrow \overset{\ominus}{C}H_2{-}N{=}\overset{\oplus}{N}$$

Diazomethane has only a small dipole moment, 1·4D, and this is in accord with the postulated structure. Electron diffraction shows it to be a linear molecule with C—N bond length 1·34 ± 0·05 Å and N—N bond length 1·13 ± 0·04 Å (*H. Boersch*, Monatsh., 1935, **65**, 331).

It is interesting to note that an early structure proposed for diazoalkanes was that containing a three-membered unsaturated ring,

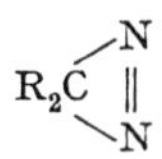

(*T. Curtius*, J. pr. Chem., 1889, **39**, 107). Compounds of this type have recently been prepared, including the cyclic isomer of diazomethane, and have been shown to be quite different in physical and chemical properties from isomeric diazoalkanes:

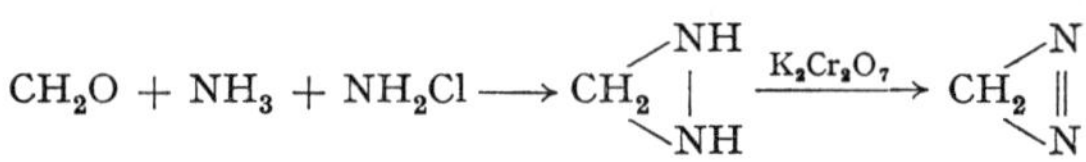

(*E. Schmitz* and *R. Ohme*, Ber., 1961, **94**, 2166; Tetrahedron Letters, 1961, 612).

Preparation of diazoalkanes. Treatment of simple aliphatic amines with nitrous acid does not give rise to stable diazonium compounds (cf. p. 125). However certain negatively substituted amines, *e.g.* esters of α-amino-acids, α-amino-nitriles etc., do give stable diazo-compounds in this way, and the first aliphatic diazo-compound to be prepared, diazoacetic ester, was made by action of nitrous acid on glycine ester hydrochloride:

$$CH_2(NH_2)CO_2C_2H_5 + HONO \longrightarrow CHN_2{\cdot}CO_2C_2H_5 + 2H_2O$$

(*T. Curtius*, Ber., 1883, **16**, 2230).

Diazomethane was first prepared by the action of bases on *N*-methyl-*N*-nitrosourethane; this reaction involves the intermediate production of an aliphatic diazotate;

$$C_2H_5O\cdot CO\cdot N(CH_3)NO \xrightarrow{C_2H_5OK} CH_3\cdot N{:}N\cdot OK,C_2H_5OH$$

$$\longrightarrow CH_2N_2 + KOH + C_2H_5OH$$

(*A. Hantzsch* and *M. Lehmann*, Ber., 1902, 35, 897).

A more conveniently prepared starting material is *N*-nitrosomethylurea which is also decomposed by bases to give diazomethane. Recently the very stable *N*-methyl-*N*-nitroso-*p*-toluenesulphonamide, p-$CH_3\cdot C_6H_4\cdot SO_2\cdot N(NO)CH_3$, has been recommended as a convenient precursor for diazomethane, since it can be stored for long periods at ordinary temperatures without decomposing (*T. J. de Boer* and *H. J. Backer*, Org. Syntheses, 1956, 36, 16). Diazomethane is usually prepared in solution in an ethereal solvent, but gaseous diazomethane can be made by reaction between chloroform, hydrazine, and potassium hydroxide:

$$CHCl_3 + NH_2\cdot NH_2 + 3KOH \longrightarrow CH_2N_2 + 3KCl + 3H_2O$$

(*H. Staudinger* and *O. Kupfer*, Ber., 1912, 45, 501), or by reaction between *N,N'*-dimethyl-*N,N'*-dinitroso-oxamide and sodium butoxide:

$$(CONH\cdot CH_3)_2 \xrightarrow{N_2O_4} [CON(NO)CH_3]_2 \longrightarrow CH_2N_2$$

(*H. Reimlinger*, Ber., 1961, 94, 2547).

Homologues of diazomethane can be divided into primary diazoalkanes, $RCHN_2$, and secondary diazoalkanes, R_2CN_2. The former class can be prepared in the following ways.

(1) Addition of a primary alkylamine to mesityl oxide gives an aminoketone which can be nitrosated. Decomposition of the nitroso-derivative by bases gives the diazoalkane:

$$(CH_3)_2C{:}CH\cdot CO\cdot CH_3 + R\cdot CH_2NH_2 \longrightarrow (CH_3)_2C(NH\cdot CH_2R)\cdot CH_2\cdot CO\cdot CH_3$$

$$\xrightarrow{HONO} (CH_3)_2C\begin{matrix}\diagup N(NO)\cdot CH_2R\\ \diagdown CH_2\cdot CO\cdot CH_3\end{matrix} \xrightarrow{CH_3ONa}$$

$$R\cdot CHN_2 + (CH_3)_2C{:}CH\cdot CO\cdot CH_3 + H_2O$$

(*J. Kenner et al.*, J. chem. Soc., 1933, 363; 1935, 286; 1937, 1551; *C. E. Redemann et al.*, Org. Syntheses Coll. Vol. III, 1955, 244).

(2) 1-Alkyl-3-nitro-1-nitrosoguanidines give diazoalkanes when they are treated with potassium hydroxide:

$$CH_3\cdot CH_2\cdot CH_2\cdot N(NO)\cdot C({:}NH)NH\cdot NO_2 \xrightarrow{KOH} CH_3\cdot CH_2\cdot CHN_2$$

(*A. F. McKay et al.*, Canad. J. Res., 1950, **28B**, 683).

(3) Base cleavage of *N*-alkyl-*N*-nitroso-oxamides, which are stable crystalline solids, gives diazoalkanes:

$$[CON(NO)CH_2\cdot CH_3]_2 \longrightarrow CH_3\cdot CHN_2$$

(*H. Reimlinger*, Ber., 1961, **94**, 2547).

(4) Secondary diazoalkanes, R_2CN_2, are prepared almost exclusively by oxidation of ketonehydrazones by mercuric oxide:

$$R_2C{:}N\cdot NH_2 + HgO \longrightarrow R_2CN_2 + Hg + H_2O$$

(*H. Staudinger* and *A. Gaule*, Ber., 1916, **49**, 1897). (For full reviews of methods of preparing diazoalkanes, and for their reactions, see *C. D. Gutsche*, Org. Reactions, 1954, **8**, 364; *R. Huisgen*, Angew. Chem., 1955, **67**, 439; *H. Zollinger*, "Azo and Diazo Chemistry", Interscience Publishers Ltd., London, 1961).

Properties and reactions. Diazoalkanes are usually yellow gases or low-boiling liquids which are toxic and which may decompose explosively. They can usually be handled safely in ethereal solutions, and most of the reactions discussed here have been carried out under these conditions.

An inspection of the resonance forms contributing to the structure of diazomethane would suggest that it should react primarily as a nucleophile. This is indeed the case, and many reactions of this type are known. There are also a few reactions in which it acts as an electrophile, and recently a good deal of work has been carried out on the reactions in which diazoalkanes are decomposed to give a derivative of methylene as an intermediate.

Nucleophilic reactions. (1) *Protonic acids* react with diazoalkanes and decompose them. The products are virtually the same as those from the reaction of a primary amine and nitrous acid, and probably derive from an identical intermediate diazonium ion (cf. p. 125):

$$R_2CN_2 \xrightarrow{H^{\oplus}} R_2CH\cdot\overset{\oplus}{N}{:}N \longrightarrow \text{Products}$$
$$R_2CH\cdot NH_2 \xrightarrow{HONO} R_2CH\cdot\overset{\oplus}{N}{:}N$$

(2) *Reactions with acidic hydrogen atoms.* This is probably the best known and most widely used reaction of diazoalkanes, and especially diazomethane. Carboxylic acids are rapidly and quantitatively converted into methyl esters, and phenols into methyl ethers, by diazomethane. Alcohols also give methyl ethers when they are treated with diazomethane in the presence of Lewis acids (*E. Müller* and *W. Rundel*, Angew. Chem., 1958, **70**, 105; *J. D. Roberts et al.*, J. Amer. chem. Soc., 1958, **80**, 2584). Compounds which exist as keto-enol tautomers usually give enol ethers when treated with diazoalkanes.

(3) *Reactions with aldehydes and ketones* (reviewed by *C. D. Gutsche*, Org. Reactions, 1954, **8**, 364).

With aldehydes and acyclic ketones diazomethane reacts in general to give mixtures of three products—two homologous carbonyl compounds and an epoxide.

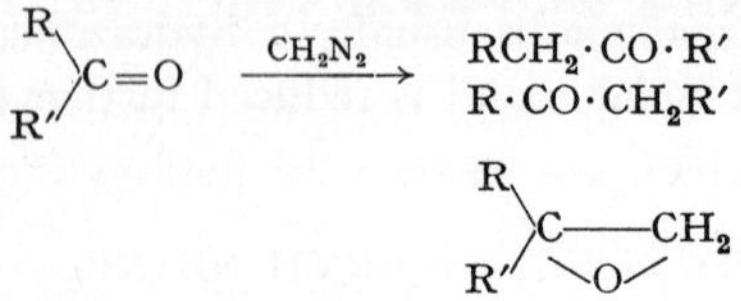

The greater the electrophilic character of the carbonyl compound the faster the reaction and, generally, the greater the proportion of epoxide formed.

With cyclic ketones ring enlargement occurs. Cyclohexanone gives a mixture of cycloheptanone and, by further reaction of this, cyclo-octanone (*E. Mosettig* and *A. Burger*, J. Amer. chem. Soc., 1930, **52**, 3456).

(4) *Reactions with carboxylic acid halides.* Diazomethane reacts with carboxylic acid chlorides to give diazoketones:

$$RCOCl + CH_2N_2 \longrightarrow RCO{\cdot}CHN_2 + HCl$$

valuable synthetic intermediates, particularly in the production of longer-chain carboxylic acids (Arndt-Eistert homologation, see *W. E. Bachmann* and *W. S. Struve*, Org. Reactions, 1942, **1**, 38).

(5) *Reactions with unsaturated compounds.* Diazoalkanes add to olefinic and acetylenic compounds to give, initially pyrazolines or pyrazoles:

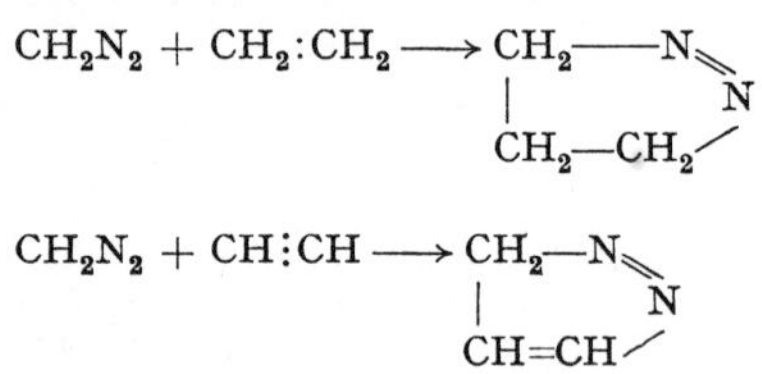

(*H. von Pechmann*, Ber., 1898, **31**, 2950; *H. Reimlinger*, Ber., 1959, **92**, 970). Other unsaturated systems, including ketens, nitriles, isocyanates, react in a similar way, but sometimes the intermediate product is unstable and loses nitrogen to give a final product containing a three-membered ring.

Electrophilic reactions. (6) *Diazoalkanes behave as electrophilic reagents towards phosphines,* and yield phosphazines by an addition reaction:

$$R_3P{:} + N_2CR_2' \longrightarrow R_3P{:}N{\cdot}N{:}CR'_2$$

(*H. Staudinger* and *J. Meyer*, Helv., 1919, **2**, 619; 1921, **5**, 75).

(7) *Reactions involving methylene intermediates (carbenes).* It has been known for some time that diazoalkanes break down on being heated or on

being irradiated to give derivative of methylenes, CH_2:. The reactions of this type of intermediate are discussed in Vol. IA (see pp. 400 and 512).

Miscellaneous reactions. (8) *Reduction of diazoalkanes.* The first product of reduction of a diazoalkane is usually a hydrazone, RR′C:N·NH_2, but sometimes this is not isolable and is reduced further to a hydrazine derivative or an amine:

$$RR'CN_2 \longrightarrow RR'C{:}N\cdot NH_2 \longrightarrow RR'CH\cdot NH\cdot NH_2 \longrightarrow RR'CH\cdot NH_2$$

(9) *Reactions with non-metallic, metalloidal, and metallic halides.* Diazoalkanes react with a wide variety of halogenated compounds to give products containing halogenoalkyl groups. These reactions have been most fully examined for diazomethane, and the examples below indicate the scope of the reaction:

$$SOCl_2 \xrightarrow{CH_2N_2} SO(CH_2Cl)_2$$

$$COCl_2 \xrightarrow{CH_2N_2} CO(CH_2Cl)_2$$

(*B. C. Saunders et al.*, J. chem. Soc., 1948, 699):

$$CCl_4 \xrightarrow{CH_2N_2} C(CH_2Cl)_4$$

(*W. H. Urry et al.*, J. Amer. chem. Soc., 1952, 74, 5822; 1954, 76, 2594):

$$AsCl_3 \xrightarrow{CH_3\cdot CHN_2} (CH_3CHCl)_2AsCl$$

$$PCl_3 \xrightarrow{CH_2N_2} CH_2ClPCl_2$$

and reactions giving Sb, Bi, Si, Sn, Pb derivatives (*A. Ya. Yakubovich et al.*, Zhur. Obshchei Khim., 1952, 22, 1528, 1534, 1783, 1788, 2060):

$$HgCl_2 \xrightarrow{CH_2N_2} CH_2Cl\cdot HgCl$$

(*L. Hellerman* and *M. D. Newman*, J. Amer. chem. Soc., 1932, 54, 2859):

$$GeCl_4 \xrightarrow{CH_2N_2} CH_2ClGeCl_3$$

$$(CH_3)_2SnCl_2 \xrightarrow{CH_2N_2} (CH_3)_2(CH_2Cl)SnCl$$

(*D. Seyferth* and *E. G. Rochow*, *ibid.*, 1955, 77, 907, 1302):

$$ZnI_2 \xrightarrow{CH_2N_2} CH_2I\cdot ZnI \xrightarrow{CH_2N_2} (CH_2I)_2Zn$$

(*G. Wittig* and *K. Schwarzenbach*, Angew. Chem., 1959, 71, 652).

(10) *Polymer formation.* The action of boron compounds or transition metal compounds on lower diazoalkanes and some aryl-substituted diazoalkanes leads to linear polymers of high molecular weight with the elimination of nitrogen (see *e.g., J. Feltzin, A. J. Restaino* and *R. B. Mesrobian,* J. Amer. chem. Soc., 1955, 77, 206; *C. E. H. Bawn, A. Ledwith* and *P. Matthies,* J. Polymer Sci., 1958, 33, 21).

Diazomethane, CH_2N_2, m.p. —145°; b.p. —24°; *diazoethane,* b.p. —19° to —17°/89·5 mm.; 1-*diazopropane,* b.p. —8°/41·5 mm.; 2-*diazopropane,* b.p. —31°/14 mm.; 1-*diazobutane,* b.p. —5·5° to —3·5°/26 mm.; 1-*diazo-2-methylpropane,* $(CH_3)_2CH{\cdot}CHN_2$, b.p. —1° to +1°/32 mm.

(vii) Alkyl azides

Much less is known about alkyl azides, $R{\cdot}N_3$, than about their aromatic counterparts. Their possible application as explosives has attracted some interest, but information about their chemistry is still scanty (see *J. H. Boyer* and *F. C. Canter,* Chem. Reviews, 1954, 54, 1).

Preparation. Alkyl azides are prepared by reaction between an alkali-metal azide and an alkyl halide or dialkyl sulphate in a sealed tube or, better, in a high-boiling ether as solvent:

$$NaN_3 + C_4H_9Br \longrightarrow C_4H_9{\cdot}N_3 + NaBr$$

(*E. Lieber, Tai Siang Chao* and *C. N. Ramachandra Rao,* J. org. Chem., 1957, 22, 238).

In alkyl azides the N_3 group is linear and it is best represented as a resonance hybrid:

$$R{-}\overset{\ominus}{N}{-}\overset{\oplus}{N}{\equiv}N \longleftrightarrow R{-}N{=}\overset{\oplus}{N}{=}\overset{\ominus}{N}$$

Properties and reactions. Alkyl azides are decomposed by heating (sometimes explosively) or photolytically. In the vapour phase above 200° methyl azide smoothly decomposes to yield hydrazoic acid and ethylene, which is presumably formed by dimerization of the methylene first formed:

$$CH_3N_3 \longrightarrow HN_3 + CH_2{:} \qquad 2CH_2{:} \longrightarrow CH_2{:}CH_2$$

Alkyl azides are not easily decomposed by bases; acid decomposition may give hydrazoic acid but the reaction is usually complex:

$$CH_3CH_2{\cdot}N_3 \xrightarrow{H_2SO_4} CH_3{\cdot}CHO,\ HCHO,\ NH_3,\ N_2,\ CH_3{\cdot}NH_2 \text{ etc.}$$

Alkyl azides are smoothly reduced, either by hydrogen plus a catalyst or by a wide variety of reducing agents, to primary amines.

Although only a few cases have so far been reported it seems that alkyl azides will add either to acetylenic triple bonds or to activated double bonds yielding triazole derivatives:

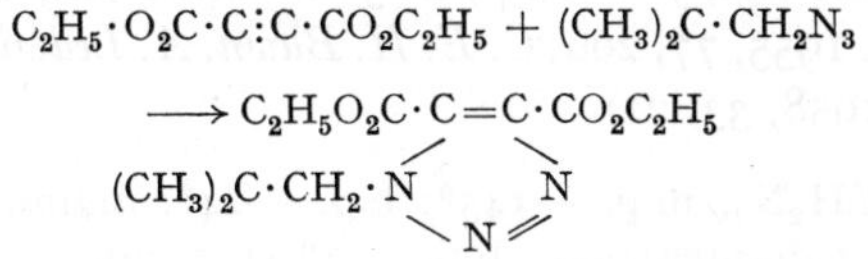

(*P. A. S. Smith, J. M. Clegg* and *J. Lakritz, ibid.*, 1958, **23**, 1595).

Methyl azide, $CH_3 \cdot N_3$, b.p. 20°; *ethyl azide*, b.p. 49°; *vinyl azide*, $CH_2:CH \cdot N_3$, b.p. 26°; *propyl azide*, b.p. 58°/357 mm.; *butyl azide*, b.p. 71°/225 mm.; n-*amyl azide*, b.p. 63°/95 mm.; *decyl azide*, b.p. 67°/0·6 mm.

(viii) Dialkyltriazenes (diazoaminoalkanes)

Only one dialkyltriazene has been prepared and investigated fully.

Reaction between methyl azide and methylmagnesium iodide, followed by hydrolysis, gave **dimethyltriazene,** $CH_3 \cdot N:N \cdot NHCH_3$, m.p. —12°, b.p. 92°, which was isolated via its copper complex (*O. Dimroth*, Ber., 1906, **39**, 3905). It is a weakly basic liquid, readily decomposed by acids:

$$CH_3 \cdot N:N \cdot NHCH_3 + 2HCl \longrightarrow CH_3Cl + N_2 + CH_3\overset{\oplus}{N}H_3\overset{\ominus}{Cl}$$

It forms silver and copper complexes, and reacts with phenyl isocyanate to form a crystalline derivative:

$$CH_3 \cdot N:N \cdot NHCH_3 + C_6H_5 \cdot NCO \longrightarrow CH_3 \cdot N:N \cdot N(CH_3) \cdot CO \cdot NHC_6H_5$$

(ix) Tetra-alkyl-2-tetrazenes

Preparation. These compounds, of formula $R_2N \cdot N:N \cdot NR_2$, are generally prepared by oxidation of *as*-dialkylhydrazines with mercuric oxide:

$$2(CH_3)_2N \cdot NH_2 + 2HgO \longrightarrow (CH_3)_2N \cdot N:N \cdot N(CH_3)_2 + 2Hg + 2H_2O$$

(For preparative details see *W. E. Bull, J. A. Seaton,* and *L. F. Audrieth,* J. Amer. chem. Soc., 1958, **80**, 2516). Reaction between nitrous oxide, amines and organometallic compounds can also yield tetrazenes:

$$2(C_2H_5)_2NH \xrightarrow{C_6H_5Li,\ N_2O} (C_2H_5)_2N \cdot N:N \cdot N(C_2H_5)_2$$

(*R. Meier*, Ber., 1953, **86**, 1483). Oxidation of *as*-dialkylhydrazines with halogens in acid solution gives a stable intermediate ion two of which, when treated with a base, combine to a tetrazene:

$$R_2N\cdot NH_2 \xrightarrow{Cl_2,\,H^{\oplus}} R_2\overset{\oplus}{N}{:}NH \xrightarrow{OH^{\ominus}} R_2N\cdot N{:}N\cdot NR_2$$

(*W. R. McBride* and *H. W. Kruse*, J. Amer. chem. Soc., 1957, **79**, 572); a number of mixed alkyltetrazenes have been prepared by a modification of this method (*W. R. McBride* and *E. M. Bens*, *ibid.*, 1959, **81**, 5546).

Properties and reactions. Tetra-alkyl-2-tetrazenes are unstable liquids which may decompose explosively when they are heated. They are weakly basic and form stable picrates and oxalates, as well as complexes with cadmium and mercuric halides. Pyrolysis or photolysis of tetramethyltetrazene at low pressures gives moderate yields of tetramethylhydrazine:

$$(CH_3)_2N\cdot N{:}N\cdot N(CH_3)_2 \longrightarrow N_2 + (CH_3)_2N\cdot N(CH_3)_2 \quad (40\%)$$

(*J. S. Watson*, J. chem. Soc., 1956, 3677). It has been claimed that tetra-alkyltetrazenes are useful accelerators for low-temperature polymerization (*J. C. McGowan* and *L. Seed*, B.P., 821,142; C.A., 1960, **54**, 2812).

Tetramethyl-2-tetrazene, $(CH_3)_2N\cdot N{:}N\cdot N(CH_3)_2$, b.p. 74°/110 mm.; *tetraethyltetrazene*, b.p. 79°/12 mm.; *tetra-allyltetrazene*, b.p. 113°; 1,4-*dibutyl*-1,4-*dimethyltetrazene*, b.p. 119°/18 mm.

(x) Hexa-alkyltetrazanes

The first stable members of this series, $R_2N\cdot NR\cdot NR\cdot NR_2$, containing a chain of four saturated nitrogen atoms, have been isolated recently as by-products from the photolysis or pyrolysis of fluorinated alkylazo-alkanes:

$$CF_3\cdot N{:}N\cdot CF_3 \longrightarrow N_2 + C_2F_6 + (CF_3)_2N\cdot N(CF_3)\cdot N(CF_3)\cdot N(CF_3)_2$$

(*J. R. Dacey* and *D. M. Young*, J. Chem. Phys., 1955, **23**, 1302; *G. O. Pritchard, H. O. Pritchard, H. I. Schiff*, and *A. F. Trotman-Dickenson*, Trans. Faraday Soc., 1956, **52**, 849). They are relatively volatile liquids which are remarkably stable to heat, decomposing only above 350° (*B. A. Ginsberg et al.*, Doklady Akad. Nauk S.S.S.R., 1962, **142**, 88).

Hexakis-trifluoromethyltetrazane, $(CF_3)_2N\cdot N(CF_3)\cdot N(CF_3)\cdot N(CF_3)_2$, b.p. 100°; 1,2,4-*trimethyl*-1,3,4-*tristrifluoromethyltetrazane*, $(CF_3)(CH_3)N\cdot N(CH_3)\cdot N(CF_3)\cdot N(CF_3)(CH_3)$, b.p. 75°/135 mm.; 1,2,4-*trisdifluoromethyl*-1,3,4-*tristrifluoromethyltetrazane*, b.p. 133°.

Chapter 7

Aliphatic Organometallic and Organometalloidal Compounds

C. B. MILNE and A. N. WRIGHT

1. Introduction

This chapter describes those aliphatic compounds which contain a bond between a carbon atom and a metal or metalloid. Transition metal compounds which contain olefins and acetylenes bonded to the metal are also discussed. The metals and metalloids include all elements in Groups I to VIII except hydrogen, carbon, nitrogen, oxygen, sulphur and the halogens*.

(i) History and development of organometallic chemistry

R. Bunsen's work (1837–1843) on cacodyl, later shown to be tetramethyldiarsine, $(CH_3)_2As—As(CH_3)_2$, is commonly regarded as the beginning of organometallic chemistry. However, it was *E. Frankland* who laid the foundations of this subject with his classic researches on the zinc alkyls (J. chem. Soc., 1848-9, **2**, 263; Ann., 1849, **71**, 171) which he later extended to other metals (Ann., 1853, **85**, 329). Zinc alkyls played an important part in synthetic organic chemistry and also in the preparation of organic compounds of other metals and metalloids. In 1900 *V. Grignard* (Compt. rend., 1900, **130**, 1322) showed that reactive organomagnesium compounds are obtained from alkyl halides and metallic magnesium in the presence of ethers. These Grignard reagents were much easier to prepare and to handle than the zinc alkyls.

P. Ehrlich's discovery (1909) that some organoarsenic compounds had therapeutic properties accelerated further research; of the 12,000 metal-carbon compounds known in 1937, about 8,000 were derived from arsenic. In 1920, T. Midgley found that tetraethyl-lead had a profound effect on the "knocking" phenomenon in petrol engines which led eventually to its manufacture on a

* The *Richter* classification, on which this book is based, is not very satisfactory for the treatment of organo-metallic compounds. It would be a distinct advantage to deal fully with, *e.g.*, cyclopentadienylmetal compounds and metal aryls alongside the aliphatic compounds and for this reason, some mention of alicyclic and aromatic metal compounds will be found in this chapter.

very large scale as an "antiknock compound". The mid-1940s saw the initiation of H. C. Brown's work on boron alkyls leading to the hydroboration process, while in 1951 T. J. Kealy and P. L. Pauson and S. A. Miller and his co-workers opened up a completely new field of organometallic chemistry with regard to transition metals with the discovery of biscyclopentadienyliron (ferrocene). The organic chemistry of the transition metals is advancing very rapidly at the present time and has already greatly increased our knowledge of the structure of simple transition metal compounds, *e.g.* Zeise's salt. More recently, K. Ziegler (1953) showed that aluminium alkyls, in conjunction with titanium tetrachloride, are catalysts for the polymerisation of olefins; this discovery already forms the basis of important commercial processes for the manufacture of polyolefins, *e.g.* polypropylene, and along with it may be mentioned the importance of organic cobalt compounds in hydroformylation reactions and of palladium compounds in the autoxidation of olefins to aldehydes and ketones (Hoechst).

(ii) Nomenclature

Organometallic compounds are now designated by the names of the organic radicals which are linked to the metal, followed by the name of the metal *e.g.* diethylzinc, $(C_2H_5)_2Zn$; ethylmagnesium chloride, C_2H_5MgCl. The metal can also be considered as a substituent in mixed compounds *e.g.* $ClHgC_6H_4CO_2H$ is chloromercuribenzoic acid. Earlier they were called metal alkyls, which is still a useful way of classification. The nomenclatures for boron, silicon and germanium derivatives and the system adopted for organophosphorus compounds are discussed on pp. 192, 208 and 219. The I.U.P.A.C. nomenclature for transition metal complexes is outlined in the introduction to the section on transition metals (p. 247).

(iii) Carbon-metal bonding

The nature of the carbon-metal bond has a large influence on the reactivity and properties of organometallic compounds. Four extreme types of bonds will be considered here.

(*a*) *Ionic bonds*. The most electropositive metals, sodium, potassium, rubidium and caesium, form metal-carbon bonds which are so highly polarized that they may be considered as ionic, $RCH_2^{\ominus}M^{\oplus}$. These compounds have a typically salt-like character; they are generally non-volatile solids, insoluble in hydrocarbons and they decompose before melting.

(*b*) *Covalent bonds*. The less electropositive metals form normal *localized covalent bonds,* usually referred to as sigma (σ) bonds, although they are usually polarized $RCH_2^{\delta-}$—$M^{\delta+}$ (see Vol. 1A, pp. 113 and 333 *et seq.*)

L. Pauling has assigned electronegativity values to the elements and it is possible to estimate the degree of ionic character of any covalent bond between two atoms A and B from the relationship:

$$\text{amount of ionic character} = 1 - e^{-\frac{1}{4}(x_A - x_B)^2}$$

where x_A and x_B are the electronegativities of the bonded atoms A and B. (*L. Pauling*, "The Nature of the Chemical Bond", 3rd Edn., Cornell University Press, New York, 1960, p. 98). *N. B. Hannay* and *C. P. Smyth* (J. Amer. chem. Soc., 1946, **68**, 171) give the alternative formula:

$$\text{amount of ionic character} = 0{\cdot}16(x_A - x_B) + 0{\cdot}035(x_A - x_B)^2$$

Pauling's complete electronegativity scale is given in Table 1

TABLE 1

ELECTRONEGATIVITIES OF THE ELEMENTS

IA	IIA	IIIA	IVA	VA	VIA	VIIA	VIII			IB	IIB	IIIB	IVB	VB	VIB	VIIB
H																
2·1																
Li	Be											B	C	N	O	F
1·0	1·5											2·0	2·5	3·0	3·5	4·0
Na	Mg											Al	Si	P	S	Cl
0·9	1·2											1·5	1·8	2·1	2·5	3·0
K	Ca	Sc	Ti	V	Cr	Mn	Fe	Co	Ni	Cu	Zn	Ga	Ge	As	Se	Br
0·8	1·0	1·3	1·5	1·6	1·6	1·5	1·8	1·8	1·8	1·9	1·6	1·6	1·8	2·0	2·4	2·8
Rb	Sr	Y	Zr	Nb	Mo	Tc	Ru	Rh	Pd	Ag	Cd	In	Sn	Sb	Te	I
0·8	1·0	1·2	1·4	1·6	1·8	1·9	2·2	2·2	2·2	1·9	1·7	1·7	1·8	1·9	2·1	2·5
Cs	Ba	La-Lu	Hf	Ta	W	Re	Os	Ir	Pt	Au	Hg	Tl	Pb	Bi	Po	At
0·7	0·9	1·1–1·2	1·3	1·5	1·7	1·9	2·2	2·2	2·2	2·4	1·9	1·8	1·8	1·9	2·0	2·2
Fr	Ra	Ac	Th	Pa	U	Np-No										
0·7	0·9	1·1	1·3	1·5	1·7	1·3										

(Reproduced by courtesy of Prof. *L. Pauling*, "The Nature of the Chemical Bond", 3rd Edn., p. 93).

The values in Table 1 refer to the common oxidation states of the elements; there is a variation in electronegativity with valence state (*M. Haïssinsky*, J. phys. Radium, 1946, **7**, 7; *W. Gordy* and *W. J. O. Thomas*, J. chem. Phys., 1956, **24**, 439). In many cases chemical reactivity is related to bond polarity although this does not apply to the transition metals.

The concept of electronegativity which also owes much to the work of *R. S. Mulliken* (J. chem. Phys., 1934, **2**, 782; J. Chim. phys., 1949, **46**, 497) has been very widely used although it is not without defects (*H. O. Pritchard* and *H. A. Skinner*, Chem. Reviews, 1955, **55**, 745). The present trend is to try to refine the electronegativity theory and to give it a clearer physical concept (*R. P. Iczkowski* and *J. L. Margrave*, J. Amer. chem. Soc., 1961, **83**, 3547).

The *s*- and *p*-type orbitals involved in covalent bonding (see Vol. 1A, p. 350 *et seq.*) may undergo *hybridization* to form new orbitals with high directional character (*C. A. Coulson*, "Valence", Chap. VIII, Clarendon Press, Oxford, 1952). It is possible to have much overlapping of atomic orbitals of different atoms when these highly directional orbitals are used and this results in strong chemical bonds. Transition metals do not easily form stable carbon-metal σ bonds although the bond can be stabilized by highly electronegative substituents or by the use of suitable ligands (*J. W. Richardson*, "Organometallic Chemistry", A.C.S. Monograph No. 147, Reinhold, New York, 1960, p. 1). This topic is dealt with more fully in the section on transition metals (p. 246).

(c) *Pi bonds.* Transition metals form stable compounds with unsaturated organic compounds such as olefins, dienes, acetylenes and aromatic hydrocarbons. The nature of the bonding of these to the metal, a problem of long standing, was clarified by *M. J. S. Dewar* (Bull. Soc. chim. Fr., 1951, **18**, C79) and *J. Chatt* and *L. A. Duncanson* (J. chem. Soc., 1953, 2939) who proposed a detailed picture of the bonding in Zeise's salt, $K[PtCl_3(C_2H_4)]$. This involves the formation of a

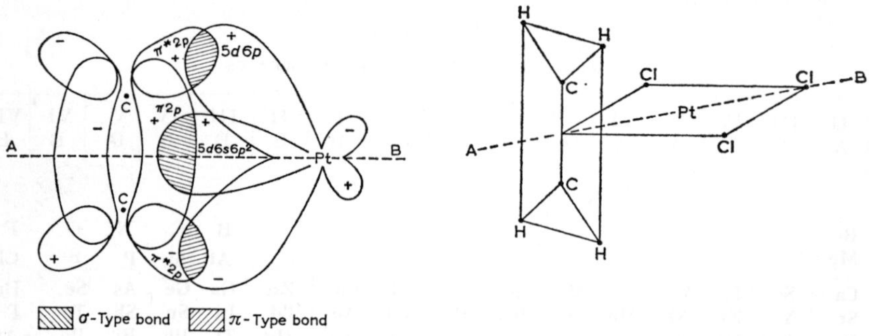

Fig. 1. Bonding in Zeise's Salt. (Reproduced with modifications by courtesy of *Prof. Chatt* and the Editor of the Journal of the Chemical Society).

σ-bond by donation of $\pi 2p$ electrons from the olefin double bond into the unfilled $5d6s6p^2$ orbital of the platinum together with a *π bond* by overlap of the vacant anti-bonding π^*2p orbital of the olefin with the filled $5d6p$ orbital of the platinum. The latter is referred to as back donation of electrons from the filled orbitals of the metal to the anti-bonding orbitals of the olefin. Evidence for the existence of this back donation has been obtained from X-ray data (*P. R. H. Alderman, P. G. Owston* and *J. M. Rowe*, Acta Cryst., 1960, **13**, 149) and low dipole moments (*J. Chatt, R. G. Guy* and *L. A. Duncanson*, J. chem. Soc., 1961, 827). In Zeise's salt the olefin molecule is symmetrically bonded at right angles to the plane containing the Pt—Cl bonds.

The interaction of π-orbitals with unfilled *d*-orbitals in the formation of organo-transition metal compounds is described as *pi(π)-bonding*. The metal is not attached to any particular carbon atom but the whole group or ring is bonded uniformly. Arene (*i.e.* aromatic hydrocarbon) complexes and most cyclopentadiene derivatives of the transition metals are of this type (*G. E. Coates* and *F. Glockling*, "Organometallic Chemistry", A.C.S. Monograph No. 147, Reinhold, New York, 1960, p. 426; *H. Zeiss*, *ibid.*, p. 380; *P. L. Pauson*, *ibid.*, p. 346).

(*d*) *Electron-deficient bonds.* The structure of a number of organometallic compounds can only be explained by assuming the formation of bonds containing less than two electrons. Such bonds are found only where a metal or metalloid atom with more low energy orbitals than valence electrons is combined with atoms or groups which do not contain unshared electron pairs (*R. E. Rundle,* J. phys. Chem., 1957, **61**, 45). Structures of this type are known as *bridge structures,* electron-deficient bonds being indicated by dotted lines. Alkyls of the lighter metals in Groups II and III contain electron-deficient bonds in which carbon atoms form the bridge.

The elements are shown in the periodic table below (Table 2). This has been marked to indicate the three main types of chemical bonds formed by the metals and metalloids.

TABLE 2

PERIODIC TABLE OF THE ELEMENTS

I A	II A	III A	IV A	V A	VI A	VII A	VIII			I B	II B	III B	IV B	V B	VI B	VII B	0
H																	He
Li	**Be**											**B**	C	N	O	F	Ne
Na	**Mg**											**Al**	**Si**	**P**	S	Cl	Ar
K	**Ca**	**Sc**	**Ti**	**V**	**Cr**	**Mn**	**Fe**	**Co**	**Ni**	**Cu**	**Zn**	**Ga**	**Ge**	**As**	**Se**	Br	Kr
Rb	**Sr**	**Y**	**Zr**	**Nb**	**Mo**	**Tc**	**Ru**	**Rh**	**Pd**	**Ag**	**Cd**	**In**	**Sn**	**Sb**	**Te**	I	Xe
Cs	**Ba**	**La***	**Hf**	**Ta**	**W**	**Re**	**Os**	**Ir**	**Pt**	**Au**	**Hg**	**Tl**	**Pb**	**Bi**	**Po**	At	Rn
Fr	**Ra**	**Ac†**															

* Lanthanides rare earths	**Ce**	**Pr**	**Nd**	**Pm**	**Sm**	**Eu**	**Gd**	**Tb**	**Dy**	**Ho**	**Er**	**Tm**	**Yb**	**Lu**
† Actinides	**Th**	**Pa**	**U**	**Np**	**Pu**	**Am**	**Cm**	**Bk**	**Cf**	**Es**	**Fm**	**Md**	**No**	

Light type: elements not classed as metals or metalloids.
Heavy type: metals and metalloids.
—·—·—· metal-carbon bonds highly ionic in character.
———— transition metals; organic derivatives contain mainly d-orbital bonding.
-------- metal-carbon bonds mainly covalent.

(*iv*) *General methods of preparation of organometallic compounds*

A brief description of the most important general preparative methods which are useful in the aliphatic series is given below. A detailed account of this subject has been given by *R. G. Jones* and *H. Gilman* (Chem. Reviews, 1954, 54, 835). More recent preparative procedures are described by *J. Eisch* and *Gilman* (Adv. in inorg. Chem. and Radiochem., 1960, **2**, 61).

Nearly all the metal alkyls can be prepared by one or more of the following reactions.

1. *Metal with alkyl halide*:

$$M + RX \longrightarrow RMX$$

$$\text{or } 2M + RX \longrightarrow RM + MX$$

This is one of the most fundamental reactions in organometallic chemistry since it can be used directly for metals of Group IA, IIA, zinc and mercury or indirectly for the preparation of other organometallic compounds (See 4).

2. *Metal alloy with alkyl halide*. When the metal reacts only slowly, or not at all, with an alkyl halide (*e.g.* metals of Groups IVB and VB) it is convenient to use a metal alloy—usually with sodium

$$RX + M/Na \longrightarrow RM + NaX$$

Organoaluminium compounds (R_2AlX) have been prepared using an aluminium/magnesium alloy.

3. *Metal exchange*:

$$M + RM' \rightleftharpoons RM + M' \text{ (M more reactive than M')}$$

This is a very general reaction and can be used for most metals. Reactions of this type are generally reversible but the equilibrium normally lies very far to one side. It is used to prepare a *more* reactive organometallic compound from a readily accessible *less* reactive one. The latter is usually a dialkylmercury.

4. *Alkyl-halogen exchange*:

$$RM + M'X \rightleftharpoons RM' + MX \text{ (RM more reactive than RM')}$$

This reaction is useful for all types of organometallic compounds including the metalloids and transition metal complexes. Grignard reagents, lithium, zinc and aluminium alkyls (RM) are easily prepared, very reactive and very useful for the preparation of less reactive organometallic compounds.

5. *Metal hydrides with olefins*. Addition of olefin to a metal hydride is a newer procedure which is particularly useful for the preparation of aluminium, boron and silicon compounds although it has also been used with hydrides of Group IA metals and those of germanium, tin, lead and Group VB metals:

$$AlH_3 + 3CH_2{=}CH_2 \rightleftharpoons Al(CH_2CH_3)_3$$

The olefin should have a terminal double bond ($RCH{=}CH_2$) to react readily with an Al—H bond but boron hydrides will add to internal double bonds.

A related reaction is the transalkylation of tri-isobutylaluminium, tri-isobutylborane and tri-isobutylgallium with higher olefins. The metal hydride bond is formed by thermal decomposition of the tri-isobutyl derivative, which then reacts with olefin to give the higher metal alkyls in good yield:

$$Al[CH_2CH(CH_3)_2]_3 + 3RCH{=}CH_2 \longrightarrow Al(CH_2CH_2R)_3 + 3CH_2{=}C(CH_3)_2$$

6. *Equilibration reaction.* Alkyls of some metals in Groups III, IV and V readily undergo interconversion reactions with the corresponding metal halides. An equilibrium is set up whose position is determined by the relative proportion of the reactants:

$$AlR_3 + AlRCl_2 \rightleftharpoons 2AlR_2Cl$$

$$R_4Si + SiCl_4 \rightleftharpoons R_3SiCl + R_2SiCl_2 + RSiCl_3$$

$$Me_3Bi + 2BiCl_3 \rightleftharpoons 3MeBiCl_2$$

7. *Metal halides with diazomethane.* Reaction of diazomethane with many metal halides gives the corresponding chloromethyl derivatives:

$$GeCl_4 + CH_2N_2 \longrightarrow Cl_3GeCH_2Cl + N_2$$

The net effect is the insertion of a methylene group and the reaction is known as *methylenation*. It has a wide scope and halides of aluminium, mercury, zinc, magnesium, boron, thallium and metals of Groups IV and VB are known to react (*D. Seyferth*, Chem. Reviews, 1955, 55, 1155).

8. *Electrolytic synthesis.* A metal which is susceptible to attack by alkyl radicals can be converted to the corresponding organometallic compound if it is used as the anode for the electrolysis of the complex $NaF \cdot 2AlR_3$. Aluminium is deposited at the cathode while the liberated radicals attack the anode. Triethyl- and tripropyl-antimony and tetraethyltin are readily prepared by this process. Unfortunately the alkyls of many metals (Pb, Hg, Zn, Cd, Bi, Tl) are partially or completely decomposed but if the electrolysis is carried out under vacuum the metal alkyl can be recovered before decomposition occurs (*K. Ziegler*, "Organometallic Chemistry", A.C.S. Monograph No. 147, Reinhold, New York, 1960, p. 251). The preparation of magnesium dialkyls by the electrolysis of alkali metal tetra-alkylaluminium solutions using a magnesium anode has commercial possibilities (*P. Kobetz* and *R. C. Pinkerton*, U.S.P., 3,028,319/1962).

(v) General physical properties of organometallic compounds

These depend chiefly on the metal since the polarity of the carbon metal bond has a large effect on the melting point and volatility of the compounds. The boiling points of metal alkyls belonging to B sub-groups II to VI increase in a regular manner as the size of the organic group increases. Alkyls of the more electropositive metals are often highly associated and an increase in the size of the alkyl group can hinder association and thus increase volatility. Physical properties are also influenced by the symmetry of the molecule.

(vi) Reactivity of metal–carbon bonds

H. Gilman and others have determined the relative reactivities of organometallic compounds by comparing the rates of addition to unsaturated bonds ("Organic Chemistry", 2nd Edn., Vol. 1, Wiley, New York, 1943, p. 520). The reactivity of the organometallic compound rises with increasing atomic number in sub-groups IA to IIIA, but falls in the corresponding B sub-groups. In groups V to VII the situation is reversed; activity falls with increasing atomic number of the parent elements in the A sub-groups, but rises in the B sub-groups.

Chemical reactivity can be related in many cases to the electronegativity difference or polarity of the carbon–metal bond. With some reservations it is the electronegativity value of the metal which determines the reactivity of the organometallic compound. However, other factors such as steric hindrance or substitution of the alkyl group by highly electronegative elements can have a profound influence. An extreme example is the per-

TABLE 3

IONIC CHARACTER OF SOME CARBON-METAL BONDS

Bond	Metal group	$x_A—x_B$	% ionic character
C—Na	IA	1·6	47
C—Cs	IA	1·8	55
C—Mg	IIA	1·3	34
C—Ba	IIA	1·6	47
C—Zn	IIB	0·9	18
C—Hg	IIB	0·6	9
C—B	IIIB	0·5	6
C—Al	IIIB	1·0	22
C—Ge	IVB	0·7	12
C—Sn	IVB	0·7	12
C—Pb	IVB	0·7	12
C—P	VB	0·4	4
C—Sb	VB	0·6	9

fluoroalkyl derivatives (Review: *R. E. Banks* and *R. N. Haszeldine*, Adv. in inorg. Chem. and Radiochem., 1961, 3, 338). Table 3 lists the percentage ionic character of various carbon-metal bonds which have been calculated from Pauling's electronegativity values (Table 1). This table is in accord with the empirical reactivity series of *Gilman* (*loc. cit.*).

E. G. Rochow, D. T. Hurd and *R. N. Lewis* have divided organometallic compounds into various classes of reactivity according to the electronegativity value of the metal; the lower the electronegativity, the wider the range of reactions of its organometallic compounds. Thus alkyl derivatives of metals with electronegativity values above 1·5 (zinc, cadmium, mercury, Groups IVB and VB) do not react with carbon dioxide. Above 1·7 (mercury, thallium, boron, Groups IV and VB) there is no reaction with acid chlorides or water ("The Chemistry of Organometallic Compounds", Wiley, New York, 1957, p. 24).

(vii) Reactions of metal and metalloid alkyls

Some of these have been dealt with in the section on the preparation of organometallic compounds. Reactions with inorganic reagents will be considered here.

(1) *Hydrogen*. Hydrogenolysis of carbon–metal bonds occurs readily with Group IA metal alkyls at atmospheric pressures of hydrogen and in the absence of catalysts (*Gilman, A. L. Jacoby* and *H. Ludeman*, J. Amer. chem. Soc., 1938, 60, 2336). The general reaction is:

$$RM + H_2 \longrightarrow RH + MH$$

Under more severe conditions (200°, 125 atmospheres and a nickel catalyst) the same reaction occurs with dibutylzinc:

$$(C_4H_9)_2Zn + 2H_2 \longrightarrow 2C_4H_{10} + ZnH_2$$

Tetraheptyl-lead is reduced to metallic lead and *n*-tetradecane (*W. H. Zartman* and *H. Adkins*, *ibid.*, 1932, 54, 3398):

$$(C_7H_{15})_4Pb + 2H_2 \longrightarrow 2C_{14}H_{30} + Pb$$

Aluminium alkyls also undergo hydrogenolysis reactions:

$$2(C_2H_5)_3Al + Al + 3/2\ H_2 \longrightarrow 3(C_2H_5)_2AlH$$

$$Al(C_2H_5)_3 + H_2 \longrightarrow (C_2H_5)_2AlH + C_2H_6$$

Few reactions of hydrogen with transition metal compounds are known. Dimethylmanganese is reduced in tetrahydrofuran to the metal and methane, while dicyclopentadienyldimethyltitanium reacts instantaneously with the formation of dicyclopentadienyltitanium and methane. Methyltitanium trichloride is not reduced (*K. Clauss* and *H. Bestian*, Ann., 1962, 654, 8).

2. *Oxygen.* Alkyls of metals with electronegativity values less than 1·7 oxidise in air. With the more reactive metals (in particular Group IA, IIA, aluminium, zinc and boron—although the last is relatively much more electronegative) and where the alkyl groups are small, the oxidation reaction is sufficiently exothermic to cause ignition. Under controlled conditions the usual reaction is insertion of oxygen between the carbon and metal atoms with the formation of an alkoxide:

$$R_3M \longrightarrow R_2MOR \longrightarrow RM(OR)_2 \longrightarrow M(OR)_3$$

These reactions become progressively less facile as oxidation proceeds.

3. *Water or mineral acids.* Alkyls of Groups IA, IIA and IIIA metals react extremely vigorously (sometimes explosively so) with water or mineral acid to form the corresponding alkane and metal hydroxide. Less reactive metal alkyls (those of mercury, silicon, tin—electronegativities greater than 1·7) react very slowly or not at all with water. If several bonds are involved cleavage may take place by a stepwise process.

4. *Halogens.* Halogens react extremely vigorously with alkyls of Groups IA and IIA metals and little work has been reported. Other metal alkyls form the corresponding metal halide and alkyl halide:

$$RM + X_2 \longrightarrow RX + MX$$

Stepwise cleavage can occur under controlled conditions, iodine being the most favoured halogen for these reactions. In some mixed metal alkyl aryls (M = Ge, Sn, Pb) the aryl group is preferentially eliminated:

$$R_2Ar_2Pb + 2I_2 \longrightarrow R_2PbI_2 + 2ArI$$

5. *Metals.* Two reactions can occur depending on whether the metal alkyl is simple (R_3M) or mixed (R_2MX). Alkylmetal halides are reduced by active metals such as lithium, sodium, potassium or sodium/potassium alloy:

$$2RAlX_2 + 3Na \longrightarrow R_2AlX + Al + 3NaX$$

Displacement reactions occur with simple organometallic compounds if a more reactive metal is added to a metal alkyl containing a less reactive metal:

$$R_2Hg \xrightarrow{Na} 2RNa + Na/Hg \text{ amalgam}$$

6. *Metal halides.* This reaction has been described on p. 170:

$$BiCl_3 + 3RMgX \longrightarrow R_3Bi + 3MgClX$$

2. Group IA: Lithium, Sodium, Potassium, Rubidium, Caesium

The carbon–metal bonds formed by the alkali metals have a high degree of ionic character. Alkali metal alkyls can be grouped into two classes.

(1) Most lithium alkyls. Lithium is the least electropositive of the alkali metals and its compounds with the possible exception of methyl-lithium, possess properties more characteristic of covalent compounds. Propyl-lithium and the higher alkyls are liquids, soluble in hydrocarbons. They have very low volatilities because they are associated, probably by some sort of electron-deficient bonding.

(2) Derivatives of sodium, potassium, rubidium and caesium, which behave in many ways as salts. They are solids, insoluble in organic solvents, decompose without melting when heated and they dissociate to form conducting solutions in such solvents as dimethylzinc and trimethylaluminium, with which they form complex ions. Methyl-lithium may also belong to this class.

The relative chemical reactivities of organometallic compounds in Group IA increase in the order RLi < RNa < RK ~ RRb < RCs (*Gilman, loc. cit.*). When dibenzofuran is metallated with the ethyl derivatives of lithium, sodium and potassium, the relative amounts of metallation products are 1, 46 and 139 respectively. The preparation, properties and reactions of alkali metal alkyls have been reviewed (*A. A. Morton*, Chem. Reviews, 1944, 35, 1; *E. A. Braude*, Prog. in org. Chem., 1955, 3, 172).

(a) Lithium

Although the lithium derivatives in class (1) are less polar than those of class (2) they are still very reactive. Most are easily prepared and for these reasons they are widely used in synthetic organic chemistry (*G. Wittig*, "Newer Methods of Preparative Organic Chemistry", Interscience, New York, 1948, p. 571). Lithium derivatives can offer certain advantages over Grignard reagents (*vide infra*); they are more reactive, the working up of the reaction mixture is often cleaner and they can sometimes be prepared when the corresponding Grignard reagents either cannot be formed, or are formed in low yield only. The last is particularly true for some halogenated aromatic compounds (*Gilman, E. A. Zoellner* and *W. M. Selby*, J. Amer. chem. Soc., 1933, **55**, 1252) and vinyl halides using diethyl ether as solvent. The reactions of organolithium compounds are generally similar to those of the corresponding Grignard reagents although there are some cases where the more reactive lithium compounds undergo different reactions, *e.g.*, unlike the magnesium derivatives, they will add to conjugated double bonds, certain phenylated ethylenes and to aromatic hydrocarbons such as anthracene:

$$(C_6H_5)_2C{=}CH_2 + C_4H_9Li \longrightarrow (C_6H_5)_2C(Li)CH_2C_4H_9$$

Treatment of a Grignard reagent with carbon dioxide gives the corresponding carboxylic acid in good yield but the higher reactivity of the lithium derivative may allow the addition of a second molecule of organolithium compound to the lithium salt of the acid giving a ketone as the major product (*Gilman* and *P. R. Van Ess*, *ibid.*, 1933, **55**, 1258):

$$RLi \xrightarrow{CO_2} RCOOLi \xrightarrow{RLi} \underset{OLi\ \ OLi}{RCR} \xrightarrow{H_2O} RCOR$$

Preparation. (1) Alkyl-lithium compounds are generally prepared by reaction between lithium metal and an alkyl halide (method 1, p. 170), using a hydrocarbon or ethereal solvent and an atmosphere of dry nitrogen or argon (see, *e.g. E. Terres et al.*, Erdöl und Kohle, 1960, **13**, 84; *T. D. Perrine* and *H. Rapoport*, Anal. Chem., 1948, **20**, 635).

Alkyl chlorides are preferable to alkyl bromides; alkyl iodides (except methyl iodide) cannot be used since they react too quickly with the resultant alkyl-lithium. The concentrations of solutions of *n*-butyl-lithium can be determined by a double titration method (*Gilman* and *A. H. Haubein*, J. Amer. chem. Soc., 1944, **66**, 1515, but see also *C. W. Kamienski* and *D. L. Esmay*, J. org. Chem., 1960, **25**, 115). Lithium alkyls attack diethyl ether and they are even less stable in tetrahydrofuran. Working temperatures recommended for the latter solvent are 0° for methyl-lithium and —30° for butyl-lithium (*Gilman* and *B. J. Gaj*, J. org. Chem., 1957, **22**, 1165).

(2) Organolithium compounds which cannot be prepared directly may be formed by a halogen–metal exchange reaction in which the lithium atom becomes attached to the more electronegative radical. Butyl-lithium is normally used for this purpose because of its high reactivity and the low electronegativity of the butyl radical. (Reviews: *R. G. Jones* and *Gilman*, Org. Reactions, 1951, **6**, 339; *Gilman* and *J. W. Morton*, *ibid.*, 1954, **8**, 258; *D. L. Esmay*, "Metal-Organic Compounds", A.C.S. Series No. 23, p. 46, Amer. chem. Soc., 1959).

Alkyl-lithium compounds are strongly associated in the liquid state and may remain so even in the vapour state. The association appears to take place through the formation of multicentre electron-deficient bonds of the type which occur in $Al_2(CH_3)_6$ (*R. West* and *W. Glaze,* J. Amer. chem. Soc., 1961, 83, 3580).

A useful bibliography, with annual supplements, on the use of organo-lithium compounds in organic synthesis is published by the Lithium Corporation of America, 500 Fifth Avenue, New York 36.

Methyl-lithium is precipitated when ethyl-lithium is added to a benzene solution of dimethylmercury (*W. Schlenk* and *J. Holtz,* Ber., 1917, **50,** 262) or methyl iodide (*T. L. Brown* and *M. T. Rogers,* J. Amer. chem. Soc , 1957, **79,** 1859). Solutions of methyl-lithium in diethyl ether are relatively stable and can be estimated by titration with acid.

Ethyl-lithium, m.p. 95°. Its infra-red and N.M.R. spectra suggest that it exists as a single species in hydrocarbon solution, probably a hexamer. In the vapour state the predominant species are the hexamer and tetramer (*Brown et al., ibid.,* 1962, **84,** 1371; J. phys. Chem., 1961, **65,** 1380).

n-*Butyl-lithium,* a liquid which can be distilled at 80–90°/10^{-4} mm., can be prepared from butyl chloride in pentane or hexane at room temperature, at which deterioration of butyl-lithium in these solvents is not appreciable. In benzene and cyclohexane solution the hexamer, $(BuLi)_6$, is the predominant species (*D. Margerison* and *J. P. Newport,* Trans. Faraday Soc., 1963, **59,** 2058), but in ethereal solution *n*-butyl-lithium exists as a solvated dimer $(BuLi)_2$, Et_2O (*Z. K. Cheema et al.,* J. Amer. chem. Soc., 1963, **85,** 3517). It is now an important industrial chemical as a catalyst for the preparation of stereospecific poly-isoprene and polybutadiene.

tert-*Butyl-lithium,* has been obtained by sublimation at 70–80°/0·1 mm., as a colourless crystalline solid, pyrophoric in air, which does not melt but de-composes at temperatures above 140°. In hexane and benzene solution it is tetrameric (*M. Weiner, G. Vogel* and *R. West,* Inorg. Chem., 1962, **1,** 654).

The direct preparation of *benzyl-lithium* in the normal way gives mainly the coupling product, bibenzyl, but solutions of benzyl-lithium can be obtained by adding lithium metal to an alkyl benzyl ether in an ether tetrahydrofuran mixture (*Gilman* and *G. L. Schwebke,* J. org. Chem., 1962, **27,** 4259).

Vinyl-lithium, a solid which decomposes on heating, is insoluble in petrol and violently pyrophoric. Unlike ethyl-lithium it forms fairly stable ethereal solutions (*E. C. Juenge* and *D. Seyferth, ibid.,* 1961, **26,** 563). It can be prepared directly, in 60 % yield, from vinyl chloride and lithium containing 2 % sodium (*R. West* and *W. H. Glaze, ibid.,* 1961, **26,** 2096) or by a transmetallation reaction involving phenyl-lithium and tetravinyltin (*D. Seyferth* and *M. A. Weiner,* J. Amer. chem. Soc., 1961, **83,** 3583). *Allyl-lithium* and *methallyl-lithium* have also been prepared by a transmetallation reaction. The former is insoluble in pentane and violently pyrophoric in air (*idem.,* J. org. Chem., 1961, **26,** 4797; *J. J. Eisch* and *A. M. Jacobs, ibid.,* 1963, **28,** 2145).

(b) Other metals of Group IA

Alkyls of class (2) are the most reactive of all organometallic compounds. They may be regarded as salts, *e.g.* n-$C_3H_7CH_2^{\ominus}Na^{\oplus}$, and their extremely high reactivity is due to the localization of the greater part of a negative charge on one carbon atom. All of them are spontaneously inflammable in air and they attack almost all organic compounds except the paraffin hydrocarbons.

Preparation. (1) Products free from alkyl halides can be prepared by an exchange reaction with the appropriate alkyl mercury (*Schlenk* and *Holtz, loc. cit.*).

(2) Most alkylsodium compounds are conveniently prepared by the reaction of the appropriate alkyl chloride with finely dispersed sodium (*A. A. Morton et al.*, J. Amer. chem. Soc., 1950, **72**, 3785).

(3) By reaction of sodium hydride with the appropriate olefin.

The use of these materials in metalation reactions has been reviewed (*R. A. Benkeser et al.*, Chem. Reviews, 1957, **57**, 867).

For a detailed review of organo-sodium and -potassium compounds see *M. Schlosser*, Angew. Chem. internat. Edn., 1964, **3**, 278, 362.

Alkyl derivatives of the heavier alkali metals are less thermally stable than those of lithium.

Methylsodium and **methylpotassium** decompose at 200°. The reaction is complex but is believed to be mainly:

$$8CH_3Na \longrightarrow 6CH_4 + Na_2C_2 + 6Na$$

Higher alkyls are even less stable; *ethylsodium* and *amylsodium* decompose rapidly below 100°. Pyrolysis of the latter is complex; both pentane and pentene are formed (*Morton* and *E. J. Lanpher*, J. org. Chem., 1956, **21**, 93). Amylsodium and sodium isopropoxide form the Alfin catalyst which is used in the preparation of synthetic rubbers from butadiene and styrene (*R. A. Stewart* and *H. L. Williams*, Ind. Eng. Chem., 1953, **45**, 173).

Benzylpotassium can be prepared by direct reaction of the metal with toluene in the presence of sodium oxide:

$$2C_6H_5CH_3 + 2K + Na_2O \longrightarrow 2C_6H_5CH_2K + NaH + NaOH$$

Sodium does not react under these conditions (*C. E. Claff* and *Morton*, J. org. Chem., 1955, **20**, 440).

Vinylpotassium (and *vinylsodium*) can be made from the metals and vinyl chloride using tetrahydrofuran as solvent (*R. G. Anderson, M. B. Silverman* and *D. M. Ritter, ibid.*, 1958, **23**, 750).

3. Group IIA: Beryllium, Magnesium, Calcium, Strontium, Barium
Group IIB: Zinc, Cadmium, Mercury

Alkyls of Group II metals vary widely from the very reactive, non-volatile types to the inert covalent types. The decrease in chemical reactivity follows the increase in the electronegativity of the metal; Ba, 0·9; Ca, 1·0; Sr, 1·0; Mg, 1·2; Be, 1·5; Zn, 1·6; Cd, 1·7; Hg, 1·9.

The most important derivatives of the Group II metals are the Grignard reagents. They owe their great utility in synthesis to the fact that they are easily formed in high yields from metallic magnesium, are soluble and stable in ether and are very reactive. Alkyls of metals which are less electronegative than magnesium are more difficult to handle while those of metals which are more electronegative are comparatively unreactive.

The lighter elements of Group II (and III) metals tend to increase their coordination number to four by the formation of donor complexes or electron-deficient bonds. For a list of all the organo-compounds of barium, calcium, strontium and beryllium described up to 1961 see *G. A. Balueva* and *S. T. Yoffe* (Russian Chem. Reviews, 1962, **31**, 439).

(a) Beryllium

Preparation. (1) By the action of a Grignard reagent on beryllium chloride.

(2) By the action of the corresponding organomercury compound on metallic beryllium.

(3) By addition of an olefin to beryllium hydride, alkyl hydride or dialkyl (*K. Ziegler*, B.P., 763,824/1956).

Little is known about organoberyllium halides. *R. E. Dessy* has suggested that there is no evidence for structures of the type RBeX (J. Amer. chem. Soc., 1960, **82**, 1580 but see *Gilman* and *Schulze*, *ibid.*, 1927, **49**, 2904).

Dimethylberyllium, a crystalline solid, sublimes at 130°/5 mm. It is spontaneously inflammable in air and carbon dioxide and reacts violently with water. Its structure has been shown by X-ray analysis to consist of polymeric chains in which beryllium atoms are attached tetrahedrally to four carbon atoms:

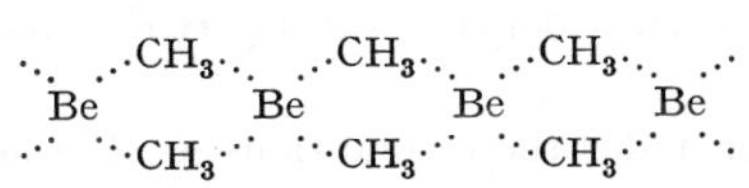

Dimethylberyllium is an electron-deficient molecule and readily forms complexes with donor solvents such as *tert*-amines, *tert*-phosphines and ethers. Trimethylamine can break down the polymer structure to form the coordination compound *trimethylamine-dimethylberyllium*, $Me_2\overset{\ominus}{Be}:\overset{\oplus}{N}Me_3$, m.p. 36°, which is associated in the liquid and solid states but is monomeric in the vapour state (*G. E. Coates, F. Glockling* and *N. D. Huck*, J. chem. Soc., 1952, 4496, 4512; *Coates* and *Huck*, *ibid.*, p. 4501). Dimethylberyllium reacts also with mono- and di-alkylamines; dimethylamine gives a trimeric cyclic compound $(MeBe{\cdot}NMe_2)_3$. In the vapour state dimethylberyllium is a mixture of monomer, dimer and trimer. It is practically insoluble in benzene.

Diethylberyllium b.p. 63°/0·3 mm.; 93–95°/4 mm., is associated but it is easily soluble in benzene. Its boiling point, suggests that it is dimeric (*J. Goubeau* and *B. Rodewald*, Z. anorg. Chem., 1949, **258**, 162).

Di-isopropylberyllium m.p. —9·5°, b.p. 280° (extrapolated) is dimeric in benzene solution. Above 50° it decomposes with evolution of propene leaving polymeric isopropylberyllium hydride (*Coates* and *Glockling*, J. chem. Soc., 1954, 22).

*Di-*tert*-butylberyllium*, a clear, colourless mobile liquid, f.p. —16°, may have a monomeric structure. It is less stable than the isopropyl derivative and decomposes at room temperature (*idem.*, *ibid.*, 1954, 2526; *E. L. Head, C. E. Holley* and *S. W. Rabideau*, J. Amer. chem. Soc., 1957, **79**, 3687.

(b) Magnesium

Magnesium alkyls are of two types (*i*) the dialkyls, R_2Mg, which are relatively unimportant and have not been studied in detail, and (*ii*) the Grignard reagents, usually represented as RMgX. *D. O. Cowan* and *H. S. Mosher* have compared the two types in addition, reduction and enolization reactions (J. org. Chem., 1962, **27**, 1). No mixed alkyls RMgR′ have been reported.

(i) Magnesium dialkyls, R_2Mg

Preparation. (1) By the action of mercury alkyls on magnesium preferably in the presence of ether (*W. Schlenk*, Ber., 1931, **64**, 734).

(2) More conveniently by adding dioxan to an ethereal solution of a Grignard reagent; magnesium halides are precipitated and the dialkylmagnesium is left in solution (*C. R. Noller* and *W. R. White*, J. Amer. chem. Soc., 1937, **59**, 1354; *W. Strohmeier* and *F. Seifert*, Ber., 1961, **94**, 2356).

(3) Recent processes involve the addition of olefins to magnesium or magnesium hydride in the presence of hydrogen at 60–120°, preferably in the presence of pre-formed alkyl (*C. Randaccio*, Ital. Pat., 548,183/1956; *S. M. Biltzer* and *T. H. Pearson*, U.S.P., 2,959,625/1960).

(4) The electrolysis of an alkali metal tetra-alkylaluminium solution using a magnesium anode liberates the low melting complex $MgR_2{\cdot}2AlR_3$. Vacuum distillation removes the volatile aluminium alkyl leaving the magnesium

dialkyl as a crystalline solid (*P. Kobetz* and *A. C. Pinkerton*, U.S.P., 3,208,319/1962).

Dimethylmagnesium and higher homologues have been isolated in the pure state. All are thermally stable, white, crystalline solids, insoluble in hydrocarbons and, apart from dimethylmagnesium, completely non-volatile. They are very reactive and ignite spontaneously in air.

(ii) Organomagnesium halides, RMgX

These compounds, discovered by Grignard in 1900, have since been used in an enormous number of reactions; a comprehensive account of their preparation, properties and reactions is given by *M. S. Kharasch* and *O. Reinmuth* ("Grignard Reactions of Nonmetallic Substances", Constable, London, 1954). A complete compilation of all Grignard reactions recorded in the literature up to the beginning of 1948 has been made (*S. T. Yoffe* and *A. N. Nesmeyanov*, "Handbook of Magnesium-Organic Compounds", 3 vols., Pergamon Press, London 1957).

Preparation. (1) The following method from magnesium and an alkyl halide is applicable to the preparation of most normal Grignard reagents, all chemicals and apparatus being previously dried as Grignard reagents are sensitive to air and moisture.

The required amount of magnesium turnings, plus 5% excess, is added to the reaction flask and covered with dry ether. A small amount of the halide is run in without stirring, to form a high local concentration. At this stage there is usually an induction period due to the presence of traces of moisture and to the coating of oxide on the magnesium turnings. Methods of initiating reaction include addition of one or two crystals of iodine, or magnesium activated by heating with iodine, or in difficult cases, addition of a pre-formed Grignard reagent. Warming may be necessary to start the reaction. Although formation of an organomagnesium halide is strongly exothermic the reaction is usually slow until it is well established. From this point it is often very fast and it is important not to add too much halide in the initial stage. When the reaction has begun to subside, the rest of the halide, in ether solution, is added at such a rate as to maintain reflux. When the addition is complete the reaction mixture is heated under reflux for a further 15–30 minutes. Suitable techniques for the preparation of Grignard reagents from gaseous or unreactive halides are described by *Kharasch* and *Reinmuth* (*loc. cit.*).

It is preferable to use an inert atmosphere during the preparation and subsequent reaction of the Grignard reagent and oxygen-free nitrogen is usually recommended. Grignard reagents prepared under helium or argon are reported to be clear and almost colourless solutions. Higher yields are also obtained and there is little or no contamination by coupling (*F. H. Owens, R. P. Fellman* and *F. E. Zimmerman*, J. org. Chem., 1960, **25**, 1808).

Iodides are the most reactive halides but generally give the lowest yields. Chlorides give the highest yields but are the most difficult to initiate; bromides are intermediate in both respects. Normal alkyl halides give better yields than secondary halides, while tertiary halides give the lowest yields.

The use of tetrahydrofuran as solvent in place of ether promotes the formation of Grignard reagents from relatively unreactive halides. The preparation of *alkenylmagnesium halides* proceeds smoothly in tetrahydrofuran. Vinylmagnesium halides are comparable in reactivity to the saturated Grignard reagents and as they can be easily made they will probably replace vinyl-lithium compounds (*H. E. Ramsden, et al.*, J. org. Chem., 1957, **22**, 1602; *H. K. Reimschuessel, ibid.*, 1960, **25**, 2256). The use of vinylmagnesium compounds in synthesis has been reviewed (*H. Normant*, Adv. in org. Chem., 1960, **2**, 1).

(2) A new way of preparing Grignard reagents has been developed in which titanium tetrachloride catalyses the exchange reaction between alkylmagnesium halides and terminal olefins:

$$C_3H_7MgBr + RCH = CH_2 \xrightarrow{TiCl_4} RCH_2CH_2MgBr + C_3H_6$$

This method is useful for the preparation of Grignard reagents where the olefin is more readily available than the corresponding halide. Yields from 20–60% are obtained (*G. D. Cooper* and *H. L. Finkbeiner*, J. org. Chem., 1961, **26**, 4779; 1962, **27**, 1493, 3395).

Di-Grignard reagents, $XMg(CH_2)_nMgX$, are known, but for a satisfactory yield n must be greater than 3 (*I. T. Millar* and *H. Heaney*, Quart. Reviews, 1957, **11**, 109). Ethylene dibromide gives ethylene while propylene dibromide undergoes a complex reaction.

Estimation. The classical methods for the quantitative determination of the concentration of Grignard reagent give either high results (*e.g.* hydrolysis and titration of magnesium hydroxide; *Volhard* halide estimation) or low ones (colorimetry using Michler's ketone). *R. D'Hollander* and *M. Anteunis* (Bull. Soc. chim. Belg., 1963, **72**, 77) have developed two more accurate methods: (a) reaction of Grignard reagent with excess benzophenone and measurement of absorbance at 333 mμ and (b) reaction with a known amount of acetone and subsequent determination of the excess volumetrically. Method (a) is faster and more convenient than method (b) but it cannot be used with *tert-* and some *sec-*alkylmagnesium halides because 1:4 addition and dehydration cause high absorbances.

Constitution. Grignard reagents are usually represented by the simple formula RMgX but their structure is much more complex and it is not yet completely resolved. They contain ether, probably bound by co-ordination to the magnesium, and may be associated. An equilibrium of the type:

$$2RMgX \rightleftharpoons R_2Mg + MgX_2$$

has been postulated as existing in ethereal solutions. The conflicting and apparently irreconcilable views on the constitution of the Grignard reagent have been summarised by *R. E. Dessy, S. E. I. Green* and *R. M. Salinger* (Tetrahedron Letters, 1964, p. 1369), who report that contrary to what they had previously reported, statistical exchange of magnesium occurs between diethylmagnesium and magnesium bromide labelled with radioactive metal. *Anteunis* (J. org. Chem., 1962, **27**, 596) has claimed that R_2Mg is not an intermediate reactive species in Grignard reactions, while other workers, using tetrahydrofuran as solvent in place of ether, have postulated that the ethyl in ethylmagnesium bromide can react from either $(C_2H_5)_2Mg \cdot MgBr_2$ or $[(C_2H_5)_2Mg]_2$ (*S. J. Storfer* and *E. I. Becker, ibid.*, 1962, **27**, 1868). More recently, *A. C. Ashby* and *W. E. Becker* have found that in tetrahydrofuran the Grignard reagent is monomeric and they conclude that, in this solvent, alkyl exchange does occur and that the predominant species is RMgX (J. Amer. chem. Soc., 1963, **85**, 118). The same applies in ether and there is no evidence of equilibrium with magnesium bromide

$$R_2Mg + MgBr_2 \rightleftharpoons R_2Mg \cdot MgBr_2$$

(*A. D. Vreugdenhil* and *C. Blomberg*, Rec. Trav. chim., 1963, **82**, 453, 461). X-ray studies on crystals of phenylmagnesium bromide di-etherate suggest that the formula is $PhMgBr \cdot 2(C_4H_{10}O)$ (*G. D. Stucky* and *R. E. Rundle*, J. Amer. chem. Soc., 1963, **85**, 1002). Proton magnetic resonance spectra of ethereal solutions of methyl, ethyl and propyl Grignard reagents closely resemble those of the corresponding dialkyls. These results support the view that Grignard reagents are better represented as $R_2Mg \cdot MgX_2$ rather than as RMgX (*D. F. Evans* and *J. P. Maher*, J. chem. Soc., 1962, 5125; *G. Fraenkel, D. G. Adams* and *J. Williams*, Tetrahedron Letters, 1963, 767).

Other evidence suggests that Grignard reagents are associated in solution to form polymeric chains involving R_2Mg and MgX_2 but *not* RMgX. The degree of polymerisation depends on the solvent (*A. Kirrmann, R. Hamelin* and *S. Hayes*, Bull. Soc. chim. Fr., 1963, 1395).

Reactions of organomagnesium compounds. Reactions of Grignard reagents fall broadly into three classes: (1) splitting of single bonds, (2) addition to multiple bonds and (3) miscellaneous, including oxidation, reduction, radical and "abnormal" reactions. Some typical reactions are given below.

(1) *Splitting of single bonds.* (a) *Reaction with compounds containing active hydrogen atoms.* The corresponding hydrocarbon RH is formed from such compounds as ROH, RSH, RNH_2, R_2NH, RC≡CH, RCO_2H, *e.g.*

$$RMgX + AH \longrightarrow RH + AMgX$$

This is the basis of the Zerewitinoff method for the estimation of active hydrogen; the product is treated with excess methylmagnesium iodide and the volume of methane liberated is measured.

(b) *Reaction with alkyl or allyl halides and alkyl sulphates.* This reaction is usually written:

$$RMgX + R'X \longrightarrow RR' + MgX_2$$

but side reactions such as elimination of HX may occur especially with secondary and tertiary halides. The best yields are obtained when R′X has a relatively polar carbon–halogen bond and does not easily lose a β-hydrogen atom. Allyl halides are particularly suitable. Better yields of the desired hydrocarbon are obtained if alkyl sulphates (or sulphonates) are used instead of the halides. A ratio of two molecules of ester to one of Grignard reagent is recommended.

(c) *Reaction with halogens.* Halogens react with Grignard reagents to form the corresponding alkyl halides:

$$RMgX + X'_2 \longrightarrow RX' + MgXX'$$

(d) *Reaction with metal or metalloid halides.* This is a general method for the synthesis of organo-derivatives of metals or metalloids which are more electronegative than magnesium:

$$PCl_3 + 3RMgX \longrightarrow PR_3 + 3MgXCl$$

$$SnCl_4 + 4RMgX \longrightarrow SnR_4 + 4MgXCl$$

Reaction with phosphoryl chloride proceeds stepwise giving, after hydrolysis, the phosphonic acid, $RPO(OH)_2$, phosphinic acid, $R_2PO(OH)$ and phosphine oxide, R_3PO. Only two chlorine atoms are readily replaceable in thiophosphoryl chloride and the main product is the tetra-alkyldiphosphine disulphide $R_2SP\cdot PSR_2$ (*P. J. Christen, L. M. Van der Linde* and *F. N. Hooge,* Rec. Trav. chim., 1959, 78, 161; *H. Niebergall* and *B. Langenfeld*, Ber., 1962, 95, 64).

(2) *Addition to multiple bonds.* Grignard reagents add to most types of unsymmetrical double or triple bonds but not to unsaturated carbon–carbon bonds. The R unit becomes attached to the more electropositive atom.

(a) *Carbonyl compounds.* Secondary or tertiary alcohols are formed by addition to aldehydes or ketones followed by hydrolysis of the reaction products:

$$RR'CO + R''MgX \longrightarrow RR'R''COMgX \xrightarrow{H_2O} RR'R''COH$$

Primary alcohols are formed from formaldehyde. Esters react in two stages; firstly by addition to the carbonyl group followed by splitting out of ROMgX, then the ketone thus formed reacts further to form a tertiary alcohol:

$$RCO_2R' + R''MgX \longrightarrow RCOR'' + R'OMgX$$

$$RCOR'' + R''MgX \longrightarrow RR''_2COH$$

Formic esters (R=H) yield secondary alcohols. Acid chlorides and anhydrides react in a similar manner to give the corresponding tertiary alcohols. Aliphatic γ-lactones give 1,4-glycols:

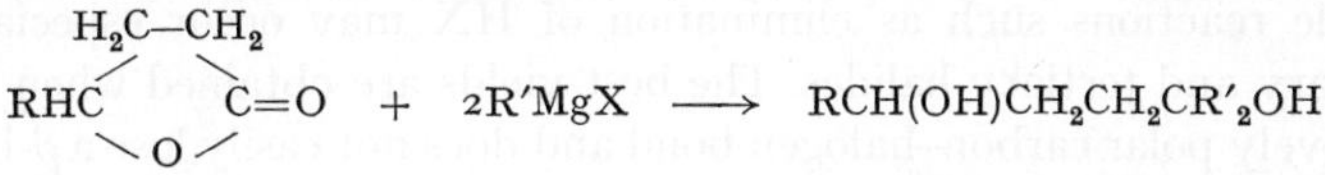

Keten itself is polymerized by alkylmagnesium halides but aromatic ketens react normally to form the corresponding enol or ketone. Many α,β-unsaturated ketones undergo abnormal addition at the 1,4 positions instead of the "normal" 1,2 positions:

$$\underset{1}{O}=\underset{2}{C}-\underset{3}{C}=\underset{4}{C}-$$

Instead of the expected unsaturated tertiary alcohol a saturated ketone is formed:

$$CH_3COCH=CHCH_3 + C_2H_5MgBr \longrightarrow CH_3C(OMgX)=CHCH(C_2H_5)CH_3 \longrightarrow CH_3COCH_2CH(C_2H_5)CH_3$$

Most α,β-unsaturated aldehydes react normally but α,β-unsaturated esters also undergo 1,4 addition.

The reaction of Grignard reagents with amides is complex. Ketones, nitriles, amines and imines have been reported among the reaction products.

Carbon dioxide readily carbonates Grignard reagents forming the salts of the corresponding carboxylic acids:

$$RMgX + CO_2 \longrightarrow RCO_2MgX \longrightarrow RCO_2H$$

Almost quantitative yields are obtained if the reagent is poured on to solid carbon dioxide.

Carbon monoxide gives two types of products. Aryl and *tert*-alkylmagnesium halides absorb one mole of carbon monoxide per mole:

$$2(CH_3)_3C\cdot MgX + 2CO \longrightarrow 2(CH_3)_3C\cdot COMgX \longrightarrow (CH_3)_3C\cdot CO\cdot CHOH\cdot C(CH_3)_3$$

Primary alkyl reagents absorb half this amount of carbon monoxide and produce olefins formed by dehydration of secondary alcohols:

$$2RCH_2MgX + CO \longrightarrow RCH_2C(MgX)(OMgX)CH_2R \longrightarrow RCH=CHCH_2R$$

Secondary alkylmagnesium halides undergo both reactions.

(b) *Double-bonded sulphur compounds.* Carbon disulphide reacts like carbon dioxide to form salts of dithiocarboxylic acids, RCSSMg. Sulphur dioxide similarly gives sulphinic acids, RSOOH. Both these products are capable of undergoing further reaction with excess Grignard reagent. Carbonyl sulphide appears to give a mixture of tertiary alcohol and thiocarboxylic acid. Dialkyl sulphoxides and dialkyl sulphides are formed when alkylmagnesium halides are treated with thionyl chloride. Sulphuryl chloride is reported to form sulphinic acids.

(c) *Nitriles.* Grignard reagents react with nitriles to form metal derivatives of ketimines which on acid hydrolysis give the corresponding ketones. In some cases it is possible to isolate the ketimine:

$$RC{\equiv}N + R'MgX \longrightarrow RR'C{=}NMgX \xrightarrow{H_2O} RR'C{=}NH \xrightarrow{acid} RR'CO$$

Nitriles are less reactive than aldehydes, ketones and esters and prolonged heating or a higher boiling solvent is usually required. Nitriles which have a labile hydrogen atom attached to the α-carbon atom are capable of undergoing a reaction analogous to a ketone enolization. In these cases the nitrile acts as a "pseudo-acid":

$$CH_3C{\equiv}N \rightleftharpoons CH_2{=}C{=}NH + RMgX \longrightarrow CH_2{=}C{=}NMgX + RH$$

This reaction occurs most readily with the lower aliphatic nitriles, and also leads to polymerization. α-Amino substituted nitriles react as if the cyano group was a halogen:

$$(CH_3)_2NCH_2CN + R'MgX \longrightarrow (CH_3)_2NCH_2R' + MgCNX$$

Cyanohydrins yield α-hydroxyketones when treated with two equivalents of Grignard reagent.

(d) *Nitro compounds.* The reaction with aliphatic nitro compounds gives disubstituted hydroxylamines but reduction to amine also occurs.

(e) *Nitroso compounds.* Little work has been carried out using alkyl Grignard reagents but the normal reaction appears to be formation of a hydroxylamine derivative with secondary reduction reactions.

(f) *Epoxides.* Ethylene oxide adds readily to Grignard reagents to give a mixture of the halohydrin $XCH_2{\cdot}CH_2OH$ and a primary alcohol two carbon numbers higher in a homologous series:

$$(CH_2)_2O + RMgX \longrightarrow RCH_2{\cdot}CH_2OMgX \longrightarrow RCH_2{\cdot}CH_2OH$$

The relative yields depend on the reaction conditions; at low temperatures and short contact times the halohydrin is the main product. Elevated temperatures or excess ethylene oxide favour the formation of the alcohol (*R. C. Huston* and *C. C. Langham*, J. org. Chem., 1947, **12**, 90; *E. E. Dreger*, Org. Syntheses, Coll. Vol. 1, 2nd Edn., Wiley, New York, 1941). Similar reactions occur with propylene oxide (*M. S. Kharasch et al.*, J. Amer. chem.

Soc., 1961, 83, 3229). Ethylene and propylene oxides have no tendency to rearrange but isomerization has been observed with more complex epoxides (*N. G. Gaylord* and *E. I. Becker*, Chem. Reviews, 1951, 49, 413).

(3) *Miscellaneous.* (a) *Oxidation reactions.* Oxygen reacts with Grignard reagents to form the corresponding alcohol.

$$RMgX + O_2 \longrightarrow ROOMgX \xrightarrow{RMgX} 2ROMgX \xrightarrow{H_2O} 2ROH$$

If the reagent is added slowly to a solution of oxygen in ether at —70° the hydroperoxide, ROOH, can be obtained in good yield. With sulphur, selenium and tellurium, mercaptans, sulphides and disulphides and their analogues are formed.

(b) *Reduction reactions.* Reduction of a ketone to the secondary alcohol usually occurs when the ketone or the Grignard reagent is highly branched. Aldehydes give normal addition products unless both reagents are branched. The normal addition and the reduction reactions may be competitive, *e.g.* with cyclohexanone:

$$(CH_2)_5CO + n\text{-}C_3H_7MgI \longrightarrow (CH_2)_5CHOH\ (50\%) + (CH_2)_5C(n\text{-}C_3H_7)OH\ (25\%)$$

(c) *Radical reactions.* Addition reactions involving Grignard reagents are normally ionic but some reactions proceed by a free radical mechanism. Homolytic dissociation of the R—MgX bond occurs in the presence of a transition metal halide giving mainly coupling products.

$$C_2H_5MgBr \xrightarrow{AgBr} C_4H_{10}\ (50\%) + C_2H_6\ (25\%) + C_2H_4\ (25\%)$$

This type of reaction has also been postulated to involve the decomposition of an unstable transition metal alkyl. With stoichiometric amounts or excess cobaltous bromide the free radicals produced by decomposition of the alkylmagnesium halide undergo disproportionation or hydrogen abstraction reactions. Ethylmagnesium bromide gave 37% ethane, 63% ethylene and 0·7% butane (*W. B. Smith,* J. org. Chem., 1961, 26, 4206).

(d) *Abnormal reactions.* The coupling, reduction and 1,4 addition reactions can be termed "abnormal". Others include the allylic rearrangement of substituted allylmagnesium halides. Butenylmagnesium halides undergo addition and condensation reactions in the 3-position to form α-methallyl derivatives:

$$CH_3\cdot CH{=}CH\cdot CH_2MgX \longrightarrow [CH_3\cdot CH(MgX)\cdot CH{=}CH_2] \xrightarrow{(CH_3)_2CO} (CH_3)_2C(OH)\cdot \underset{\displaystyle CH_3}{\underset{|}{CH}}\cdot CH{=}CH_2\ (81\%)$$

Organomagnesium halides in hydrocarbon solvents. Unsolvated organomagnesium halides have been prepared in hydrocarbon media from alkyl halides and magnesium in yields of up to 95 %. The lower alkyl halides (CH_3, C_2H_5), which react less readily than the higher ones, form suspensions while the latter are markedly soluble in iso-octane, dodecane and tetralin (*L. I. Zakharkin, O. Yu. Okhlobystin* and *B. N. Strunin*, Tetrahedron Letters, 1962, 631). These compounds are linearly associated and frequently correspond to the empirical formula R_3Mg_2X. They are much more reactive for the polymerization of vinyl monomers than ethereal Grignard reagents (*D. Bryce-Smith* and *G. O. Cox*, J. chem. Soc., 1961, 1175; *E. T. Blues* and *Bryce-Smith*, Chem. and Ind., 1960, 1533; *Bryce-Smith*, Bull. Soc. chim. Fr., 1963, 1418).

(c) Calcium, strontium and barium

Little work has been reported on the properties of these metal alkyls but they appear to resemble those of lithium. Complexes with diethylzinc ($MZnEt_4$, M = Ca, Sr, Ba) have been studied by *H. Gilman* (J. Amer. chem. Soc., 1945, 67, 520) who showed that activity decreased in the order Ba > Sr > Ca > Mg. (Review: *G. A. Balueva* and *S. T. Yoffe*, Russian Chem. Reviews, 1962, 31, 439).

Dialkyls. Dimethyl derivatives can be prepared from the metal and methyl iodide in pyridine. These compounds are unchanged in vacuum up to 400°. All three hydrolyse readily and become incandescent if they are exposed to oxygen or carbon dioxide (*D. A. Payne* and *R. T. Sanderson*, J. Amer. chem. Soc., 1958, 80, 5324).

Organocalcium halides can be prepared in ethereal solutions from metallic calcium (preferably containing a small amount of magnesium as a promotor) and an alkyl halide, preferably bromide or iodide. They resemble organolithium compounds rather than Grignard reagents and reaction with carbon dioxide tends to give more of the diketone, RCOR, than the corresponding carboxylic acid (*Bryce-Smith* and *A. C. Skinner*, J. chem. Soc., 1963, 577).

(d) Zinc

Dialkyls

Preparation. (1) By reaction of zinc, or better, a zinc–copper couple, with an alkyl iodide to give the alkylzinc iodide, which disproportionates on heating to the dialkylzinc (*C. R. Noller*, Org. Syntheses, Coll. Vol. II, p. 184):

$$2Zn + 2C_2H_5I \longrightarrow 2C_2H_5ZnI \longrightarrow ZnI_2 + (C_2H_5)_2Zn$$

(2) Either from zinc and a dialkylmercury or from zinc halides and alkyls

of lithium, magnesium or aluminium (*S. M. Biltzer* and *T. H. Pearson*, U.S.P., 2,969,381/1961).

Properties. Zinc dialkyls are colourless covalent liquids, in which the zinc–carbon bonds are co-linear, so that the molecules are non-polar and not associated. (Physical properties are given in Table 4, p. 192). They react rapidly with air and water and readily with acid chlorides and tertiary halides. They are chemically less reactive than the analogous magnesium compounds, being almost inert to cyanides, ketones and esters, and can be handled in an atmosphere of carbon dioxide. They are of value in synthesising compounds whose functional groups are attacked by Grignard reagents, *e.g.* ketones.

Alkylzinc halides

These are described as intermediates (above) but rarely have they been isolated. They are probably structurally RZnX (rather than $R_2Zn{\cdot}ZnX_2$) since they give unsymmetrical zinc dialkyls when treated with Grignard reagents (*E. Krause* and *W. Fromm*, Ber., 1926, **59**, 931).

Diazomethane reacts with zinc iodide to give *iodomethylzinc iodide* (*G. Wittig* and *K. Schwarzenbach*, Ann., 1961, **650**, 1) which decomposes thermally with the formation of "methylene" (carbene); the latter reacts with olefins to give cyclopropane derivatives (see Vol. IA, p. 400 and *H. E. Simmons* and *R. D. Smith*, J. Amer. chem. Soc., 1959, **81**, 4256):

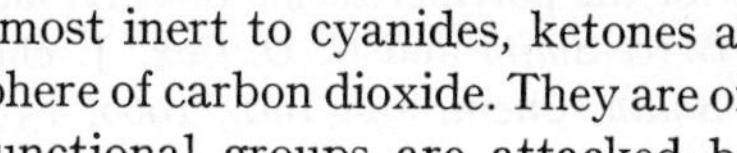

Anionic complexes. Although diethylzinc has a very low electrical conductivity, zinc dialkyls dissolve alkali metal alkyls, giving conducting solutions containing anions such as $ZnR_3^{\ominus}$ and $ZnR_4^{\ominus\ominus}$ (*D. T. Hurd*, J. org. Chem., 1948, **13**, 711).

(e) Cadmium

Dialkyls

Preparation. (1) By addition of powdered anhydrous cadmium chloride to a refluxing solution of a Grignard reagent. This solution is normally used in synthesis *e.g.* of ketones from acid chlorides (*D. A. Shirley*, Org. Reactions, 1954, **8**, 28) but the dialkylcadmium can be isolated by distillation *in vacuo*.

(2) From the reaction of cadmium acetate with aluminium alkyls (*Biltzer* and *Pearson, loc. cit.*).

Properties. They are very similar to those of zinc alkyls (cf. Table 4,

p. 192), but they are less reactive to air and water. Only dimethylcadmium is stable, all the higher homologues decomposing to cadmium and hydrocarbons, slowly in the cold but often violently on heating.

Cadmium dialkyls are more useful than zinc alkyls in the synthesis of ketones from acid chlorides (*Shirley, loc. cit.*) because cadmium chloride is not hygroscopic; zinc alkyls are preferable in some cases, however, *e.g.* in introducing secondary or tertiary alkyl groups. *J. Kollonitsch* (Nature, 1960, **188**, 140) showed that purified cadmium dialkyls react slowly with acid chlorides unless magnesium salts are present. Carboxylic anhydrides also react with cadmium alkyls (*F. Tatibouët* and *P. Freon*, Bull. Soc. chim. Fr., 1963, 1496).

(f) Mercury

Alkylmercury compounds, dialkyls and alkylmercury halides, have been extensively studied since the low electronegativity and low oxygen-affinity of the metal makes them very stable to air and water. They are extremely toxic by skin-contact and by inhalation if volatile. They have been used in pharmacy and as seed-dressings, although their use is decreasing. The separate existence of both R_2Hg and RHgX is established but no compounds containing Hg—Hg bonds are known and their existence seems unlikely.

Preparation. (1) Mercuric chloride reacts with Grignard or lithium reagents in ethers giving either the *monoalkylmercury halide* or the *dialkylmercury*, depending on the ratio of the reactants. *Gilman* and *R. E. Brown* (J. Amer. chem. Soc., 1930, **52**, 3314) showed that excess of Grignard reagent is unnecessary for the preparation of the latter. Aluminium and boron alkyls can also be used. Mercurous chloride reacts to give a dialkylmercury and metallic mercury (*C. S. Marvel* and *V. L. Gould, ibid.*, 1922, **44**, 153).

(2) Mercury reacts with some alkyl halides, especially methyl, benzyl and allyl iodides, giving alkylmercury halides. Sunlight, ethyl acetate and use of a sodium amalgam accelerate the reaction.

(3) Alkylmercury halides can be prepared from the dialkylmercury and halogen or mercuric halide.

(4) *α-Substituted alkylmercury compounds* have been obtained by the reaction of mercuric chloride with diazomethane, giving chloromethyl derivatives (*L. Hellerman* and *M. D. Newman, ibid.*, 1932, **54**, 2859), and with methylenetriphenylphosphorane (*S. O. Grim* and *D. Seyferth*, Chem. and Ind., 1959, 849).

(5) Dimethylmercury can be prepared by electrolysis of methylmercuric acetate in aqueous pyridine at room temperature. At —80°, an intermediate forms on the cathode: it has a formula (CH_3Hg) and is best regarded as an

organic metal, since it has none of the properties of a free radical and has a metallic appearance and conductivity (*B. G. Gowenlock et al.*, J. chem. Soc., 1957, 2114; 1958, 535).

(*i*) *Mercury dialkyls*

Properties and reactions. Mercury dialkyls are monomeric liquids (for physical properties see Table 4, p. 192), which often decompose on standing at room temperature giving mercury and hydrocarbons. At high temperatures the decomposition appears to involve the formation of two free alkyl radicals and zerovalent mercury rather than an alkyl mercuri radical RHg· (*M. E. Russell* and *R. B. Bernstein*, J. chem. Phys., 1959, **30**, 607, 613). They fail to react in many organic Grignard reactions, but react slowly with acid chlorides, and readily with many more electropositive metals, giving mercury and the metal alkyl.

Bistrifluoromethylmercury, $(CF_3)_2Hg$, m.p. 163°, is obtained as a white crystalline solid from mercuric oxide and tris(trifluoromethyl)phosphine. It is slightly soluble in water, the solutions conducting electricity due to ionization of the hydrate $(CF_3)_2Hg \cdot OH_2$ (*J. E. Griffiths* and *A. B. Burg*, J. Amer. chem. Soc., 1960, **82**, 5759).

(*ii*) *Alkylmercury oxides, hydroxides and halides*

Alkylmercury halides are also largely covalent, being derived from the weakly basic hydroxides. The Hg—C bond appears to be less reactive than in the dialkyls, since methylmercuric chloride is unchanged when distilled in concentrated hydrochloric acid vapour. They are appreciably volatile and therefore also hazardous to handle.

Methylmercury hydroxide dehydrates readily to *methylmercury oxide* $(CH_3Hg)_2O$, which reacts with methylmercuric bromide to give a salt $(CH_3Hg)_3O^{\oplus}Br^{\ominus}$ (*D. Grdenic* and *F. Zado*, Croatica Chemica Acta, 1957, **29**, 425).

Addition compounds. Mercuric salts, especially mercuric acetate, react readily with many olefinic compounds:

$$C_2H_4 + Hg(NO_3)_2 + NaOH \longrightarrow HOCH_2CH_2HgNO_3 + NaNO_3$$

Alkali is added to neutralise the liberated acid. The products are normally well-defined, stable solids but by-products such as $O(C_2H_4HgX)_2$ may form also. The addition compounds often behave as olefin complexes, regenerating olefin on treatment with acid or cyanide, but equally on reduction or iodination they behave as alkyls (*J. Chatt*, Chem. Reviews, 1951, **48**, 7). Their formulation as alkyls seems definite in view of their proton resonance spectra (*F. A. Cotton* and *J. R. Leto*, J. Amer. chem. Soc., 1958, **80**, 4823).

Mercuric chloride in hydrochloric acid adds readily to acetylene giving trans-

2-chlorovinylmercuric chloride, m.p. 124° (*A. N. Nesmeyanov*, Bull. Acad. Sci., U.S.S.R., 1945, 239 etc.) but in neutral solution acetaldehyde derivatives are formed. In alkali, the main products are acetylides.

TABLE 4

PHYSICAL PROPERTIES OF SOME GROUP IIB METAL ALKYLS

R	ZnR_2		CdR_2		HgR_2	
	m.p.	b.p.	m.p.	b.p.	m.p.	b.p.
CH_3	—29°	44°	— 2°	106°	—154°	92°
C_2H_5	—30°	117°	—21°	64°/19·5 mm.	—	159°
n-C_3H_7	—83°	154°	—83°	84°/21·5 mm.	—	189°
n-C_4H_9	—58°	193°	—48°	104°/12·5 mm.	—	117°/18 mm.

4. Group IIIB: Boron, Aluminium, Gallium, Indium, Thallium

All these elements form well-defined alkyl compounds, showing marked variations in reactivities. The alkyls of boron and aluminium have been extensively studied as a result of industrial interest in their manufacture and use in synthesis and as catalysts in polymerization. Their great tendency to form co-ordination compounds has led also to work on electron-deficient molecules such as $Al_2(CH_3)_6$, where a carbon atom acts as a bridge. Boron alkyls are not associated, unless two hydride bridges can be formed, but it is possible that alkyl groups can bridge boron atoms to aluminium, at least in intermediates (*R. Köster* and *G. Bruno*, Ann., 1960, **629**, 89). The tendency to form co-ordination compounds decreases sharply from boron to gallium; indium and thallium have little tendency to do this.

The chemical reactivity depends partly on the electronegativities (see Table 1, p. 167); boron, 2·0; aluminium, 1·5; gallium, 1·6; indium, 1·7; thallium, 1·8, and partly on the tendency towards co-ordination especially in reaction with oxygen.

(a) Boron

Boron forms compounds with one, two or three alkyl groups joined to a boron atom. (For a review of their chemistry see *W. Gerrard*, "The Organic Chemistry of Boron", Academic Press, London, 1961; *M. F. Lappert*, Chem. Reviews, 1956, **56**, 959).

Nomenclature. In the definitive I.U.P.A.C. rules for inorganic chemistry for 1963, boron hydride, BH_3, is designated *(mono)borane*, hence the alkyl derivatives are mono-, di- and tri-alkylboranes. The term *borine* (to agree with *phosphine*) has been widely used also.

The trivial name *diborane* is retained for the hydride B_2H_6 and higher hydrides and their derivatives are named systematically, *e.g.* B_4H_{10} is tetraborondecahydride.

(i) Trialkylboranes

Preparation. Alkyl groups can be attached to boron in two chief ways.

(1) From an organometallic compound, normally a Grignard reagent, and a boron halide, especially boron trifluoride or its etherate (*H. C. Brown*, J. Amer. chem. Soc., 1945, **67**, 374):

$$BF_3 \cdot (C_2H_5)_2O + 3RMgX \longrightarrow R_3B + 3MgFX + (C_2H_5)_2O$$

Boron trichloride or orthoborate esters can also be used. Aluminium alkyls are especially useful since they are readily available, avoid the use of large volumes of solvent and give good yields of trialkylboranes, especially from boron trichloride: ortho- and meta-borates also react (*Köster*, Ann., 1958, **618**, 31; *L. H. Long* and *A. C. Sanhueza*, Chem. and Ind., 1961, 588). Orthoborates, often with less reactive alkyls *e.g.* dimethylzinc, are preferable in preparing partially substituted compounds such as methylboron dichloride (*E. Wiberg* and *W. Ruschmann*, Ber., 1937, **70**, 1538).

(2) By the addition of an olefin to a boron hydride, *i.e.* a B—H bond. This reaction, often called *hydroboration*, is strongly catalysed by ether solvents, especially 2,5,8-trioxanonane.

In a typical reaction, diborane is prepared *in situ* from sodium borohydride and aluminium chloride in the ether, and the olefin added at room-temperature. Alternatively, diborane can be bubbled into a mixture of olefin and ether. In either case yields can exceed 85% (*H. C. Brown* and *B. C. Subba Rao*, J. org. Chem., 1957, **22**, 1136, 1137, J. Amer. chem. Soc., 1959, **81**, 6423 *et seq.*; *Brown*, "Organometallic Chemistry" A.C.S. Monograph No. 147, Reinhold, New York, 1960, p. 150).

Addition is faster to terminal olefins, *e.g.* in a mixture pent-1-ene reacts in preference to pent-2-ene. The latter, by itself, reacts readily to give a mixture of isomers, as is shown by oxidation to pentan-2-ol and pentan-3-ol. However, such mixed boranes isomerise on heating, giving pure tri-*n*-alkylboranes. It is therefore unlikely that tri-isopropylborane or tri-*tert*-butylborane have been prepared pure: tri-*sec*-butylborane can be prepared (from *sec*-butylmagnesium bromide and boron trifluoride), but it isomerises to tri-*n*-butylborane on refluxing (*G. F. Hennion et al.*, J. Amer. chem. Soc., 1957, **79**, 5190).

The hydroboration mechanism probably involves addition of a polar $\overset{\delta+}{B}$—$\overset{\delta-}{H}$ bond to olefin *via* a four-centre transition state: this rationalizes the normal preference of the boron for the terminal position on the chain:

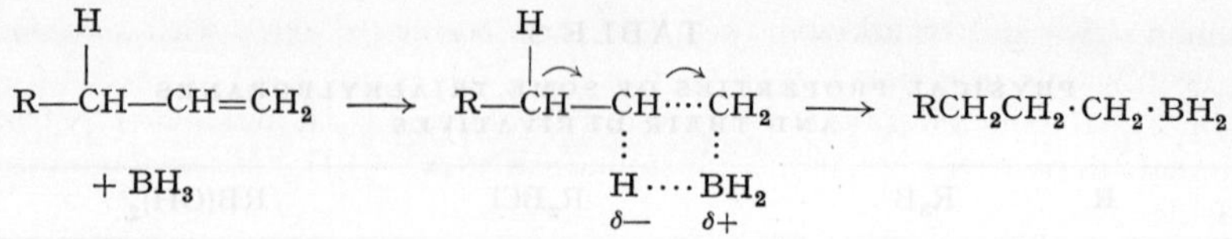

(*H. C. Brown*, Tetrahedron, 1961, **12**, 117).

At high temperatures, the tri-*n*-alkylborane is in equilibrium with the dialkylborane and the corresponding olefin. Therefore it is possible to prepare a higher alkyl compound from a higher olefin and a lower borane, the alkyl groups of which are displaced (*Köster, loc. cit.*):

$$3\ n\text{-}C_8H_{17}CH{=}CH_2 + [(CH_3)_2CHCH_2]_3B \longrightarrow (n\text{-}C_8H_{17}CH_2CH_2)_3B + (CH_3)_2C{=}CH_2$$

Heterocyclic compounds are obtained from dienes, *e.g.* butadiene and diborane give:

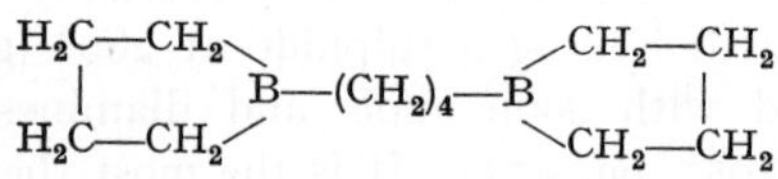

(*Köster*, Angew. Chem., 1959, **71**, 520). Some cyclic compounds are produced also when pentyl- and hexyl-boranes are heated alone.

Diborane adds readily to non-terminal acetylenes, but with terminal acetylene groups it tends to add two B—H groups; single addition occurs only with the less reactive bis-(3-methyl-2-butyl)borane (*Brown* and *G. Zweifel*, J. Amer. chem. Soc., 1961, **83**, 1241, 3834).

(3) Boron halides do not normally add to olefins, but *H. R. Arnold* (U.S.P., 2,402,589 and 2,402,590/1946) obtained β-chlorovinylboron dichloride from boron trichloride and acetylene at 200°. Trifluoromethylboron difluoride, CF_3BF_2, has been obtained by the action of boron trifluoride on the reaction product from potassium di-*n*-butylborane and trifluoromethyl iodide in ether (*A. B. Burg et al.*, J. Amer. chem. Soc., 1961, **83**, 250).

(4) Halogenomethylboron compounds have been prepared using diazomethane (*J. Goubeau* and *K. H. Rohwedder*, Ann., 1957, **604**, 168; *H. S. Turner*, Chem. and Ind., 1958, 1405).

Properties and reactions. Trialkylboranes are volatile non-associated liquids or gases (for b.ps. and m.ps. of the lower members see Table 5), which form stable complexes with electron-donors like trimethylamine; these complexes afford a convenient means of storing the more volatile compounds since the alkylborane can be regenerated with hydrogen chloride. The stability of amine-complexes is decided mainly by the steric effects of both the boron and nitrogen substituents (*H. C. Brown et al.*, J. Amer. chem. Soc., 1948, **70**, 2878, etc.).

TABLE 5

PHYSICAL PROPERTIES OF SOME TRIALKYLBORANES AND THEIR DERIVATIVES

R	R_3B		R_2BCl		$RB(OH)_2$
	m.p.	b.p.	m.p.	b.p.	m.p.
CH_3	—160°	—22°	—	5°	—
C_2H_5	— 93°	95°	— 85°	79°	—
n-C_3H_7	— 66°	165°	—125°	127°	107°
n-C_4H_9	—	91°/9 mm.	—	173°	94°
iso-C_4H_9	—	68°/7 mm.	—	—	112°
sec-C_4H_9	—	60°/2·5 mm.	—	—	—

Trialkylboranes are normally inert to water, trimethylborane being only partly hydrolysed at 180°; triallylborane is unusually reactive. Trimethylborane also reacts with hydrogen sulphide at 280° (giving methane and boron sulphide) and with some diols and diamines (*Goubeau* and *D. Ulmschneider*, Ber., 1957, **90**, 2733). It is the most thermally stable boron alkyl, decomposing only at 400° to give methane, hydrogen and polymers $[CH_2—B(CH_3)]_n$ including a cyclic tetramer (*Goubeau* and *R. Epple*, *ibid.*, 1957, **90**, 171). Other boranes lose olefin and form less completely alkylated boranes containing B—H bonds.

The lower trialkylboranes inflame in air; controlled oxidation in solution gives initially a *peroxide*, which is converted into a *dialkyl alkylboronate*:

$$R_3B + O_2 \longrightarrow [R_3B \leftarrow O_2] \longrightarrow R_2BOOR \longrightarrow RB(OR)_2$$

The peroxide can sometimes be isolated *e.g.* *sec*-$C_4H_9B(O_2$*sec*-$C_4H_9)_2$ (*S. B. Mirviss*, J. Amer. chem. Soc., 1961, **83**, 3051). The rate of autoxidation is highest for the trialkylboranes, dialkyl and monoalkyl compounds being increasingly stable in air. Trivinylborane is stable in air at room temperature (*T. D. Parsons, M. Silverman* and *D. M. Ritter*, *ibid.*, 1957, **79**, 5091).

Since hydroboration is a convenient, mild reaction, it offers routes from olefins via boranes to *alcohols* (on oxidation with alkaline hydrogen peroxide), to *hydrocarbons* (the olefin dimer is formed with silver nitrate, or the saturated hydrocarbon is formed with carboxylic acids) or to other *organometallic compounds*.

Trialkylboranes react with halogens, hydrogen halides and boron halides at about 150°, giving *mono-* and *di-alkylboron halides*. Boric oxide reacts reversibly, giving *alkylboroxines* $(RBO)_3$. Aldehydes are reduced to *esters* of *dialkylborinic acids* (*H. Meerwein et al.*, J. pr. Chem., 1936, **147**, 226):

$$C_6H_5CHO + (C_2H_5)_3B \longrightarrow C_6H_5CH_2OB(C_2H_5)_2 + C_2H_4$$

Trialkylboranes are converted by lithium alkyls into salt-like *boron-lithium alkyls, e.g.* $LiB(CH_3)_4$, which are probably ionic, and are much less stable to water than the corresponding aryls.

(ii) Alkyl- and dialkylboranes; alkylboron hydrides

Preparation. (1) By the equilibration of trialkylboranes and diborane (*H. I. Schlesinger* and *A. O. Walker*, J. Amer. chem. Soc., 1935, **57**, 621):

$$B_2H_6 + (CH_3)_3B \longrightarrow (B_2H_2)H_{4-n}\,(CH_3)_n$$

(2) By the pyrolysis of long chain boranes.

(3) By partial alkylation of a borohydride with an aluminium trialkyl (*L. H. Long* and *A. C. Sanhueza*, Chem. and Ind., 1961, 588).

Some *alkyl derivatives* of higher boranes have been prepared. *Long (loc. cit.)* obtained trimethylpentaboron hexahydride, $B_5H_6(CH_3)_3$ and tetramethyl pentaboron pentahydride, $B_5H_5(CH_3)_4$ as by-products of the preparation of alkylboron hydrides, and derivatives of decaborane of formula $B_{10}H_{14-n}(C_2H_5)_n$ where n = 1, 2, 3, 4 have been obtained by *R. L. Williams* and co-workers (J. chem. Soc., 1960, 424, 430). 6-Benzyldecaborane has been prepared from sodium decaboronate and benzyl bromide (*R. J. F. Palchak, J. H. Norman* and *R. E. Williams*, J. Amer. chem. Soc., 1961, **83**, 3380).

The alkylboron hydrides equilibrate readily. Pentamethyldiborane is unknown and tetramethyldiborane, $[(CH_3)_2BH]_2$, reverts readily to diborane and trimethylborane. They react with ammonia to form heterocyclic *B*-alkylborazenes, $[RBNR']_3$, six-membered boron-nitrogen ring compounds (*H. I. Schlesinger, L. Horwitz* and *A. B. Burg*, J. Amer. chem. Soc., 1936, **58**, 409; *M. F. Hawthorne, ibid.*, 1961, **83**, 833):

$$CH_3B_2H_5 + NH_3 \xrightarrow{200^0} (CH_3)_nB_3N_3H_{(6-n)} \text{ where } n=1,2,3$$

$$3RBH_2\cdot N(CH_3)_3 + 3NH_3 \xrightarrow[\text{diglyme}]{100^0} B_3R_3N_3H_3 + 3N(CH_3)_3 + 6H_2$$

(iii) Alkylboron halides, RBX_2 and R_2BX

Preparation. (1) By treating trialkylboranes with halogen or hydrogen halide.

(2) Preferably by equilibrating a mixture of boron halide and trialkylborane at 200°.

(3) By partial alkylation of a boron trihalide or orthoborate ester with a zinc or aluminium alkyl (*B. Lengyel* and *B. Csakvari*, Z. anorg. Chem., 1963, **322**, 103).

(4) Monoalkylboron dihalides can also be obtained from a trialkylboroxole

and a boron trihalide (*P. A. McCusker, E. C. Ashby* and *H. S. Makowsky*, J. Amer. chem. Soc., 1957, **79**, 5182):

$$[RBO]_3 + BCl_3 \longrightarrow RBCl_2 + B_2O_3$$

Amine Derivatives. Di-alkylboron halides form complexes with primary or secondary amines which on treating with *e.g.* triethylamine, lose hydrogen halide giving *borazenes*; the Raman spectrum of the complex, $(CH_3)_2{}^{\ominus}B \cdot N^{\oplus}(CH_3)_2$, suggests considerable double bonding between the B and N atoms (*H. J. Becher* and *Goubeau*, Z. anorg. Chem., 1952, **268**, 133). Alkylboron dihalides and primary amines give cyclic *borazenes*, $[RBNR']_3$, on dehydrohalogenation. These compounds contain a six-membered heterocyclic ring of boron and nitrogen atoms isoelectronic with benzene. (For a review of borazene chemistry see *J. C. Sheldon* and *B. C. Smith*, Quart. Reviews, 1960, **14**, 200).

The cyclic *phosphinoborine*, $[(CH_3)_2{}^{\ominus}B \cdot P^{\oplus}(CH_3)_2]_3$, m.p. 334°, prepared from dimethylphosphine, $(CH_3)_2PH$, and dimethylboron bromide, $(CH_3)_2BBr$, in the presence of trimethylamine, is extremely resistant to acid hydrolysis and unaffected by air (*A. B. Burg* and *R. I. Wagner*, J. Amer. chem. Soc., 1953, **75**, 3872).

(iv) Alkylboronic acids, $RB(OH)_2$, *and dialkylborinic acids,* R_2BOH

Preparation. (1) From trialkylboranes by controlled oxidation or by equilibration with trimethyl borate, followed by hydrolysis (*R. Köster*, Angew. Chem., 1959, **71**, 31).

(2) By partial alkylation of a boron halide or orthoborate ester with a Grignard or other alkylmetal compound.

They are weak acids which dehydrate readily below 100° giving *boroxoles*, $[RBO]_3$, and *borinic anhydrides*, $(R_2B)_2O$. These acids are much less sensitive to oxidation than the trialkylboranes, the alkylboronic acids being only slowly oxidised by air.

(b) Aluminium

Aluminium alkyls are more reactive than those of the other elements in Group III particularly with air and water. Several are now commercially available and they are becoming important reagents in large scale processes. Suitable precautions should be taken when quantities of these alkyls are being handled (*J. E. Knap et al.*, Ind. Eng. Chem., 1957, **49**, 874; *E. D. Zeratsky*, Petroleum Refiner, 1961, **40**, (8), 145). For a review of the preparation and chemistry of organo-aluminium compounds see *K. Ziegler*, "Organometallic Chemistry", A.C.S. Monograph 147, Reinhold, 1960, p. 194; Bull. Soc. chim. Fr., 1963, 1456.

Aluminium forms three types of organometallic compounds containing aluminium–carbon bonds, viz. (a) aluminium alkyls, (b) alkylaluminium halides and (c) alkylaluminium hydrides.

(i) Preparation of organo-aluminium compounds

(a) *Aluminium trialkyls.*

(1) Reaction of alkyl halides with aluminium/magnesium (70:30) alloys gives primarily dialkylaluminium halides which can be reduced to the trialkyl derivatives by treatment with sodium (*A. V. Grosse* and *J. M. Mavity*, J. org. Chem., 1940, 5, 106):

$$3R_2AlX + 3Na \longrightarrow 2AlR_3 + 3NaX + Al$$

Excess of sodium must be avoided or compounds of the type $NaAlR_4$ are formed. Alloys containing 57–65 % magnesium are reported to give mainly aluminium trialkyls (*Ziegler*, B.P., 767,400/1957).

(2) Disproportionation of fluoride complexes at 200–300° provides another way of converting dialkylaluminium chlorides to aluminium trialkyls (*Ziegler* and *R. Köster*, Ann., 1957, 608, 1):

$$R_2AlCl + 2NaF \longrightarrow Na(R_2AlF_2) + NaCl$$

$$3Na(R_2AlF_2) \longrightarrow Na_3AlF_6 + 2R_3Al$$

(3) Lower aluminium alkyls (except trimethylaluminium) can be prepared on a large scale by direct synthesis from aluminium, hydrogen and the appropriate olefin, preferably in the presence of preformed aluminium alkyl (*H. E. Redman*, U.S.P., 2,787,626/1957; *Ziegler*, B.P., 770,707/1957).

(4) Alkyls can also be made by addition of an α-olefin to an Al—H bond (*Ziegler*, B.P., 763,824/1956). Primary aluminium alkyl bonds are formed in this reaction; this is not an anti-Markovnikov addition since the aluminium–hydrogen bond is polarized $Al^{\delta+}$—$H^{\delta-}$. Ease of addition decreases in the order $CH_2{=}CH_2 > RCH{=}CH_2 > RR'C{=}CH_2$. Cyclo-olefins and olefins with an internal double bond react very slowly (*Ziegler et al.*, Ann., 1954, 589, 91). These additions are reversible:

$$\text{al-H} + RCH{=}CH_2 \underset{\text{heat}}{\rightleftharpoons} \text{al}CH_2CH_2R \qquad \text{al} = {}^1/_3Al$$

(5) If tri-isobutylaluminium is heated with an α-olefin, isobutene is liberated and the corresponding aluminium trialkyl is formed (*Ziegler et al.*, Angew. Chem., 1955, 67, 424):

$$[(CH_3)_2CHCH_2]_3Al + 3\,RCH{=}CH_2 \longrightarrow (RCH_2CH_2)_3Al + 3CH_2{=}C(CH_3)_2$$

(b) *Alkylaluminium halides*

(1) From alkyl halides (RX) and aluminium which react to give "sesqui-halides"-mixtures of R_2AlX and $RAlX_2$. The appropriate amount of sodium will convert the sesquihalide into the dialkyl derivative (*Grosse* and *Mavity*, *loc. cit.*):

$$2R_3Al_2X_3 + 3Na \longrightarrow 3R_2AlX + Al + 3NaX$$

(2) Mixtures of aluminium derivatives which are not adjacent in the series R_3Al, R_2AlX, $RAlX_2$, AlX_3 rapidly disproportionate providing a convenient method of interconversion:

$$R_2AlX \underset{AlR_3}{\overset{AlX_3}{\rightleftarrows}} RAlX_2$$

(3) Alkylaluminium dichlorides can be obtained from the reaction of a dialkylaluminium chloride with free chlorine (*H. Coates, W. H. Hunter* and *B. Topley*, B.P., 718,198/1954).

(c) *Alkylaluminium hydrides*

(1) Prepared best on a large scale directly from metallic aluminium, olefin and excess hydrogen.

(2) Hydrogenolysis of triethylaluminium gives diethylaluminium hydride in good yield (*H. E. Podall, H. E. Petree* and *J. R. Zietz*, J. org. Chem., 1959, **24**, 1222):

$$R_3Al + H_2 \longrightarrow R_2AlH + RH$$

(3) Dialkylaluminium hydrides are formed from the reaction of dialkyl-aluminium halides with lithium or sodium hydride (*Ziegler et al.*, Ann., 1954, **589**, 91):

$$(C_2H_5)_2AlCl + LiH \longrightarrow (C_2H_5)_2AlH + LiCl$$

(4) Aluminium trialkyls lose olefin on moderate heating to give dialkyl-aluminium hydrides. Dissociation is particularly easy if the alkyl is branched as in triisobutylaluminium which loses isobutene at 150°:

$$[(CH_3)_2CH{\cdot}CH_2]_3Al \longrightarrow [(CH_3)_2CH{\cdot}CH_2]_2AlH + CH_2{=}C(CH_3)_2$$

(d) *Aluminium alkyls containing other substituents*

The halogen atom in alkylaluminium halides can be replaced by other groups; R_2AlY, where Y=H, D, F, OR and —C≡C—R, have been formed from the corresponding sodio derivatives. Partial decomposition of aluminium trialkyls with alcohols, phenols, thiols, amines, acids etc. also gives rise to substituted derivatives:

$$AlR_3 + R'_2NH \longrightarrow R_2AlNR'_2 + RH$$

(ii) Properties of organo-aluminium compounds

The physical constants of the lower alkyls and their derivatives are recorded in Table 6.

TABLE 6

PHYSICAL CONSTANTS OF SOME ALUMINIUM ALKYLS

		b.p.	m.p.	d^{20}
AlR_3				
	R=Me	126^0	15^0	0·75
	Et	69^0/2 mm.	—46^0	0·84
	n-Pr	250^0	—107^0	0·82
	iso-Bu	73^0/5 mm.	4^0	0·785
R_2AlCl				
	R=Me	119^0	—45^0	—
	Et	125^0/50 mm.	—	0·97
	iso-Bu	152^0/10 mm.	—40^0	0·91
$RAlCl_2$				
	R=Me	100^0/100 mm.	73^0	—
	Et	100^0/30 mm. (194^0)	22^0	d^{25} 1·32
	iso-Bu	105^0/5 mm.	—30^0	1·12
R_2AlH				
	R=Me	154^0	—	—
	Et	55^0/0·001 mm.	—	0·81

The lower alkyls are spontaneously inflammable liquids which react violently with water and alcohols. Tri-isobutylaluminium and the higher alkyls are readily oxidised in air but they are not spontaneously inflammable. Trimethyl- and triethyl-aluminium are dimeric and a bridge structure with electron deficient bonds has been suggested:

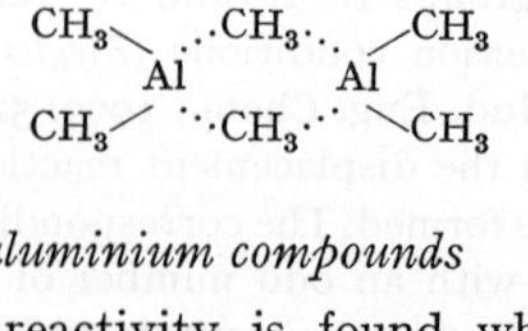

(iii) Reactions of organo-aluminium compounds

The highest chemical reactivity is found when all three aluminium valencies are bound to carbon or to carbon and hydrogen. Compounds of the type $AlR_{3-n}X_n$ (n = 0, 1, 2, X = halogen) are electron deficient and form stable coordination compounds with donor molecules. Complex stability decreases in the order $Me_3N > Me_3P > Me_2O > Me_2S$. Anionic complexes are formed by the addition of alkali metal halides, hydrides, aryls and alkyls *e.g.* $Na^{\oplus}[Al(C_2H_5)_4]^{\ominus}$. Some 1:2 complexes such as $NaF \cdot 2Al(C_2H_5)_3$ also exist.

(1) *Addition to Al—H and Al—C bonds.* Unlike Grignard reagents, aluminium trialkyls do not replace the active hydrogen atoms in acetylene and monosubstituted acetylenes but these compounds add easily to diethylaluminium hydride at room temperature:

$$(C_2H_5)_2AlH + CH{\equiv}C \cdot C_2H_5 \longrightarrow (C_2H_5)_2AlCH{=}CH \cdot C_2H_5$$

Disubstituted acetylenes also add to an Al—H bond; after hydrolysis the corresponding *cis*-olefin is formed (*G. Wilke* and *H. Müller*, Ber., 1956, **89**, 444):

$$CH_3 \cdot CH_2 \cdot C{\equiv}C \cdot CH_2 \cdot CH_3 + R_2AlH \longrightarrow CH_3 \cdot CH_2 \cdot C(AlR_2){=}CH \cdot CH_2 \cdot CH_3$$
$$\xrightarrow{H_2O} CH_3 \cdot CH_2 \cdot CH{=}CH \cdot CH_2 \cdot CH_3$$

Straight chain α-olefins add to an Al—C bond so that the aluminium atom becomes attached to a primary carbon atom ("Aufbau" reaction):

$$RCH_2al + R'CH{=}CH_2 \longrightarrow RCH_2 \cdot CHR'CH_2al \quad al = 1/3Al$$

With propylene and the higher olefins this reaction goes no further because olefin is readily split off at this stage by disproportionation or displacement reactions:

$$RCH_2 \cdot CHR' \cdot CH_2al \longrightarrow RCH_2 \cdot CR'{=}CH_2 + alH$$

$$RCH_2 \cdot CHR' \cdot CH_2al + R'CH{=}CH_2 \longrightarrow RCH_2 \cdot CR'{=}CH_2 + R'CH_2 \cdot CH_2al$$

Further reaction then takes place and catalytic quantities of aluminium alkyl will dimerize an α-olefin.

Ethylene gives but-1-ene in the presence of colloidal nickel, otherwise it does not split off at the dimer stage but it adds to triethylaluminium to form AlRR'R″ where the lengths of the groups R, R' and R″ vary about a statistical mean which depends on the reaction conditions (*Ziegler*, Angew. Chem., 1960, **72**, 829; *W. A. Skinner et al.*, Ind. Eng. Chem., 1960, **52**, 695). Addition of a nickel salt at this stage catalyses the displacement reaction and olefins with an even number of carbon atoms are formed. The corresponding paraffins are liberated by hydrolysis. Hydrocarbons with an odd number of carbon atoms are prepared from tri-*n*-propylaluminium since trimethylaluminium does not undergo the growth reaction.

Primary alcohols are formed by oxidation of aluminium trialkyls:

$$AlR_3 + 1\tfrac{1}{2}O_2 \longrightarrow Al(OR)_3 \xrightarrow{H_2O} 3ROH + Al(OH)_3$$

The third atom of oxygen is absorbed much more slowly than the first two. This is a convenient way of making primary alcohols from α-olefins and it forms the basis of the "Alfol" process for the preparation of long chain alcohols from ethylene via the "Aufbau" reaction (*G. E. Hinds*, Oil Gas J., 1959, 57, [16] 96). This reaction does not go to completion with sulphur or selenium and $(C_2H_5)_2AlSC_2H_5$ and $(C_2H_5)_2AlSeC_2H_5$ are the main reaction products.

(2) *Aluminium alkyls as reducing agents.* Tri-isobutylaluminium and dialkylaluminium hydrides are powerful reducing agents. Aldehydes, ketones and esters are reduced to alcohols, and nitriles to aldehydes or primary amines. At low temperatures (—70°), the reduction of carboxylic esters with di-isobutylaluminium hydride stops at the intermediate stage to give aldehydes in good yield (*L. I. Zakharkin* and *I. M. Khorlina*, Tetrahedron Letters, 1962, 619). Normally only the Al—H bond is involved in the reduction but tri-isobutylaluminium acts a a mixture of isobutene and di-isobutylaluminium hydride (*K. Ziegler*, B.P., 803,178/1958; *Ziegler et al.*, Ann., 1959, **623**, 9; *A. E. G. Miller, J. W. Bliss* and *L. H. Schwartzman*, J. org. Chem., 1959, **24**, 627). Triethylaluminium will reduce halogenated aldehydes only, *e.g.*:

$$Al(C_2H_5)_3 + 3CCl_3\cdot CHO \longrightarrow 3C_2H_4 + (Cl_3C\cdot CH_2O)_3Al$$

Aluminium trialkyls and dialkyl hydrides have been used as reducing agents in the preparation of manganese, chromium, molybdenum and tungsten carbonyls from the appropriate metal acetate or chloride and carbon monoxide (*H. E. Podall, J. H. Dunn* and *H. Shapiro*, J. Amer. chem. Soc., 1960, **82**, 1325).

(3) *Aluminium alkyls as alkylating reagents.* Suitable compounds of boron, silicon, antimony and arsenic, alkylate readily in the presence of aluminium alkyls. Phosphorus trichloride forms trialkylphosphines, or alkylphosphonous dichlorides, $RPCl_2$, if the trichloride is in excess (*Zakharkin* and *O. Yu. Okhlobystin*, Doklady Akad. Nauk S.S.S.R., 1957, **116**, 236). Diethylzinc, trimethylgallium and tetra-ethylgermane have been obtained by this method (*Idem*, J. gen. Chem., 1961, **31**, 3417). The preparation of zinc, cadmium and mercury alkyls has been reported (*S. M. Biltzer* and *T. H. Pearson*, U.S.P., 2,969,381/1961). Aluminium trialkyls and dialkyl-halides rapidly alkylate tin(IV) chlorides in high yields (*W. K. Johnson*, J. org. Chem., 1960, **25**, 2253). Tetraethyl-lead has been prepared from di- and tetra-valent lead salts using triethylaluminium; lead metal is formed as a by-product

in both cases (*F. W. Frey* and *S. E. Cook*, J. Amer. chem. Soc., 1960, **82**, 530).

(4) *Other reactions of aluminium alkyls.* Dialkylaluminium chlorides and alkylaluminium dichlorides form *ketones* with acid chlorides (*H. Adkins* and *C. Scanley*, J. Amer. chem. Soc., 1951, **73**, 2854):

$$R_2AlCl + R'COCl \longrightarrow RAlCl_2 + RCOR'$$

$$RAlCl_2 + R'COCl \longrightarrow AlCl_3 + RCOR'$$

Only one alkyl group is used in the first reaction because the two products form a stable complex which is unreactive towards acid chlorides. In the second reaction, aluminium chloride forms the more stable complex and the dichloride is able to react completely. Aluminium trialkyls cannot be used in this reaction.

Aluminium salts of *carboxylic acids* are formed from the reaction of aluminium trialkyls with carbon dioxide but only one Al—C bond reacts under normal conditions.

Sulphur dioxide adds readily to all three bonds to give *sulphinic acid* derivatives (*Ziegler et al.*, Ann., 1960, **629**, 251):

$$AlR_3 + 3SO_2 \longrightarrow Al(OSOR)_3$$

Aluminium alkyl halides have been claimed as *catalysts* for *isomerization* (*J. M. Mavity*, U.S.P., 2,388,428/1945), *hydrogenation* (*J. A. Ridgway*, Ind. Eng. Chem., 1958, **50**, 1139) and *alkylation* (*I. V. Nicolescu*, *M. Iovu*, and *G. I. Nikishin*, Izvest. Akad. Nauk, S.S.S.R., Otdel. Khim. Nauk, 1960, 94).

(5) *Aluminium alkyls as polymerization catalysts.* Under certain conditions aluminium alkyls and titanium halides can cyclopolymerize acetylene and substituted acetylenes to form the corresponding substituted benzenes in good yield (*E. F. Lutz*, J. Amer. chem. Soc., 1961, **83**, 2551; *B. Franzus*, *P. J. Canterino* and *R. A. Wickliffe*, *ibid.*, 1959, **81**, 1514). Butadiene has been cyclopolymerized by similar catalysts, the main product being *trans-trans-cis*-1,5,9-cyclododecatriene (*G. Wilke*, J. Polymer Sci., 1959, **38**, 45).

High molecular weight products are normally formed when olefins and dienes are polymerized by catalysts prepared from aluminium alkyls and a metal halide, usually titanium trichloride or tetrachloride. Certain catalysts will polymerize olefins in a stereoregular manner (*J. K. Stille*, Chem. Reviews, 1958, **58**, 541; *W. Cooper*, "Progress in High Polymers", Vol. 1, p. 279, Heywood and Co., London, 1961).

(c) Gallium

(i) Gallium alkyls

Preparation. (1) By reaction of gallium(III) halides with Grignard reagents (*L. M. Dennis* and *W. Patnode*, J. Amer. chem. Soc., 1932, 54, 182).

(2) By a displacement reaction between gallium metal and mercury alkyls (*G. E. Coates*, J. chem. Soc., 1951, 2003; *J. P. Oliver* and *L. G. Stevens*, J. inorg. nuclear Chem., 1962, 24, 953).

(3) Reaction of gallium(III) halides with zinc alkyls.

Two new methods have been developed in which the gallium alkyls are conveniently prepared from the readily accessible aluminium alkyls.

(4) A transalkylation reaction between gallium (III) halides and an aluminium alkyl.

(5) A displacement reaction using tri-isobutylgallium (prepared by method 4) and a higher olefin (*J. J. Eisch*, J. Amer. chem. Soc., 1962, 84, 3605).

In method (4) it is necessary to use a 3:1 molar ratio of aluminium alkyl to gallium halide and to add three equivalents of potassium chloride to prevent the formation of gallium/aluminium complexes of the type $Ga[AlR_2Cl_2]_3$:

$$GaCl_3 + 3AlR_3 + 3KCl \longrightarrow GaR_3 + 3K[AlR_2Cl_2]$$

Although gallium alkyls form complexes with potassium fluoride, $K[GaR_3F]$, unlike the aluminium alkyls they do not complex with sodium fluoride and so gallium alkyls can be freed from aluminium alkyls by distillation over sodium fluoride.

The last method is an excellent route to the higher non-distillable gallium alkyls:

$$Ga[CH_2{\cdot}CH(CH_3)_2]_3 + RCH{=}CH_2 \xrightarrow{150\text{–}160^0} Ga(CH_2{\cdot}CH_2R)_3 + 3CH_2{=}C(CH_3)_2$$

Properties. Gallium alkyls have similar properties to the aluminium alkyls but they are markedly less reactive. Besides the displacement reaction above, gallium alkyls undergo a growth reaction with ethylene although the necessary reaction conditions are much more severe:

$$Ga(C_2H_5)_3 + 3nCH_2{=}CH_2 \xrightarrow[100\text{ ats.}]{170^0} Ga[(CH_2CH_2)_nC_2H_5]_3$$

The alkyl groups can be successively hydrolysed by water at room temperature, water at 100⁰ and warm dilute sulphuric acid. Gallium alkyls burn in air but under controlled oxidation conditions an alkoxydialkylgallium, $ROGaR_2$, is formed which is stable towards further oxidation at ordinary temperatures. These materials have a characteristically sweet odour (*Eisch, ibid.*, 1962, 84, 3830).

Trimethylgallium, $(CH_3)_3Ga$, can be prepared by fractional distillation of a mixture of gallium, dimethylmercury and methylmercuric chloride (*G. E. Coates*, J. chem. Soc., 1951, 2003), or from gallium chloride and dimethylzinc (*E. Wiberg, T. Johannsen* and *O. Stecher*, Z. anorg. Chem., 1943, **251**, 114). It has m.p. —15·8°, b.p. 56°, and explodes in air and water, but careful hydrolysis in ether gives the *oxide* $[(CH_3)_2Ga]_2O$, and an amphoteric *hydroxide*, $[(CH_3)_2GaOH]_4$, which is stable up to 150°, when it loses methane to give $(CH_3GaO)_n$. Halogens, acid chlorides and benzyl chloride react rapidly giving gallium halides. Trimethylgallium gives a complex *etherate* $(CH_3)_3Ga \cdot O(C_2H_5)_2$, m.p. —76°, b.p. 98·3°, which dissociates more readily than that of aluminium. It also forms complexes with amines, *e.g. trimethylgallium-ammonia* $(CH_3)_3Ga \cdot NH_3$, m.p. 31°, which on pyrolysis gives methane and $(CH_3GaNH_2)_2$ and $(CH_3GaNH)_n$ (*Coates, loc. cit.*).

Triethylgallium, m.p. —82·3°, b.p. 143°, from gallium and diethylmercury, is associated in solutions, but monomeric in the vapour state. It burns in air with a purple flame (*A. W. Laubengayer* and *W. F. Gilliam*, J. Amer. chem. Soc., 1941, **63**, 477).

Tripropylgallium, b.p. 184°, can be prepared from propylmagnesium chloride and gallium(III) chloride under oxygen-free and absolutely anhydrous conditions. Ether-free tripropylgallium is obtained by distillation at normal pressure. It decomposes violently on contact with water (*H. Hartmann* and *H. Lutsche*, Naturwiss., 1961, **48**, 601).

Trivinylgallium, m.p. —7 to —9°, a liquid which polymerizes above 70°, has been prepared from divinylmercury and an excess of gallium metal at room temperature. Hydrolysis by water or dilute mineral acids produces ethylene (*J. P. Oliver* and *L. G. Stevens*, J. inorg. nuclear Chem., 1961, **19**, 378).

(ii) Alkylgallium halides

Alkylgallium halides are obtained from trialkylgallium and hydrogen chloride or, more usually, gallium(III) halides (*Eisch, loc. cit.*):

$$2Ga(C_2H_5)_3 + GaCl_3 \longrightarrow 3Ga(C_2H_5)_2Cl$$

Dimethylgallium chloride, $(CH_3)_2GaCl$, m.p. 45·5° and **methylgallium dichloride,** CH_3GaCl_2, m.p. 46° are dimeric and form amine and phosphine complexes. (*C. A. Kraus* and *F. E. Toonder*, Proc. nat. Acad. Sci., 1933, **19**, 392). Bromides and iodides are also known as well as several higher alkylgallium halides.

(iii) Alkylgallium hydrides

The alkylgallium hydrides are obtained from trialkylgalliums by reduction with hydrogen in a flow discharge (*Wiberg et al., loc. cit.*) or by an exchange reaction between a dialkylgallium halide and a dialkylaluminium hydride in the presence of potassium chloride (*Eisch, loc. cit.*):

$$Ga(C_2H_5)_2Cl + Al(C_2H_5)_2H \xrightarrow{KCl} Ga(C_2H_5)_2H + K[Al(C_2H_5)_2Cl_2]$$

Dimethylgallium hydride, $(CH_3)_4Ga_2H_2$, b.p. 172° (extrap., dec. 130°) is dimeric; with triethylamine it gives gallium hydride and trimethylgallium-triethylamine, $(CH_3)_3Ga \cdot N(C_2H_5)_3$. *Diethylgallium hydride* is similar and readily adds to both olefins and acetylenes at moderate temperatures giving unsymmetrical gallium alkyls. The latter disporoportionate on distillation.

(d) Indium

Indium alkyls can be prepared by most of the methods described for gallium alkyls. Their chemical properties are very similar although the indium alkyls are less reactive and they do not undergo the growth reaction with ethylene. The displacement method for the preparation of higher indium alkyls is less successful since they are unstable at the temperatures required for the evolution of isobutene. Complexes do not appear to be formed with sodium or potassium fluorides (*Eisch*, J. Amer. chem. Soc., 1962, **84**, 3605, 3830).

Indium alkyls are colourless liquids, soluble in organic solvents and sensitive to air, light and water. They are less stable than gallium alkyls (*Hartmann* and *Lutsche*, Naturwiss., 1962, **49**, 182).

Trimethylindium, $(CH_3)_3In$, from dimethylmercury and indium at 100°, has m.p. 88·4°, b.p. 135·8° dec. and sublimes *in vacuo* (*L. M. Dennis et al.*, *ibid.*, 1934, **56**, 1047). It is monomeric in the vapour state but tetrameric in benzene solution (*Laubengayer* and *Gilliam*, *loc. cit.*). The crystal structure is a polymeric network of tetramer units (*E. L. Amma* and *R. E. Rundle*, J. Amer. chem. Soc., 1958, **80**, 4141). It forms a weak complex with ether (*F. Runge et al.*, Z. anorg. Chem., 1951, **267**, 39). Runge has also prepared *triethylindium* $(C_2H_5)_3In$, m.p. —32°, b.p. 144°, 51–54°/3 mm., *tri*-n-*propylindium*, $(C_3H_7)_3In$, m.p. —51°; b.p. 178° and *tri*-n-*nonylindium*, $(n\text{-}C_9H_{19})_3In$, which do not form ether complexes. Triethylindium and *tri-isobutylindium*, b.p. 65–66°/2 mm., are best prepared from indium(III) chloride and the trialkylaluminium in the presence of potassium chloride followed by redistillation over potassium fluoride.

Trimethylindium burns in air but slow oxidation at —78° gives *dimethylindium oxide* $[(CH_3)_2In]_2O$. Cold water removes two alkyl groups, giving *methylindium dihydroxide*, $CH_3In(OH)_2$. Under the same conditions triethylindium gives *diethylindium hydroxide*, $(C_2H_5)_2InOH$ but at 90° indium hydroxide is formed. Ethanol gives *diethylindium ethoxide* $(C_2H_5)_2InOC_2H_5$.

(e) Thallium

Thallium(III) alkyls, TlR_3, resemble the trialkyls of gallium and indium. Thallium also forms an exceptionally stable series R_2TlX (X = halogen), the $R_2Tl^{\oplus}$ ion being isoelectronic with the R_2Hg compounds. Thallium(I) forms a stable cyclopentadienyl derivative C_5H_5Tl.

Thallium trialkyls are normally prepared from dialkylthallium halides, thallic or thallous halides, and lithium reagents (*H. Gilman* and *R. G. Jones*, J. Amer. chem. Soc., 1946, 68, 517), although *H. P. A. Groll* (*ibid.*, 1930, 52, 2998) obtained *triethylthallium*, $(C_2H_5)_3Tl$, from ethyl chloride and a thallium-sodium alloy. They are yellow liquids which decompose at about 90°, and oxidize in air. They do not form stable co-ordination compounds. Hydrolysis gives dialkylthallium hydroxides.

Dialkylthallium halides are formed from thallic halides and Grignard reagents or zinc alkyls, together with thallous salts, from which they are difficult to separate. They are inert to water and air and are fairly stable to heat. Silver oxide converts the halides to the strongly basic *hydroxides* from which many other salts can be obtained (see *E. Krause* and *A. von Grosse*, in bibliography p. 277).

Mono-alkylthallium dihalides are prepared from the dialkyls by equilibration with thallic halides (*Gilman* and *Jones*, J. Amer. chem. Soc., 1939, 61, 1513; *N. N. Melnikov* and *G. P. Gracheva*, J. gen. Chem. U.S.S.R., 1935, 5, 1786). They are relatively more water soluble, but tend to decompose to thallous halides and alkyl halides:

$$RTlX_2 \longrightarrow RX + TlX$$

5. Group IVB: Silicon, Germanium, Tin, Lead

All these elements form stable alkyl derivatives. Their valency electrons are directed towards the four corners of a tetrahedron due to sp^3 orbital hybridisation as in carbon.

Boiling points of analogous compounds increase regularly with increasing atomic weight of the parent element. The reactivity of the carbon-metal bond shows a steady increase with atomic number in the series silicon to lead. This is due to the increasing bond length, and to the increasing openness to nucleophilic attack on the metal atom. Values for the electronegativities are: C, 2·5; Si, 1·8; Ge, 1·8; Sn, 1·8; Pb, 1·8.

Alkyl derivatives are chemically inert compared with those of earlier groups, but halogen and other groups attached to partially alkylated derivatives are highly reactive, especially in comparison with the analogous carbon compounds.

None of these elements is known to form multiple bonds, such as occur in olefins or ketones. Instead, polymeric structures are formed. However, there is good evidence that partial multiple bonding can occur.

Thus, although silicon is less electronegative than carbon, trisilylamine is a weaker base than trimethylamine, due to contributions from structures such as $H_3\overset{\ominus}{Si}{=}\overset{\oplus}{N}(SiH_3)_2$. This accounts for the planarity of the trisilylamine molecule (*K. Hedberg*, J. Amer. chem. Soc., 1955, **77**, 6491), and for the large Si—O—Si angle of 140°–180°, compared with the expected value of 109°.

Alkyl derivatives of divalent tin and lead are known, but are unstable and tend to polymerize or oxidize. There is evidence that divalent silicon and germanium intermediates exist, comparable to carbenes (*M. E. Volpin, Yu. D. Koreshkov* and *D. N. Kursanov*, Izvestia Akad. Nauk S.S.S.R., 1961, 1355).

Nomenclature

The nomenclature of silicon compounds recommended by the I.U.P.A.C. (Handbook for Chemical Society Authors, Special Publication No. 14, 1960, p. 115), has been adopted in most publications. Briefly, it rules that SiH_4 shall be called *silane* and that compounds Si_nH_{2n+2} shall be called *disilane, trisilane,* etc. Compounds substituted with alkyl, alkoxy or halogen groups are named as substituted silanes. Compounds with Si—O—Si groups are *siloxanes*, and with Si—N—Si groups, *silazanes*. Hydroxyl groups attached to silicon are indicated by the suffixes *ol*, *diol* and *triol*. However, a functionally substituted compound such as $(CH_3)_3SiCH_2CONHPh$ can be named in terms of the organic part, *i.e.* α-trimethylsilylacetanilide. Similar nomenclature is applicable to germanium compounds (*O. H. Johnson*, Chem. Reviews, 1951, **48**, 259).

The alkyl compounds of tin and lead are most conveniently described by the usual method for metal alkyls (see p. 166): *e.g.* $(CH_3)_4Sn$ is *tetramethyltin,* although the suffix *stannane* is sometimes used for hydrides, *e.g.* $(CH_3)_3SnH$ may be called trimethylstannane, and for compounds with tin-tin bonds such as $(CH_3)_3Sn{-}Sn(CH_3)_3$, *hexamethyldistannane*. The terms stannyl and plumbyl may be used for a trialkylmetal group; *e.g.* $(C_2H_5)_3PbNa$, triethylplumbyl sodium.

(a) Silicon

The organic chemistry of silicon has been reviewed (*C. Eaborn*, "Organosilicon Compounds", Butterworths, London, 1960).

General methods for the preparation of alkylsilanes and their derivatives. (1) Alkylsilanes are formed from a silicon halide, an alkoxysilane or a silicon hydride, by reaction with a metal alkyl, usually a lithium alkyl or a Grignard reagent. Alkyl groups are introduced progressively so that partially alkylated silicon compounds can be obtained, especially if metal alkyl is added to silicon halide in the cold. Complete alkylation of silicon tetrachloride normally requires heating:

$$SiCl_4 + 4RMgX \longrightarrow R_4Si + 4MgXCl$$

Tetra-alkoxysilanes are less reactive and normally give alkoxytrialkylsilanes in good yields. Silanes containing Si—H bonds are still less reactive, normally being unaffected by Grignard reagents, but attacked by lithium alkyls. The latter are also more reactive with sterically hindered halides.

(2) Methylsilicon chlorides are prepared industrially by passing methyl chloride over a mixture of copper and silicon at 300–400°. Methylcopper may be an intermediate. Higher alkyl halides also react but the products are complex.

(3) Silanes add to olefins and alkynes under free radical conditions, up to 99% of *n*-octylsilicon trichloride being obtainable from trichlorosilane, octene-1 and diacetyl peroxide. Compounds with sensitive functional groups can be made this way if the olefin does not polymerize readily (*L. H. Sommer, E. W. Pietrusza* and *F. C. Whitmore,* J. Amer. chem. Soc., 1948, **70**, 484):

$$X_3SiH + CH_2{=}CHR \longrightarrow X_3SiCH_2CH_2R$$

Since the silane is functioning as a chain-transfer reagent in what would otherwise be a free-radical polymerization, the silane should be present in excess and, if possible, high temperatures avoided.

Addition also occurs with some metal catalysts, probably by an ionic mechanism:

$$Cl_3Si^{\oplus}H^{\ominus} + CH_2{=}CH\cdot CO_2CH_3 \longrightarrow CH_3\cdot CH(SiCl_3)\cdot CO_2CH_3$$

Such reactions do not lead to telomerization and are not prevented by free-radical inhibitors. Some organic bases, such as pyridine or nickel chloride-pyridine complexes, are also efficient catalysts; they probably also lead to a polar reaction (*Eaborn, loc. cit.*). Alkynes can add two molecules of a silane, trichlorosilane and *n*-hexyne giving 1,2-bis(trichlorosilyl)hexane (*C. A. Burkhard, ibid.*, 1950, **72**, 1402):

$$HSiCl_3 + HC{\equiv}CC_4H_9 \longrightarrow Cl_3SiCH{=}CHC_4H_9 \longrightarrow Cl_3SiCH_2CH(SiCl_3)C_4H_9$$

(4) Methylenation of silicon chlorides with diazomethane gives chloromethylsilicon chlorides (*D. Seyferth* and *F. G. Rochow,* Inorg. Synth., 1960, **6**, 37):

$$SiCl_4 + CH_2N_2 \longrightarrow ClCH_2SiCl_3 + N_2$$

Silanes methylenate to methylsilanes by insertion of carbene (*K. A. W. Kramer* and *A. N. Wright,* J. chem. Soc., 1963, 3604).

(5) Silicon halides react with methane and some other hydrocarbons at high temperatures (*P. W. Schafer* and *G. H. Wagner,* B.P., 662,916/1951).

(6) Alkylsilanes or mixtures of these with silicon halides disproportionate on heating to 180° with aluminium chloride (*P. D. George, Sommer* and *Whitmore,* J. Amer. chem. Soc., 1955, **77**, 1677).

(*i*) *Tetra-alkylsilanes*

Physical properties. The tetra-alkylsilanes are colourless liquids, with boiling points near those of the analogous carbon compounds; the b.ps. are related approximately by the equation:

$$\text{b.p.}^0\ SiR_4 = 0{\cdot}878\ (\text{b.p.}^0\ CR_4) + 19{\cdot}8$$

(see Table 7, p. 213). Other physical properties also can be forecast (*Gilman* and *G. E. Dunn*, Chem. Reviews, 1953, **52**, 77).

Reactions. Tetra-alkylsilanes are not chemically very reactive. Thermal decomposition of tetramethylsilane at 500^0 involves the C—H bonds as well as the Si—C bonds; the products include silicon, carbon, hydrogen, methane, and methylene linked compounds such as $(CH_3)_3SiCH_2SiH(CH_3)(C_2H_5)$, (*G. Fritz* and *B. Raabe*, Z. anorg. Chem., 1956, **286**, 149; Z. Naturforsch., 1957, **12b**, 66). Tetra-alkylsilanes with longer alkyl chains are nearly as stable. Oxidation also normally attacks the organic group rather than the Si—C bond, and it is difficult to remove an unsubstituted alkyl group without destroying it.

Most reactions are by electrophilic attack. Chlorine and bromine substitute hydrogen atoms in the alkyl groups (*J. L. Speier*, J. Amer. chem. Soc., 1951, **73**, 824) but up to two alkyl groups can be removed from tetraethylsilane with iodine and aluminium iodide, or other Friedel-Crafts catalysts. Electrophilic attack on an alkyl group is accelerated by the presence of electron withdrawing substituents on the other alkyl groups, but the latter are more open to cleavage at the Si—C bond by nucleophilic reagents. Increasing chlorination of methylsilicon groups gives a progressive increase in the rate of hydrolysis, and trichloromethyl groups are removed by water. Such groups may also be attacked by lithium alkyls or by basic reagents such as sodamide or alcoholic alkali (*H. Gilman, A. G. Brook* and *L. S. Miller*, *ibid.*, 1953, **75**, 4531):

$$Cl_3CSiPh_3 + C_4H_9Li \longrightarrow [Cl_3CLi] + C_4H_9SiPh_3$$

β-Halogenoalkyl groups are readily split off, often giving the olefin and halogenosilane on heating. A similar reaction occurs with Grignard reagents, and the expected coupling product is not usually obtained (*A. D. Petrov* and *V. F. Mironov*, Izvestia Akad. Nauk S.S.S.R., 1952, 635):

$$(C_2H_5)_3SiCH_2CH_2Cl \longrightarrow (C_2H_5)_3SiCl + CH_2{=}CH_2$$

γ-Halogenoalkyl groups can give cyclopropanes (*Sommer, R. E. van Strien* and *Whitmore*, J. Amer. chem. Soc., 1949, **71**, 3056):

$$Cl_3Si(CH_2)_3Cl \xrightarrow{NaOH} Si(ONa)_4 + \overline{CH_2CH_2CH_2}$$

Substituents on carbon atoms in the δ or further-removed positions do not show abnormal properties.

The preparation and properties of silicon-carbon compounds with substituents on carbon atoms have been reviewed (*P. D. George, M. Prober* and *J. R. Elliott*, Chem. Reviews, 1956, **56**, 1065).

(ii) Alkylsilicon halides

Preparation. (1) Normally directly by the partial alkylation of a silicon halide.

(2) By halogenation of a silicon hydride or an aminosilane.

(3) From a silanol or alkoxysilane and acetyl or benzoyl chloride.

(4) Siloxanes can be converted to alkylsilicon halides with sulphuric acid and ammonium halide.

(5) Alkylsilicon halides can be interconverted by replacement with a metal salt (often of silver), the replacement then being in the order:

$$\text{SiI} \longrightarrow \text{SiS} \longrightarrow \text{SiBr} \longrightarrow \text{SiNC} \longrightarrow \text{SiCl} \longrightarrow \text{SiNCS} \longrightarrow \text{SiNCO} \longrightarrow \text{SiO} \longrightarrow \text{SiF}$$

unless the volatility of one component affects the equilibrium. Silicon-fluorine compounds, however, often give other halides with aluminium salts, perhaps owing to the high lattice energy of aluminium fluoride (*C. Eaborn*, J. chem. Soc., 1953, 494):

$$3R_3SiF + AlCl_3 \longrightarrow 3R_3SiCl + AlF_3$$

Properties. Alkylsilicon halides are highly reactive, unless hindered sterically. They react with alcohols and ethers (to give alkoxysilanes), alkali metals (which normally give a Wurtz type condensation rather than metalation), amines (forming silazanes), epoxides (forming β-halogenoethoxy derivatives) and carboxylic acids (which give mixed anhydrides like $R_3Si{\cdot}O{\cdot}CO{\cdot}R'$).

Their hydrolysis has been shown to involve an S_N2 type mechanism, in which the use of the silicon *d*-orbitals lowers the energy of formation of the intermediate pentacovalent complex. There is evidence that this may be formed both by the normal in-line attack—as with carbon compounds—or by a side attack (*L. H., Sommer* and *G. L. Baughman*, J. Amer. chem. Soc., 1961, **83**, 3346).

For some physical constants of individual alkylsilicon halides see Table 7, p. 213.

(iii) Alkylsilanes with SiH groups; alkylsilicon hydrides

Preparation. (1) Normally by reduction of the corresponding chloride with lithium aluminium hydride in ether solvents. Only complete reduction is possible, and sterically hindered halides may require heating in a high boiling ether. Other halides, alkoxysilanes and some siloxanes also react.

(2) Hydrides can also be prepared by selective alkylation of silicon halides, or halogenosilanes, or by addition of some of the silane groups to an olefin:

$$Cl_3SiH + 3RLi \longrightarrow R_3SiH + 3LiCl$$

Properties. Except for the lower monoalkylsilanes, they are not spontaneously inflammable in air. They can be oxidized to silanols, usually with benzoyl peroxide and oxygen, and lose hydrogen when treated with alkali (a reliable qualitative test). The hydrogen is replaceable by silver salts (*C. Eaborn*, J. chem. Soc., 1955, 2517):

$$R_3SiH + AgClO_4 \longrightarrow R_3SiClO_4 + Ag + \tfrac{1}{2}H_2$$

(iv) Alkylsilanes with other substituents

(1) *Silanols*, R_3SiOH, are normally prepared by hydrolysis of the chlorides in neutral, (preferably buffered) solution, to avoid condensation to siloxanes.

(2) *Silanediols*, $R_2Si(OH)_2$, are obtained similarly but no alkylsilanetriols are known.

These hydroxysilanes lose water with acid or on heating unless highly hindered sterically. They are weak acids, but more acidic than the alcohols. Acid chlorides convert them initially to halogenosilanes, which often react subsequently with the carboxylic acid formed, losing hydrogen chloride, and forming a mixed anhydride. Silanols resemble alcohols in adding to vinyl ethers and isocyanates, but they can be estimated by the Karl Fischer procedure for water (*M. F. Shostakovskii et al.*, Zh. prikl. Khimii, 1958, 31, 952; *H. Gilman and L. S. Miller*, J. Amer. chem. Soc., 1951, 73, 2367).

(3) *Disiloxanes*, $R_3SiOSiR_3$, prepared from silanols on heating or with acid, are the simplest organic compounds containing the Si—O—Si linkage (see Table 7, p. 213). They are inert to thermal and chemical attack, except under vigorous conditions.

(4) *Silicones.* The dehydration of diols gives oils with the overall formula (R_2SiO), which are called silicones: they are not structurally analogous to ketones but have the polymeric unit $\left[Si(R_2){-}O\right]_n$. The crude product of the dehydration includes cyclic compounds, such as cyclic trisiloxanes $[R_2SiO]_3$, and long chain polymers of high molecular weight.

Commercial silicones normally consist of mixtures of cyclic and linear polymers where the alkyl groups are mainly methyl or phenyl. The polymer structure and average molecular weight are determined initially by the conditions of the polymerization. The free silanol groups and part of the cyclic material are then removed by heating with hexamethyldisiloxane and alkali or sulphuric acid. This stage converts much of the low polymer to long chains, the remainder being subsequently removed by vacuum distillation, and recycled. Such oils are useful as antifoaming agents and in polishes and leather treatment. Silicones used as water repellent coatings often contain Si—H groups, introduced by the hydrolysis of mixtures containing CH_3SiHCl_2, since these lead to cross-linking and a more permanent finish. Cross-linking can also be achieved by hydrolysis of a mixture of R_2SiCl_2 and $RSiCl_3$. *Silicone resins* are obtained by carefully controlled condensation of silanediols and/or dihalides to a high polymer, often followed by partial cross-linking, especially when vinyl groups are present on the chains. *Silicone rubbers* are obtained from the resins by compounding with silica and further cross-linking. They are useful at high temperatures ("Silicones", Ed. *S. Fordham*, Newnes, London, 1960). The physical properties of silicones are relatively little affected by temperature changes, due to their low polarity and the flexibility of the Si—O—Si linkages.

TABLE 7

PHYSICAL PROPERTIES OF SOME GROUP IV METAL ALKYLS AND THEIR DERIVATIVES

Alkyl	m.p. Si	b.p.	m.p. Ge	b.p.	m.p. Sn	b.p.	m.p. Pb	b.p.
R_4M								
R=CH_3	—100°	26·5°	—88°	43°	—54°	78°	—27°	110°
C_2H_5	—83°	153°	—90°	163°	—112°	181°	—135°	78°/10
n-C_3H_7	—46°	212°	—73°	225°/746	—109°	225°	—	126°/13
n-C_4H_9	—55°	231°	—	180°/733	—97°	145°/10	—	140°/1
R_3MMR_3								
R=CH_3	14°	113°	—	—	23°	182°	38°	238° dec.
C_2H_5	—	252°	—	265°	—	160°/23	—	100°/2 dec.
R_3MCl								
R=CH_3	—57°	58°	—	115°	38°	152°	190° dec.	(subl. 187°)
C_2H_5	—	144°	—	176°	16°	210°	123° dec.	—
R_2MCl_2								
R=CH_3	—76°	70°	—22°	124°	108°	190°	155° dec.	—
C_2H_5	—	129°	—39°	175°	89°	277°	—	207° dec.
$RMCl_3$								
R=CH_3	—80°	66°	—	111°	46°	—	—	—
C_2H_5	—106°	98°	—	144°	—10°	198°	—	—
R_3MOMR_3								
R=C_2H_5	—	233°/734	—	254°	—	272°	—	120°/2 dec.

(5) *Organosilylmetallic compounds*, R_3SiM, at present known are mainly aryl compounds, although triethylsilyl-lithium, $(C_2H_5)_3SiLi$, and methyldiphenylsilyl-lithium, $(C_6H_5)_2(CH_3)SiLi$, have been prepared (*D. Wittenberg* and *H. Gilman*, Quart. Reviews, 1959, 13, 116).

(b) Germanium, tin, lead

The organic chemistry of these elements has been reviewed for organogermanium compounds by *O. H. Johnson* (Chem. Reviews, 1951, 48, 259) and *D. Quane* and *R. S. Bottei* (*ibid.*, 1963, 63, 403); for organotin compounds by *R. K. Ingham, S. D. Rosenberg* and *H. Gilman* (*ibid.*, 1960, 60, 459), *J. G. A. Luijten* and *G. J. M. van der Kerk* ("Investigations in the Field of Organotin Chemistry", Tin Research Organization, 1955) and *W. P. Neumann* (Angew. Chem., Internat. Edn., 1963, 3, 165), and for organolead compounds, by *R. W. Leeper, L. Summers* and *H. Gilman* (Chem. Reviews, 1954, 54, 101).

(i) Tetra-alkyl derivatives, R_4M

Preparation. (1) Tetra-alkyl derivatives of these elements are normally prepared from the metal chloride and a metal alkyl. Partially alkylated products are usually formed unless excess alkyl and a high boiling ether, such as di-*n*-butyl ether, are employed. Lithium alkyls are especially useful in obtaining complete substitution. Dimethylzinc is recommended in preparing tetramethylgermane (*J. H. Lengel* and *V. H. Dibeler*, J. Amer. chem. Soc., 1925, 74, 2683). Tin tetra-alkyls are conveniently made from stannic chloride and the appropriate aluminium alkyl (*W. P. Neumann*, Ann., 1962, 653, 157) and the use of triethylaluminium for the preparation of tetraethyl-lead from lead acetate has been patented (*S. M. Biltzer* and *T. H. Pearson*, U.S.P., 2,859,225/1959). In general, lead (II) halides give lead(IV) derivatives, the proposed mechanism being:

$$2CH_3MgI + PbCl_2 \longrightarrow (CH_3)_2Pb + MgI_2 + MgCl_2$$

$$3(CH_3)_2Pb \longrightarrow Pb + (CH_3)_3Pb\cdot Pb(CH_3)_3 \longrightarrow 3/2(CH_3)_4Pb + \tfrac{1}{2}Pb$$

(2) Mixed alkyls of lead can be obtained by reaction of an alkyl-lead halide with a Grignard reagent (*G. Grüttner* and *E. Krause*, Ber., 1916, 49, 1125). It is essential to use an excess of Grignard reagent and to avoid the use of acid in the working-up process or equilibration may occur. Alternatively the equilibration reaction can be used as a preparative method. A mixture of two symmetrical tetra-alkyl compounds of tin or lead rearranges when heated in the presence of Friedel-Crafts catalysts to give a random mixture

which can be separated by fractional distillation (*G. Calingaert* and *H. A. Beatty*, U.S.P., 2,270,108/1942; "Organic Chemistry", Ed. *H. Gilman*, Vol. II, 2nd Edn., Wiley, New York, 1943, p. 1806; *P. D. George, L. H. Sommer* and *F. C. Whitmore*, J. Amer. chem. Soc., 1955, 77, 1677).

(3) Germanium will react with methyl chloride at 320° (*E. G. Rochow*, J. Amer. chem. Soc., 1948, **70**, 1801); but alloys of tin and lead, with sodium (about 1:1 atomic ratio) are used for reaction with alkyl halides. This is the basis of the commercial preparation of tetraethyl-lead from ethyl chloride.

(4) Alkylstannonic and plumbonic acids can be made by the Meyer reaction (*G. Meyer*, Ber., 1883, **16**, 1442; *M. Lesbre*, Compt. rend., 1940, **210**, 535):

$$K_2SnO_2 + RI \longrightarrow RSnOOK + KI$$

(5) Alkyltin hydrides add readily to olefin bonds, without catalysis, at 60–100° to give fully alkylated derivatives:

$$R_3SnH + CH_2{=}CHR' \longrightarrow R_3SnCH_2CH_2R'$$

This is apparently a polar reaction, since quinol does not retard it; as with silanes, it is especially valuable as a means of preparing C-substituted compounds, containing nitrile, amide, ester or other groups. Allyl bromide, however, gives propene when treated with triphenyl-lead hydride (*G. J. M. van der Kerk* and *J. C. Noltes*, J. Appl. Chem., 1957, 7, 356; 1959, **9**, 106).

(6) Other syntheses of partially alkylated derivatives involve the use of diazoalkanes (*D. Seyferth*, Chem. Reviews, 1955, 55, 1155; *Lesbre* and *R. Buisson*, Bull. Soc. chim. Fr., 1957, 1204) and methylenetriphenylphosphorane (*Seyferth* and *S. O. Grim*, Chem. and Ind., 1959, 849). Trifluoromethyltrimethyltin has been prepared from hexamethylditin and trifluoromethyl iodide (*H. C. Clark* and *C. J. Willis*, J. Amer. chem. Soc., 1960, **82**, 1888).

Properties and reactions. The tetra-alkyls are stable liquids (for physical constants see Table 7, p. 213), burning in air if ignited and soluble in organic solvents but insoluble in water The lead compounds are very toxic, especially it volatile. Their stability to heat decreases with increasing atomic weight, particularly from tin to lead. Tetramethyltin decomposes only above 400° (*C. E. Waring* and *W. S. Horton*, J. Amer. chem. Soc., 1945, 67, 540), while tetramethyl-lead decomposes rapidly at 100°, probably with the formation of free alkyl radicals and lead metal. The tetra-alkyls react with halogens and hydrogen halides giving di- and tri-alkylmetal halides. The ease of cleavage of alkyl groups by iodine from tin decreases in the order $CH_3 > C_2H_5 >$ n-$C_3H_7 >$ n-$C_4H_9 >$ iso-C_5H_{11} (*Z. M. Manulkin*, J. gen. Chem. U.S.S.R., 1943, 13, 42, 46). Alkyl groups are also cleaved by phenols, carboxylic acids, sodium in liquid ammonia and by some metal alkyls. Tetravinyltin, from vinylmagnesium bromide and stannic chloride in tetrahydrofuran, is cleaved by phenyl-lithium to give vinyl-lithium and tetraphenyltin (*E. C. Juenge* and *D. Seyferth*, J. org. Chem., 1961, **26**, 563). Allyl and ethynyl groups are also unusually reactive.

(ii) Alkylhalogeno derivatives

As with silicon, these alkylmetal halides are highly reactive towards water and alcohols, and react with metal alkyls to form unsymmetrical tetra-alkyl derivatives.

The *monohalides*, R_3MX, of germanium are best prepared from the tetra-alkyl and the halogen; those of tin from the tetra-alkyl and the tin tetrahalide; and those of lead from the tetra-alkyl and hydrogen halide. Partial alkylation of tin(IV) halides with aluminium alkyls also gives compounds of the type R_3SnX (X = Br, Cl, I) in good yield (*Neumann, loc. cit.*).

Dihalides, R_2MX_2, can be prepared similarly. In addition, dihalides of germanium can be prepared from dialkyldiphenylgermane and bromine (when the phenyl groups are preferentially removed) (*A. E. Flood*, J. Amer. chem. Soc., 1932, **54**, 1663). Those of germanium and tin are obtained from a metal-copper alloy and alkyl chlorides (*A. C. Smith* and *E. G. Rochow*, *ibid.*, 1953, **75**, 4103, 4105) and dialkyl-lead dihalides are prepared from the dialkyl-lead sulphites, R_2PbSO_3, and hydrogen halides (*R. Heap*, *B. C. Saunders* and *G. J. Stacey*, J. chem. Soc., 1951, 658).

Acidification of an aqueous or acetone solution of lead tetra-alkyldinitrosyl nitrate with hydrochloric acid also gives dialkyl-lead dichloride (*B. Hetnarski* and *T. Urbanski*, Rocz. Chem., 1963, **37**, 1073):

$$[R_4Pb(NO)_2]^{2\oplus}[NO_3]_2^{\ominus} \xrightarrow{HCl} R_2PbCl_2$$

The preparation of several organotin(IV) compounds by new or more convenient methods has been described. In the series $(C_4H_9)_2SnX_2$, *dibutyltin difluoride*, m.p. 157–60°, is obtained by the action of potassium fluoride on the dichloride. Dibutyltin oxide and hydriodic acid give *dibutyltin diiodide* b.p. 124–30°/0·2 mm. (*D. L. Alleston* and *A. G. Davies*, J. chem. Soc., 1962, 2050).

The *trihalides*, RMX_3, of germanium are best prepared by heating alkyl halides with germanium(II) halides (*E. A. Flood*, J. Amer. chem. Soc., 1933, **55**, 4935):

$$GeI_2 + EtI \longrightarrow EtGeI_3$$

but partial alkylation of germanium tetrachloride is also possible (*O. H. Johnson* and *L. V. Jones*, J. org. Chem., 1952, **17**, 1172). Alkyltin trihalides are obtained by halogenation of the tetra-alkyltin, although methyltin trichloride can be obtained from stannous chloride and methyl chloride (*Smith* and *Rochow*, *loc. cit.*) or from chlorination of the stannonic acids (*E. Krause* and *K. Weinberg*, Ber., 1930, **63**, 381). The alkyl-lead trihalides are much less stable, and have not been prepared pure (*M. Lesbre*, Compt. rend., 1940, **210**, 535).

(iii) Trialkylmetal hydroxides, R_3MOH

These are formed from the halides on hydrolysis in dilute alkali. They condense on heating, giving oxides, but can be characterized in the cold. They become increasingly basic from germanium to lead, trialkyl-lead hydroxides being readily converted to salts such as sulphate, acetate or sulphide with the appropriate acid. Hydrolysis of dialkylmetal dihalides does not give diols in a pure form. The final products are *oxides* $[R_2MO]_n$, which often dissolve readily in both acid and alkali, with destruction of the polymer chain. However, intermediates in the hydrolysis have been isolated (*Alleston*, *Davies* and *B. N. Figgis*, Proc. chem. Soc., 1961, 457):

$$R_2SnX_2 \xrightarrow{OH^\ominus} R_4Sn_2X_2O \xrightarrow{OH^\ominus} R_4Sn_2X(OH)O \longrightarrow HO{-}[R_2SnO]_n{-}H$$

(iv) Alkylmetal hydrides

These are formed on reduction of the halides with lithium aluminium hydride or from the metal derivatives, *e.g.* R_3MNa, and water. Those of germanium and tin are stable, and useful in synthesis, since they will add to olefins. Organotin hydrides have also been prepared in good yield from the corresponding halides, oxides and alkoxides by reduction with dialkylaluminium hydrides or lithium aluminium hydride (*Neumann* and *H. Niermann*, Ann., 1962, **653**, 164; *W. J. Considine* and *J. J. Ventura*, Chem. and Ind., 1962, 1683).

Methyl- and *ethyl-lead hydrides*, from the corresponding chloride and lithium aluminium hydride in dimethyl ether at —78°, are unstable above about —30°. Decomposition gives metallic lead and the lead tetra-alkyl (*W. E. Becker* and *S. E. Cook*, J. Amer. chem. Soc., 1960, **82**, 6264):

$$4R_3PbH \longrightarrow 2H_2 + 3PbR_4 + Pb$$

Alkylgermanium hydrides, R_3GeH, R_2GeH_2, $RGeH_3$ are similarly prepared by reduction of the halide with lithium aluminium hydride or sodium borohydride. The hydrides will add to activated double bonds at elevated temperatures (*J. Satge*, Ann. Chim., 1961, **6**, 519). Addition to the carbonyl group of saturated aldehydes and ketones at 150° in the presence of copper powder forms trialkylalkoxygermaniums in high yield. The last compounds are also formed by reaction of alkylgermanium hydrides with alcohols (*M. Lesbre* and *Satge*, Compt. rend., 1962, 254, 1456, 4051).

(v) Alkali metal derivatives

These are normally prepared (1) by metallation of a hexa-alkyldimetal compound:

$$R_3GeGeR_3 + 2Li \longrightarrow 2R_3GeLi$$

or (2) (for tin) from a halide or dihalide and sodium:

$$R_2SnCl_2 + 4Na \longrightarrow R_2SnNa_2 + 2NaX$$

or (3) for lead from a tetra-alkyl-lead and sodium in liquid ammonia:

$$R_4Pb + 2Na + NH_3 \longrightarrow R_3PbNa + NaNH_2 + RH$$

They are highly reactive, and rarely isolated.

(vi) Compounds with metal–metal bonds

Digermanes can be prepared by refluxing trialkylgermane bromides with potassium. They are stable to air, monomeric in benzene and do not react with cold concentrated sulphuric acid. *Distannanes*, prepared similarly or in liquid ammonia, are much more reactive to air and acid. *Diplumbanes* are much less stable thermally. They are prepared from Grignard reagents and lead(II) halides in the cold, or from alkali derivatives and monohalides:

$$R_3PbLi + R_3PbX \longrightarrow R_3Pb{-}PbR_3 + LiX$$

On warming they give metal and tetra-alkyl-lead:

$$2R_6Pb_2 \longrightarrow 3R_4Pb + Pb$$

6. Group VB: Phosphorus, Arsenic, Antimony, Bismuth

Organic derivatives of these elements show a striking gradation in properties as the element becomes more metallic, from compounds of phosphorus, which are very stable, to the unstable compounds of bismuth. All Group V elements exhibit valency states of three and five. Trivalent derivatives have an unshared pair of electrons. Ease of formation of quaternary salts decreases in the order $P > As > Sb > Bi$. Basicity, and the tendency to form coordination complexes with strong electron accepting metallic compounds, decrease in the same order. Derivatives of bismuth show little or no donor character. Compounds containing metal–metal bonds are formed, but these become less stable as the atomic weight increases.

Large numbers of arsenic and antimony compounds have been synthesised in the search for compounds of therapeutic interest. More attention is now being devoted to the organic compounds of phosphorus than to those of the remaining elements.

(a) Phosphorus

The chemistry of organophosphorus compounds has been surveyed up to 1949 by *G. M. Kosolapoff* ("Organophosphorus Compounds", Wiley, New York, 1950) and more recent work has been reviewed by *P. C. Crofts* (Quart. Reviews, 1958, **12**, 341), *M. Grayson* (Chem. and Eng. News, Dec. 3rd 1962, p. 90), *E. L. Gefter* ("Organophosphorus Monomers and Polymers", Translated by *G. M. Kosolapoff*, Associated Technical Services, New Jersey, 1962) and by *J. I. G. Cadogan* (Quart. Reviews, 1962, **16**, 208).

Quinquevalent phosphorus normally has a coordination number of four with three atoms joined by single bonds to phosphorus and one double (or semi-polar) bond from phosphorus to oxygen or other bivalent atoms or groups. There are some exceptions including the quaternary phosphonium salts, $R_4P^{\oplus}X^{\ominus}$, and the phosphoranes. Trivalent phosphorus compounds are good reducing agents because of the high reactivity of the unshared pair of electrons and the ease of formation of the $P \rightarrow O$ group. They are readily oxidised by oxygen, ozone, peroxides and other oxidising agents.

The carbon-phosphorus bond is similar in strength to the carbon–carbon bond and is very resistant to oxidation and hydrolysis.

Nomenclature

Rules for the nomenclature of organophosphorus compounds have been agreed between the Chemical Society and the American Chemical Society ("Handbook for Chemical Society Authors", Special Publication No. 14, The Chemical Society, London, 1960, p. 154). The terminology is based on a number of parent structures, some of which have only a hypothetical existence. The parent structures include: H_3P phosphine, H_3PO phosphine oxide, $(HO)_3P$ phosphorous acid, $HP(OH)_2$ phosphonous acid, $H_2P{\cdot}OH$, phosphinous acid, $(HO)_3PO$ phosphoric acid, $HPO(OH)_2$ phosphonic acid, and $H_2PO{\cdot}OH$ phosphinic acid. Tervalent and quinquevalent acids end in "-ous" or "-ic", respectively. Compounds containing five covalent bonds to phosphorus are classed as "phosphoranes" after the systematic name for H_5P. Phosphonium hydroxide $H_4P^{\oplus}OH^{\ominus}$ is named as a special case.

Structures containing a carbon-phosphorus linkage formed by replacing hydrogen in an H—P bond are named by prefixing the radical name of the replacing groups to the parent name, *e.g.* $C_2H_5PO(OH)_2$ is ethylphosphonic acid.

Esters are named by prefixing (separated by a space) the radical name to the parent acid and replacing -ic or -ous by -ate or -ite respectively. Partial esters have the word "hydrogen" inserted between the other two.

Examples of this nomenclature are: $H_2PC_2H_5$ ethylphosphine, $(C_4H_9)_3PO$ tributylphosphine oxide, $P(OC_2H_5)_3$ triethyl phosphite, $C_4H_9PO(OC_2H_5)_2$ diethyl butylphosphonate, $P(OH)(OC_2H_5)_2$ diethyl hydrogen phosphite, $C_4H_9PCl_2$ butylphosphonous dichloride, $C_4H_9POCl_2$ butylphosphonic dichloride.

Unfortunately there is some confusion about compounds which can exist in the tautomeric forms $>P{-}OH \rightleftharpoons >PO{\cdot}H$ and it is not clear whether these compounds should be named after the tervalent or quinquevalent parent. This system of nomenclature is not yet universally accepted and at present its use is mainly confined to American and British literature.

Derivatives of tervalent phosphorus

(i) Phosphines

Preparation. (1) When alkyl halides are heated with phosphonium iodide in the presence of zinc oxide at 100–180°, progressive alkylation takes place and the mixture of primary, secondary and tertiary phosphines can be separated by fractional distillation (*N. Davidson* and *H. C. Brown,* J. Amer. chem. Soc., 1942, **64,** 718). The extent of alkylation can be better controlled by reaction of sodium (or potassium) hydrogen phosphides (obtained by reaction of the metal with the appropriate phosphine) with alkyl halides (*G. W. Watt* and *R. C. Thompson, ibid.*, 1948, **70,** 2295; *R. I. Wagner* and *A. B. Burg, ibid.*, 1953, **75,** 3869):

$$NaPH_2 + RX \longrightarrow NaX + RPH_2$$

$$NaPHR + R'X \longrightarrow RR'PH + NaX$$

(2) Olefins will add to phosphine in the presence of cationic catalysts to give mainly primary phosphines. Tertiary phosphines are formed in the presence of free-radical catalysts (*H. C. Brown,* U.S.P. 2,584,112/1952; *M. M. Rauhut et al.*, J. org. Chem., 1961, **26,** 5138).

(3) The best method for the preparation of primary phosphines appears to be reduction of alkylphosphonous dichlorides, alkylphosphonic dichlorides and alkylphosphonous acids or esters with lithium aluminium hydride. Secondary phosphines are similarly obtained from the corresponding dialkylphosphinous derivatives (*L. Horner, H. Hoffmann* and *P. Beck,* Ber., 1958, **91,** 1583; *F. Pass* and *H. Schindlbauer,* Monatsh., 1959, **80,** 148).

(4) A series of alkylphosphines have been obtained by the reaction of the solid complex $H_3P{\cdot}AlCl_3$ m.p. 80–83° (from phosphine and aluminium chloride) with alkyl halides. Higher yields are obtained with bromides than chlorides; iodides produce side reactions (*F. Pass, E. Steininger* and *H. Zorn,* Monatsh., 1962, **93,** 230).

(5) Tertiary phosphines containing three identical groups are prepared

from phosphorus trichloride and a Grignard reagent or organolithium compound (*C. Screttas* and *A. F. Isbell*, J. org. Chem., 1962, **27**, 2573). The yields are increased significantly if the hydrolysis step can be omitted (*R. A. Zingaro* and *R. E. McGlothlin*, J. chem. eng. Data, 1963, **8**, 226). Unsymmetrical products can be obtained (a) from the reaction of alkylphosphonous dichlorides or dialkylphosphinous chlorides with a Grignard reagent; (b) by the reduction of benzyl-containing phosphonium compounds with lithium aluminium hydride (*W. J. Bailey, S. A. Buckler* and *F. Marktscheffel, ibid.*, 1960, **25**, 1996); (c) by the alkylation of the metal salt of a secondary phosphine (*F. Pass, E. Steininger* and *H. Schindlbauer*, Monatsh., 1959, **80**, 792); (d) by the stepwise alkylation of phosphorus trichloride with tetra-alkyl-lead compounds (*L. Maier*, J. inorg. nucl. Chem., 1962, **24**, 1073):

$$3PCl_3 + PbR_4 \longrightarrow 3RPCl_2 + PbCl_2 + RCl$$

$$3RPCl_2 + PbR'_4 \longrightarrow 3RR'PCl + PbCl_2 + R'Cl$$

$$3RR'PCl + PbR''_4 \longrightarrow 3RR'R''P + PbCl_2 + R''Cl$$

(e) by the successive alkylation of methylolphosphines and subsequent dealkylation of the methylolphosphonium halide in the presence of tertiary amines (*K. A. Petrov, V. A. Parshina* and *M. B. Luzanova*, J. gen. Chem. U.S.S.R., 1962, **32**, 553, English Translation p. 542):

$$P(CH_2OH)_4Cl \xrightarrow{Et_3N} P(CH_2OH)_3 \xrightarrow{RCl} RP(CH_2OH)_3Cl \xrightarrow{Et_3N} RP(CH_2OH)_2$$

$$\xrightarrow{R'Cl} RR'P(CH_2OH)_2Cl \xrightarrow{Et_3N} RR'PCH_2OH$$

$$\xrightarrow{R''Cl} RR'R''P(CH_2OH)Cl \xrightarrow{Et_3N} RR'R''P$$

(6) *Methinophosphide*, a unique compound containing a multiple carbon-phosphorus bond, has been obtained by passing phosphine through a low intensity graphite arc. It is a very reactive colourless gas with the possible structure $H{-}C{\equiv}P$ (*T. E. Gier*, J. Amer. chem. Soc., 1961, **83**, 1769).

Reactions. The phosphines are generally obnoxious toxic materials which are readily oxidised. Primary phosphines give phosphonous acids on careful oxidation with 30% hydrogen peroxide; more vigorous conditions give only phosphonic acids. All three types add sulphur to give mono-, di-, and tri-alkylphosphine sulphide, respectively (*Zingaro* and *McGlothlin, loc. cit.*). Dialkylphosphines react with sulphur in dilute aqueous ammonium hydroxide to give salts of dialkyldithiophosphinic acids, $R_2PS(SNH_4)$ (*M. M. Rauhut, H. A. Currier* and *V. P. Wystrach*, J. org. Chem., 1961, **26**, 5133). Reaction with halogen is vigorous but with care it will replace hydrogen. Tertiary phosphines form the unstable dihalogenophosphoranes, R_3PX_2.

Primary phosphines react with ketones and aromatic aldehydes in concentrated hydrochloric acid to give oxygen transfer products:

$$RPH_2 + 2R'CHO \longrightarrow RPO(CH_2R')CH\begin{smallmatrix}OH \\ R'\end{smallmatrix}$$

The 1-hydroxy tertiary phosphine oxide is formed unless the substituents are bulky when the reaction stops at the secondary phosphine oxide, HRPO·CHR'R''. Secondary phosphines react at about 100^0 with ketones and aldehydes (except formaldehyde) in strongly acid solutions to produce the corresponding tertiary phosphine oxide, $R_2PO{\cdot}CHR'_2$. Under milder conditions the normal carbonyl addition product, $R_2P{\cdot}CR_2OH$, is formed (*M. Epstein* and *S. A. Buckler*, Tetrahedron, 1962, **18**, 1231). Phosphines are basic, the degree of basicity increasing in the order primary < secondary < tertiary; the reverse of the order which applies to the corresponding amines (*W. Henderson* and *C. A. Streuli*, J. Amer. chem. Soc., 1960, **82**, 5791).

The sequence of base strength of some tertiary phosphines towards trimethylborane is ethyldimethylphosphine > trimethylphosphine > dimethylvinylphosphine ≫ triethylphosphine > trivinylphosphine (*H. D. Kaesz* and *F. G. A. Stone*, J. Amer. chem. Soc., 1960, **82**, 6213). Alkyl halides add to form the corresponding quaternary derivatives, tertiary phosphines being the most reactive. Tertiary phosphines can be characterized by the adducts which they form with carbon disulphide, $R_3P^{\oplus}CSS^{\ominus}$ (*W. C. Davies* and *W. P. Walters*, J. chem. Soc., 1935, 1786). Di-*tert*-phosphines of the type $R_2P—(CH_2)_n—PR_2$ where n = 3, 4, 5 have also been described (*K. Issleib* and *D-W. Müller*, Ber., 1959, **92**, 3175). Bi- and tri-dentate tertiary phosphines are formed from the reaction between lithium diethylphosphide and aliphatic polychloro compounds (*W. Hewertson* and *H. R. Watson*, J. chem. Soc., 1962, 1490).

Boiling points of some of the lower phosphines are given in Table 8, p. 225).

(ii) Diphosphines

Compounds of the type $R_2P{\cdot}PR_2$ can be prepared by treating a dialkylphosphinous chloride with the sodium salt of a secondary phosphine:

$$R_2PCl + R_2PNa \longrightarrow R_2P{\cdot}PR_2 + NaCl$$

($R{=}CH_3$, *A. B. Burg*, J. Amer. chem. Soc., 1961, **83**, 2226; $R{=}C_2H_5$, C_4H_9, *K. Issleib* and *W. Seidel*, Ber., 1959, **92**, 2681) or by reduction of a tetra-alkyldiphosphine disulphide (from RMgX and $PSCl_3$) with copper or iron (*H. Niebergall* and *B. Langenfeld*, *ibid.*, 1962, **95**, 64).

(iii) Alkylphosphonous dihalides, RPX_2, *and dialkylphosphinous halides,* R_2PX

Preparation. (1) The dihalides are formed when phosphorus trichloride is partially alkylated by organocadmium derivatives (*R. B. Fox,* J. Amer. chem. Soc., 1950, 72, 4147) or aluminium alkyls, preferably complexed with ether or pyridine (*O. Yu. Okhlobystin* and *L. I. Zakharkin,* Doklady Akad. Nauk S.S.S.R., 1957, 116, 236):

$$R_3Al + 3PCl_3 \longrightarrow 3RPCl_2 + AlCl_3$$

Alkylmagnesium chlorides will give pure dialkylphosphinous chlorides providing the alkyl group is branched and the reaction temperature is kept below —25° (*N. Voskuil* and *J. F. Arens,* Rec. Trav. chim., 1963, 82, 302). If alkylmagnesium bromides are used the corresponding dialkylphosphinous bromides are formed:

$$2RMgCl + PCl_3 \longrightarrow R_2PCl + 2MgCl_2$$

$$2RMgBr + PCl_3 \longrightarrow R_2PBr + MgCl_2 + MgBrCl$$

(2) Addition of α-olefins to phosphorus trichloride under free radical conditions produces β-chloroalkylphosphonous dichlorides:

$$RCH{=}CH_2 + PCl_3 \longrightarrow RCHClCH_2PCl_2$$

(3) Treatment of alkylphosphonous diesters or dialkylphosphinous esters with phosphorus trichloride gives good yields or the corresponding di- and mono-chlorides (*M. Sander,* Ber., 1960, 93, 1220):

$$RP(OR')_2 + PCl_3 \longrightarrow RPCl_2 + (R'O)_2PCl$$

$$R_2POR' + PCl_3 \longrightarrow R_2PCl + R'OPCl_2$$

Disproportionation of a mixture of phosphonous diester and alkylphosphonous dihalide gives the phosphonous monoester monochloride in good yield (*E. Steininger,* Ber., 1962, 95, 2993):

$$RP(OR')_2 + RPCl_2 \rightleftharpoons 2RP(OR')Cl$$

(4) Alkyl halides, aluminium chloride and phosphorus trichloride, form complexes of the type $[RPCl_3]^{\oplus}[AlCl_4]^{\ominus}$. Electron donating solvents such as diethyl phthalate (DEP) form more stable complexes with the aluminium chloride and the alkyltetrahalogenophosphorane ($RPCl_4$) is liberated. In the presence of powdered antimony, the phosphorane is reduced to the corresponding alkylphosphonous dichloride in high yield. This is a good general

method for the synthesis of the more volatile alkylphosphonous dichlorides; isomerisation occurs if normal alkyl chlorides are used. (*J. L. Ferron, B. J. Perry* and *J. B. Reesor*, Nature, 1960, **188**, 227; Canad. J. chem., 1963, **41**, 2299). Alkylphosphonous dichlorides also form complexes which are similarly reduced to dialkylphosphinous chlorides (*Ferron*, Canad. J. Chem., 1961, **39**, 842):

$$RCl + R'PCl_2 + AlCl_3 \longrightarrow [RR'PCl_2]^{\oplus}[AlCl_4]^{\ominus} \xrightarrow[DEP]{Sb} RR'PCl + [AlCl_3]\,[DEP] + SbCl_2$$

This is a modification of the Kinnear-Perren synthesis (*vide infra*) in which the same complex of aluminium chloride and a tetrachlorophosphorane is formed. The complex or the halogenophosphorane can be reduced to the phosphonous dihalide with elemental phosphorus:

$$3[RPCl_3]^{\oplus}\,[AlCl_4]^{\ominus} + 2P \longrightarrow 3RPCl_2 + 2PCl_3 + 3AlCl_3$$

$$3RPCl_4 + 2P \longrightarrow 3RPCl_2 + 2PCl_3$$

(5) Good yields of dialkylphosphinous chlorides are obtained by the reaction of dry hydrogen chloride with an aminodialkylphosphine (*A. B. Burg* and *P. J. Slota*, J. Amer. chem. Soc., 1958, **80**, 1107; *Issleib* and *Seidel*, *loc. cit.*):

$$PCl_3 + (C_2H_5)_2NH \longrightarrow (C_2H_5)_2N\cdot PCl_2 \xrightarrow{RMgX} (C_2H_5)_2N\cdot PR_2 \xrightarrow{HCl} (C_2H_5)_2NH\cdot HCl + R_2PCl$$

(6) Alkylphosphonous dihalides and dialkylphosphinous halides are formed by the partial alkylation of phosphorus trichloride using lead tetra-alkyls (*Maier*, J. inorg. nucl. Chem., 1962, **24**, 1073).

(7) Methyl- and ethyl-phosphonous dichlorides have recently been obtained by direct alkylation of phosphorus trichloride with methane and ethane at 600°. The conversion is low (10–15%) but the yield based on phosphorus trichloride consumed is high (*J. A. Pianfetti* and *L. D. Quin*, J. Amer. chem. Soc., 1962, **84**, 851).

Reactions. The lower alkylphosphonous dihalides react vigorously with halogens to form alkyltetrahalogenophosphoranes:

$$RPCl_2 + Cl_2 \longrightarrow RPCl_4$$

Oxygen, sulphur and selenium add to give phosphonic dihalides and their analogues (although direct reaction with oxygen is not a satisfactory preparative method). Addition of sulphur is best carried out using aluminium chloride as catalyst. Reduction of the di- and mono-halides with lithium aluminium hydride gives the corresponding primary and secondary phosphines respectively. For a review of the preparation and reactions of alkylphosphinous dihalides see *A. W. Frank*, Chem. Reviews, 1961, 61, 389.

Boiling points of some tervalent phosphorus compounds are given in Table 8.

TABLE 8

BOILING POINTS OF SOME TERVALENT PHOSPHORUS ALKYLS AND HALIDES

R	H_2RP	HR_2P	R_3P	$RPCl_2$	R_2PCl
CH_3	—14°	25°	38°	81°	73°
C_2H_5	25°	85°	127°	112°	133°
n-C_3H_7	53°	—	103°/25 mm.	140°	100°/15 mm.
iso-C_3H_7	41°	118°	81°/22 mm.	135°	46–47°/10 mm.
n-C_4H_9	—	170°	150°/22 mm.	160°	91–2°/12 mm.

(iv) Phosphonous acids $RP(OH)_2$

Phosphonous acids and their derivatives have been reviewed (*Frank, loc. cit.*). Free phosphinous acids $R_2P{\cdot}OH$ probably exist as secondary phosphine oxides HR_2PO but they can be obtained as esters $R_2P(OR')$ (see p. 226).

Preparation. (1) Addition of olefin to sodium hypophosphite in the presence of free-radical initiators (*A. R. Stiles, D. Harman* and *F. F. Rust*, U.S.P., 2,724,718/1955):

$$RCH{=}CH_2 + H_2PO(ONa) \longrightarrow HPO(CH_2CH_2R){\cdot}ONa$$

(2) Hydrolysis of phosphonous dichlorides; improved yields and purity are obtained if aqueous ethanol is used (*G. M. Kosolapoff* and *J. S. Powell*, J. Amer. chem. Soc., 1950, 72, 4291):

$$RPCl_2 + 2H_2O \longrightarrow HRPO{\cdot}OH + 2HCl$$

(3) Hydrolysis of phosphonous esters. The first ester group is cleaved by traces of mineral acid but the second is much more resistant to hydrolysis and strong acids or bases are required.

Properties. Phosphonous acids have the "ideal" formula $RP(OH)_2$, but normally they behave as the tautomeric monobasic form $HRPO{\cdot}OH$. The acids are viscous liquids which disproportionate to phosphines and phosphonic acids when heated to 150°:

$$3RP(OH)_2 \longrightarrow RPH_2 + 2RPO(OH)_2$$

They exhibit the characteristic reducing powers of compounds containing P—H bonds. They can be reduced to primary phosphines with lithium aluminium hydride.

(v) Esters of phosphonous and phosphinous acids

Preparation. (1) Reaction of a Grignard reagent with a trialkyl phosphite or dialkyl chlorophosphite, $ClP(OR)_2$ gives a phosphonous ester. Phosphinous esters are similarly obtained from alkyl dichlorophosphites and two equivalents of the Grignard reagent (*M. Sander*, Ber., 1960, 93, 1220; *M. I. Kabachnik* and *E. N. Tsvetkov*, Doklady Akad. Nauk S.S.S.R., 1960, 135, 323):

$$2RMgX + Cl_2P(OR') \longrightarrow R_2P(OR') + 2MgXCl$$

(2) Neutral esters can also be made by addition of alcohol to the appropriate alkylphosphonous dichloride or phosphinous chloride in the presence of a tertiary base.

(3) Alkyl hydrogen phosphonites can be prepared directly (*Kosolapoff*, J. Amer. chem. Soc., 1950, 72, 4292):

$$RPCl_2 + 2R'OH \longrightarrow RP(OR')\cdot OH + R'Cl + HCl$$

Derivatives of quinquevalent phosphorus

(vi) Phosphine oxides

(a) *Primary phosphine oxides* are formed in the reaction between phosphine and ketones (*S. A. Buckler* and *M. Epstein*, Tetrahedron, 1962, 18, 1211):

$$RCOR' + PH_3 \xrightarrow{H^{\oplus}} H_2PO\cdot CHRR'$$

A better method is the controlled oxidation of primary phosphines with hydrogen peroxide in ethanol at 0^0 (*idem, ibid.*, 1221).

Primary phosphine oxides are thermally unstable; *octylphosphine oxide*, m.p. 46–8^0 undergoes substantial decomposition after one hour at room temperature. Alcohol solutions are more stable although they tend to disproportionate into primary phosphines and phosphonous acids:

$$2H_2RPO \longrightarrow H_2PR + HRPO\cdot OH$$

Primary phosphine oxides will add to olefinic double bonds in the presence of basic or free-radical catalysts to give secondary and tertiary phosphine oxides:

$$H_2RPO + 2CH_2{=}CHCN \xrightarrow{NaOCH_3} RPO\cdot(CH_2CH_2CN)_2$$

$$H_2PO\cdot(n\text{-}C_8H_{17}) \xrightarrow{C_8H_{16}} HPO\cdot(n\text{-}C_8H_{17})_2 \xrightarrow{C_8H_{16}} (n\text{-}C_8H_{17})_3PO$$

Reactions with carbonyl compounds and isocyanates are also reported (*idem, loc. cit.*).

(b) *Secondary phosphine oxides (phosphinous acids). Preparation.* (1) By the reaction of a Grignard reagent with di-*n*-butyl phosphite.

(2) By reduction of dialkylphosphinic chlorides with lithium aluminium hydride (*R. H. Williams* and *L. A. Hamilton, ibid.*, 1952, 74, 5418; 1955, 77, 3411).

(3) Hydrolysis of phosphinous esters (*Sander, loc. cit.*).

(4) Oxidation of secondary phosphines (*M. M. Rauhut et al.*, U.S.P., 2,953,596/1960).

Properties. Secondary phosphine oxides are neutral, the higher ones being moderately stable, white solids. They are readily oxidised by 30% hydrogen peroxide to the corresponding phosphinic acids. With chlorine, thionyl chloride and phosphorus pentachloride they react vigorously to form dialkylphosphinic chlorides. In virtue of the reactivity of the P—H bond the compounds react with a large number of unsaturated compounds in the presence of basic catalysts (*R. C. Miller et al.*, J. Amer. chem. Soc., 1957, 79, 424):

$$HR_2PO + R'R''CO \longrightarrow R_2PO{\cdot}C(OH)R'R''$$

The adducts from both aldehydes and ketones are thermally unstable.

(c) *Tertiary phosphine oxides. Preparation.* (1) By reaction of a lithium alkyl or a Grignard reagent with phosphorus oxychloride, a phosphonic halide, phosphinic halide or an ester of a phosphonic acid.

(2) By the oxidation of a trialkylphosphine.

(3) By decomposition of a quaternary phosphonium hydroxide:

$$R_4POH \longrightarrow R_3PO + RH$$

Properties. tert-Phosphine oxides are among the most stable of organic phosphorus compounds (see Table 9, p. 229). Pyrolysis occurs at above 300° by a cyclic mechanism, if the alkyl group contains β hydrogen atoms, to form an olefin and a secondary phosphine oxide. Oxides with no β hydrogen atoms are even more stable; trimethylphosphine oxide shows no sign of decomposition at 700°. (*W. J. Bailey, W. M. Muir* and *F. Marktscheffel*, J. org. Chem., 1962, 27, 4404). They are usually crystalline materials which are often hygroscopic and stable hydrates may be formed. For a detailed review of the preparation and reactions of secondary and tertiary phosphine oxides see *K. D. Berlin* and *G. B. Butler*, Chem. Reviews, 1960, 60, 243.

(vii) Diphosphine dioxides

Preparation. (1) Compounds of the type $R_2PO{\cdot}(CH_2)_n{\cdot}POR_2$ where n = 4, 5 or 6 have been synthesized from the phosphinic chloride R_2POCl and a di-Grignard reagent.

(2) Reaction of a glycol ditoluene-*p*-sulphonate with the Grignard reagent of a secondary phosphine oxide allows the preparation of the compounds where n = 2 and 3 (*G. M. Kosolapoff* and *R. F. Struck*, J. chem. Soc., 1959, 3950; 1961, 2423):

$$2R_2PO\cdot MgX + (p\text{-}CH_3\cdot C_6H_4\cdot SO_3)_2(CH_2)_n \longrightarrow R_2PO\cdot(CH_2)_nPOR_2 + 2p\text{-}CH_3\cdot C_6H_4\cdot SO_3\cdot MgX$$

(3) A four stage synthesis from dibutyl hydrogen phosphite to give compounds where n = 1 has been reported (*J. J. Richard et al.*, J. Amer. chem. Soc., 1961, 83, 1722).

Diphosphine disulphides, $R_2PS\cdot PSR_2$, are formed from the reaction of a Grignard reagent with thiophosphoryl chloride (*P. J. Christen, L. M. Van Der Linde* and *F. N. Hooge*, Rec. Trav. chim., 1959, 78, 161).

(viii) Alkylphosphonic dihalides, $RPOX_2$*, and dialkylphosphinic halides,* R_2POX

Preparation. (1) Phosphonic and phosphinic acids, or their esters, are converted into the corresponding acid chlorides by the action of phosphorus pentachloride or thionyl chloride:

$$RPO(OR)_2 + 2PCl_5 \longrightarrow RPOCl_2 + 2POCl_3 + 2RCl$$

$$R_2PO\cdot OH + PCl_5 \longrightarrow R_2POCl + POCl_3 + HCl$$

(2) Phosphonous and phosphinous halides may be oxidised directly but it is better to carry out the reaction in two stages. Addition of halogen forms the halogenophosphorane which is then treated with sulphur dioxide:

$$RPCl_2 \xrightarrow{Cl_2} RPCl_4 \xrightarrow{SO_2} RPOCl_2 + SOCl_2$$

(3) Direct reaction of a hydrocarbon with phosphorus trichloride in the presence of oxygen forms phosphonic dichlorides. The reaction is not selective but tertiary carbon atoms are attacked in preference to primary carbon atoms (*J. O. Clayton* and *W. L. Jensen*, J. Amer. chem. Soc., 1948, 70, 3880).

(4) *J. P. Clay* (J. org. Chem., 1951, 16, 892) showed that alkyl halides, phosphorus trichloride and aluminium chloride form a complex $[RPCl_3]^{\oplus}$-$[AlCl_4]^{\ominus}$ which can be hydrolysed under controlled conditions. *A. M. Kinnear* and *E. A. Perren* (J. chem. Soc., 1952, 3437) found that *n*-alkyl groups containing more than three carbon atoms are isomerized and the corresponding secondary alkylphosphonic dichloride is formed. The con-

ditions for hydrolysis are critical. In an improved method the complex is dissolved in diethyl phthalate and the alkyltetrahalogenophosphorane which is liberated is treated with sulphur dioxide (*J. B. Reesor et al.*, Canad. J. Chem., 1960, 38, 1416).

(5) Alkyl-phosphonic and -phosphinic halides are also formed by the partial alkylation of phosphoryl chloride with lead tetra-alkyls (*Maier*, J. inorg. nucl. Chem., 1962, **24**, 1073).

(6) Dialkylphosphinic chlorides have recently been obtained in excellent yield from the reaction of thionyl chloride with tetra-alkyldiphosphine disulphides (*K. A. Pollart* and *H. J. Harwood*, J. org. Chem., 1962, **27**, 4444):

$$R_2PS \cdot PSR_2 + 2SOCl_2 \longrightarrow 2R_2POCl + 2S + S_2Cl_2$$

The boiling points and melting points of some quinquevalent phosphorus oxides and chlorides are given in Table 9.

TABLE 9

BOILING POINTS OF QUINQUEVALENT PHOSPHORUS OXIDES AND CHLORIDES

R	R_3PO	$RPOCl_2$	R_2POCl
CH_3	(m.p. 140°)	163° (m.p. 32°)	204° (m.p. 66°)
C_2H_5	(m.p. 53°)	175°	80°/15 mm.
n-C_3H_7	(m.p. 38°)	90°/50 mm.	119°/11 mm.
iso-C_3H_7	—	82°/50 mm.	109°/12 mm.

(*ix*) *Alkylphosphonic acids*, $RPO(OH)_2$ *and dialkylphosphinic acids*, $R_2PO \cdot OH$

Work prior to 1949 has been reviewed by *Kosolapoff* (Org. Reactions, 1951, 6, 273) and more recent work on phosphonic acids by *L. D. Freedman* and *G. O. Doak* (Chem. Reviews, 1957, 57 479).

Preparation. (1) Both types of acids are easily obtained by the hydrolysis of their esters (*vide infra*) or their chlorides.

(2) Alkylphosphonic difluorides, $RPOF_2$ react with Grignard reagents at low temperatures to form a solid complex $RR'POF \cdot MgXF$, which on alkaline hydrolysis gives good yields of dialkylphosphinic acids (*T. P. Dawson* and *K. C. Kennard*, J. org. Chem., 1957, **22**, 1671).

(3) Symmetrical phosphinic acids are obtained from the reaction of dialkyl phosphites with Grignard reagents, followed by oxidation with hydrogen peroxide of the resulting secondary phosphine oxides which need not be isolated (*G. M. Kosolapoff* and *R. M. Watson*, J. Amer. chem. Soc., 1951, 73, 4101; *R. H. Williams* and *L. A. Hamilton*, *ibid.*, 1952, 74, 5418; 1955, 77, 3411). A later method using bromine as the oxidising agent is described by *Kosolapoff* and *Struck* (*loc. cit.*).

(4) *N,N*-Diethylphosphoramidic dichloride, $(C_2H_5)_2NPOCl_2$, will react with only two equivalents of a Grignard reagent to form $R_2PON(C_2H_5)_2$, which is readily hydrolysed to the corresponding phosphinic acid (*Kosolapoff*, J. Amer. chem. Soc., 1949, **71**, 369).

(5) The reaction between quinquevalent phosphorus acid chlorides and Grignard reagents does not usually go to completion owing to the formation of ether-insoluble complexes such as $R_2POCl \cdot MgXCl$ which are relatively unreactive. Hydrolysis liberates a mixture of phosphine oxide and dialkylphosphinic acid from which the latter is extracted with sodium hydroxide (*K. A. Petrov, N. K. Bliznyuk* and *V. P. Korotkova*, J. gen. Chem., U.S.S.R. 1960, **30**, 2995).

Properties. Phosphonic acids are moderately strong dibasic acids. The pK values decrease with an increase in the number of carbon atoms and in chain branching. Both acid and neutral salts may be formed and these are used as a means of purification.

Methylphosphinic acid, a strong monobasic acid (pK 2·3), is prepared by the low temperature methanolysis of methylphosphonous dichloride:

$$CH_3PCl_2 + 2CH_3OH \longrightarrow H(CH_3)PO \cdot OH + 2CH_3Cl$$

The acid, which is stable at room temperature, is oxidised by aqueous bromine to methylphosphonic acid (*D. Fiat et al.*, J. chem. Soc., 1962, 3837).

Some unsaturated derivatives of phosphonic acid prepared from bis(2-chloroethyl) vinylphosphonate have recently been described (*C. M. Welch, E. J. Gonzales* and *J. D. Guthrie*, J. org. Chem., 1961, **26**, 3270).

(x) Esters of phosphonic and phosphinic acids

Preparation. (1) *The Arbuzov reaction.* In the formation of a dialkyl alkylphosphonate from a trialkyl phosphite and an alkyl halide, the alkyl group of the latter becomes attached to the phosphorus atom, while one of the alkyl groups from the ester is removed as alkyl halide. Alkyl dialkylphosphinates are similarly obtained from dialkyl alkylphosphonites:

$$P(OR)_3 + R'X \longrightarrow R'PO(OR)_2 + RX$$

$$RP(OR')_2 + R''X \longrightarrow RR''PO(OR) + R'X$$

(2) *The Michaelis reaction.* Alkali metal derivatives of dialkyl phosphites when treated with alkyl halides give phosphonic esters:

$$(RO)_2P \cdot OH \xrightarrow{Na} (RO)_2PO \cdot Na \xrightarrow{R'X} R'PO(OR)_2$$

Phosphinic esters can be obtained from alkyl hydrogen phosphonites (*Kosolapoff*, J. Amer. chem. Soc., 1950, **72**, 4292):

$$RPCl_2 \xrightarrow{2ROH} RP(OR)OH \xrightarrow{Na} RPO(OR)Na \xrightarrow{R'X} RR'PO(OR)$$

(3) Diphenyl *n*-alkylphosphonates having large alkyl groups can be synthesized in one step in 70% yield by heating triphenyl phosphite with the appropriate alcohol at 220–250° in the presence of sodium iodide (*R. G. Laughlin*, J. org. Chem., 1962, **27**, 3644).

(xi) Alkylhalogenophosphoranes, R_nPX_{5-n}

These compounds can be formed by the action of (1) halogen on alkylphosphorus halides or (2) phosphorus pentachloride on a tertiary phosphine oxide:

$$RPX_2 + X_2 \longrightarrow RPX_4$$

$$R_2PX + X_2 \longrightarrow R_2PX_3$$

$$R_3PO + PCl_5 \longrightarrow R_3PCl_2 + POCl_3$$

Dialkyltrichlorophosphoranes, R_2PCl_3, are conveniently prepared in 80% yields by the action of chlorine on the readily available tetra-alkyldiphosphine disulphides (*W. Kuchen* and *K. Strolenberg*, Angew. Chem., 1962, **74**, 27; internat. Edn. 1962, **1**, 46):

$$R_2PS{-}PSR_2 + Cl_2 \longrightarrow 2R_2PS{\cdot}Cl$$

$$2R_2PS{\cdot}Cl + 4Cl_2 \longrightarrow 2R_2PCl_3 + 2SCl_2$$

These compounds are generally crystalline solids which are readily hydrolysed and thermally unstable, *e.g. trichlorodiethylphosphorane* and *trichlorodipropylphosphorane* decompose above 65° and 55° respectively. Pyrolysis results in cleavage of a carbon-phosphorus bond and the formation of a tervalent phosphorus compound:

$$R_3PCl_2 \longrightarrow R_2PCl + RCl$$

$$R_2PCl_3 \longrightarrow RPCl_2 + RCl$$

There is evidence that dialkyltrichlorophosphoranes have an ionic structure, $[R_2PCl_2]^{\oplus}\,[R_2PCl_4]^{\ominus}$. The corresponding *dialkyltribromophosphoranes* cannot be isolated by the method of Kuchen and Strolenberg, instead a complex of the type dialkyl tribromophosphorane dibromide R_2PBr_5 is formed; *tribromodiethyl-* and *-dipropyl-phosphorane dibromides* form orange-red needles which melt without decomposition at 103·5° and 87·5°, respectively. An ionic structure, $[R_2PBr_2]^{\oplus}Br_3^{\ominus}$ has been proposed.

Fluorophosphoranes are prepared in high yield from alkylphosphonous dichlorides or dialkylphosphinous chlorides and arsenic or antimony trifluoride (*R. Schmutzler*, Chem. and Ind., 1962, 1868):

$$3RPCl_2 + 4AsF_3 \longrightarrow 3RPF_4 + 2As + 2AsCl_3$$

$$3R_2PCl + 3SbF_3 \longrightarrow 3R_2PF_3 + 2Sb + SbCl_3$$

The products are isolated by distillation and appear to be more thermally stable than the chlorine analogues.

Methyltetrafluorophosphorane, CH_3PF_4, b.p. 10°; *ethyltetrafluorophosphorane* b.p. 32–33°; n-*butyltetrafluorophosphorane*, b.p. 85–86°.

(xii) Quaternary phosphonium compounds, R_4PX (X = halogen or hydroxyl)

The simplest preparation for halides is addition of an alkyl halide (preferably iodide) to a tertiary phosphine:

$$R_3P + RX \longrightarrow R_4P^{\oplus}X^{\ominus}$$

Moist silver oxide converts these to the hydroxides, $R_4P^{\oplus}OH^{\ominus}$, hygroscopic syrups which behave as strong bases. Thermal decomposition yields tertiary phosphine oxides:

$$R_4P^{\oplus}OH^{\ominus} \longrightarrow R_3PO + RH$$

(b) Arsenic

The properties and reactions of organo-arsenic compounds are in many cases similar to those of the corresponding phosphorus derivatives, though there is a marked decrease in stability. A large number of organo-arsenic compounds have been investigated largely because of their potential therapeutic value; others have been used as poison gases. (Short review, see *R. M. Kary*, "Metal-Organic Compounds", A.C.S. No. 23, p. 319, Amer. chem. Soc., 1959).

Derivatives of tervalent arsenic

(i) Alkylarsines

(1) *Primary and secondary arsines*, $RAsH_2$ and R_2AsH. *Preparation.* By the reduction of arsonic acids, $RAsO(OH)_2$ and arsinic acids, $R_2AsO(OH)$, respectively, using zinc and hydrochloric acid.

Properties. They are volatile, very poisonous compounds with objectionable odours. Oxidation occurs readily, the secondary arsines being spontaneously inflammable. Primary arsines are oxidized to $(RAsO)_n$ or, under aqueous conditions, to arsonic acids. Addition of methyl iodide at 100° gives the quaternary salt:

$$RAsH_2 + 3CH_3I \longrightarrow R(CH_3)_3As^{\oplus}I^{\ominus} + 2HI$$

Secondary arsines are more reactive; aqueous iodine will oxidise dimethylarsine to dimethylarsinic acid. Controlled oxidation gives first cacodyl, $(CH_3)_2As—As(CH_3)_2$, or cacodyl oxide $[(CH_3)_2As]_2O$. Primary and secondary arsines are weakly basic, their salts being completely hydrolysed in water.

Methylarsine, CH_3AsH_2, b.p. 2°; *Ethylarsine,* b.p. 36°, d^{22} 1·217. Both form crystalline *sulphates* $(RAsH_2)_2 \cdot H_2SO_4$.

Dimethylarsine, $(CH_3)_2AsH$, b.p. 35·6°/747 mm., d^{29} 1·213; *diethylarsine,* b.p. 105°.

(2) *Tertiary arsines,* R_3As. *Preparation.* Generally by the action of lithium alkyls or Grignard reagents on arsenic trichloride or alkylchloroarsines. Unsymmetrical tertiary arsines can be made by the alkylation of an alkylhalogenoarsine by a tetra-alkyl-lead (*Maier,* J. inorg. nucl. Chem., 1962, **24,** 1073).

Properties. The lower tertiary arsines are poisonous volatile liquids with extremely repulsive odours. Oxidation to the oxide, R_3AsO, takes place readily; sulphur and selenium form the corresponding sulphides, R_3AsS, and selenides, R_3AsSe. The halogens also add readily to give the quinquevalent halides, R_3AsX_2, and alkyl halides form the corresponding quaternary salts. Mercuric chloride forms stable addition products (*mercurichlorides*), $(R_3As)_2 \cdot HgCl_2$ (*W. M. Dehn,* Amer. chem. J., 1908, **40,** 88). Physical constants of many arsines are given by *W. J. Jones et al.* (J. chem. Soc., 1932, 2284); see also Table 10, p. 237.

(ii) Diarsines, $R_2As—AsR_2$

Preparation. (1) By reduction of dialkylchloroarsines, R_2AsCl, with zinc.

(2) By reduction of oxygen derivatives with phosphorous acid.

(3) By the action of secondary arsines on dialkylchloroarsines (*Dehn* and *B. B. Wilcox,* Amer. chem. J., 1908, **40,** 123):

$$R_2AsH + R'_2AsCl \longrightarrow R_2As\text{-}AsR'_2 + HCl$$

Properties. For b.ps. see Table 10, p. 237. They are foul-smelling liquids which are rapidly oxidised in air but, unlike the arsines, they form the corresponding peroxides, $R_2AsO \cdot OAsR_2$. The metal-metal bond is cleaved by halogens.

Tetramethyldiarsine, *cacodyl,* $(CH_3)_2As—As(CH_3)_2$ b.p. 170°, is spontaneously inflammable.

(iii) Polyarsines

Reduction of sodium methylarsonate, $CH_3AsO(ONa)_2$ with hypophosphorous acid gives "*arsenomethane*", $(CH_3As)_n$, which exists in two forms, a yellow liquid f.p. 12° and a red solid, m.p. ~ 180°. The X-ray diffraction pattern is compatible with five-membered rings (*J. Waser* and *V. Schomaker*, J. Amer. chem. Soc., 1945, **67**, 2014). Linear polymers have been obtained by the polymerization of arseno-methane by iodine (*M. Ya. Kraft* and *V. V. Katyshkina*, Doklady Akad. Nauk S.S.S.R., 1949, **66**, 207).

(iv) Halides, $RAsX_2$ and R_2AsX

Preparation. Both types of halides may be prepared by the following methods.

(1) Equilibration reactions between tertiary arsines and arsenic trichloride.

(2) Thermal decomposition of the appropriate quinquevalent halide:

$$R_3AsX_2 \longrightarrow R_2AsX + RX$$

(3) Direct reaction of methyl chloride or bromide with arsenic at 350–375° using a copper catalyst.

A mixture of the mono- and di-halides (about 1:2) is obtained. Methyl iodide reacts at 280° but the yields are poor (*L. Maier*, *E. G. Rochow* and *W. C. Fernelius*, J. inorg. nucl. Chem., 1961, **16**, 213).

(4) Reaction of tervalent arsenic oxides, RAsO and $(R_2As)_2O$, with halogen acids. Alkylarsonic acids can be used if they are simultaneously reduced (*A. J. Quick* and *R. Adams*, J. Amer. chem. Soc., 1922, **44**, 805).

Properties. The halides are hydrolysed under alkaline conditions. Halogens will add directly, the dihalides forming the unstable tetrahalides, $RAsX_4$, while the monohalides form the trihalides, R_2AsX_3, which are relatively more stable. Hydrogen sulphide converts the dichlorides into the corresponding sulphides, RAsS.

The dichlorides are colourless liquids with toxic and vesicant properties.

Methyldichloroarsine, CH_3AsCl_2 b.p. 132°, and *ethyldichloroarsine*, $C_2H_5AsCl_2$ b.p. 155°, were used as poison gases in World War I. Their chemistry and properties have been discussed (*K. E. Jackson*, Chem. Reviews, 1935, **17**, 251). Even more potent were the *Lewisites*: mono-, di- and tri-2-chlorovinyldichloroarsines, $ClCH{=}CH_2AsCl_2$, etc., prepared by passing acetylene into arsenic trichloride in the presence of aluminium trichloride (*K. E. Jackson* and *M. A. Jackson*, *ibid.*, 1935, **16**, 439).

Dimethylchloroarsine, $(CH_3)_2AsCl$, b.p. 109°; *dimethylbromoarsine*, b.p. 129°.

(v) Alkylarsine oxides, RAsO and $(R_2As)_2O$

The corresponding acids, $RAs(OH)_2$ and R_2AsOH, dehydrate extremely readily to the oxides. Primary oxides, prepared by hydrolysis of the dihalides, are often colourless polymeric solids of low volatility. Secondary oxides can be obtained from the primary oxides on distillation with potassium hydroxide:

$$4CH_3AsO + 6KOH \longrightarrow [(CH_3)_2As]_2O + 2As(OK)_3 + 3H_2O$$

Esters are formed from the reaction of oxides with alcohols (*G. Kamai* and *Z. L. Khisamova*, J. gen. Chem., 1960, 30, 3611):

$$(R_2As)_2O + 2R'OH \rightleftharpoons 2R_2AsOR' + H_2O$$

Methylarsine oxide, CH_3AsO, m.p. 95°, has a garlic-like smell.

Tetramethyldiarsine oxide (*cacodyl oxide*), $[(CH_3)_2As]_2O$, b.p. 120°, and *tetraethyldiarsine oxide*, b.p. 93°/9 mm., are converted to the chlorides by concentrated hydrochloric acid. They are regenerated by alkaline hydrolysis. Cacodyl oxide is oxidized in air to dimethylarsinic (cacodylic) acid, $(CH_3)_2AsO(OH)$.

Derivatives of quinquevalent arsenic

(vi) Alkyls

Pentamethylarsenic, $(CH_3)_5As$, may have been prepared by the action of methyl-lithium with trimethylarsenic dibromide at low temperatures, but the pure compound does not appear to have been isolated (*G. Wittig* and *K. Torssell*, Acta chem. Scand., 1953, 7, 1293).

(vii) Halides

(1) *Quaternary arsonium salts*, $R_4As^{\oplus}X^{\ominus}$, are obtained from tertiary arsines and alkyl iodides (*W. Steinkopf* and *G. Schwen*, Ber., 1921, 54, 1451). Moist silver oxide converts these compounds to the *tetra-alkylarsonium hydroxides*, $R_4As^{\oplus}OH^{\ominus}$, which are strongly basic, deliquescent solids.

(2) *Trialkylarsenic dihalides*, R_3AsX_2, are formed (a) when tertiary arsines are treated with halogen, (b) by the action of hydrogen halides or halogen acids on the oxides R_3AsO and (c) by addition of an alkyl halide to a trivalent monohalide, R_2AsX. The dihalides are very easily hydrolysed by water.

(3) *Dialkylarsenic trihalides*, R_2AsX_3, are obtained by addition of halogen to dialkylhalogenoarsines in carbon disulphide. *Dimethylarsenic trichloride*, $(CH_3)_2AsCl_3$, has also been prepared by the action of phosphorus pentachloride on cacodylic acid, $(CH_3)_2AsO(OH)$. It decomposes into methyldichloroarsine and methyl chloride at 40–50° and is hydrolysed by water to cacodylic acid.

(4) *Alkylarsenic tetrahalides*, $RAsX_4$, are less stable than the trihalides.

Methylarsenic tetrachloride, CH_3AsCl_4, formed by passing chlorine into a solution of methyldichloroarsine in carbon disulphide at −10°, decomposes at 0° into arsenic trichloride and methyl chloride.

(viii) Trialkylarsine oxides, R_3AsO

Trialkylarsine oxides are formed by oxidation of tertiary arsines in air or by hydrolysis of the dihalides followed by dehydration. They are very weak bases, their salts being almost completely hydrolysed by water.

(ix) Alkylarsonic acids, $RAsO(OH)_2$

Preparation. (1) From primary arsines, oxides or dihalides when oxidised in the presence of water.

(2) Conveniently, in high yields, by addition of alkyl halides to sodium arsenite (*Meyer's reaction*). (*Quick* and *Adams, loc. cit.*):

$$As(ONa)_3 + RX \longrightarrow RAsO(ONa)_2 + NaX$$

Properties. Alkylarsonic acids are colourless, crystalline solids, soluble in alcohol and water but not in ether. They are dibasic and relatively strong but weaker than the corresponding phosphonic acids (*H. J. Backer* and *C. C. Bolt,* Rec. Trav. chim., 1935, 54, 186). They are readily reduced by sulphur dioxide in the presence of hydrochloric acid to give good yields of the corresponding alkyldichloroarsine.

(x) Dialkylarsinic acids, $R_2AsO(OH)$

Preparation. (1) From an alkyldichloroarsine by dissolution in sodium hydroxide solution and subsequent treatment with an alkyl halide (*Quick* and *Adams, loc. cit.*):

$$RAsCl_2 \xrightarrow{NaOH} RAs(ONa)_2 \xrightarrow{R'X} RR'AsO(ONa) + NaX$$

(2) By hydrolysis of dialkylarsenic trihalides, R_2AsCl_3.

These acids are weaker than the arsonic acids.

Dimethylarsinic acid (cacodylic acid), $(CH_3)_2AsO(OH)$, m.p. 200°, is soluble in water and alcohol. It is extremely stable, very resistant to hydrolysis, oxidation and reduction. In strongly alkaline solution it is tribasic, forming $(CH_3)_2As(ONa)_3$.

(c) Antimony

Antimony forms organic derivatives in which the metal covalency is three (R_3Sb), four ($R_4Sb^{\oplus}X^{\ominus}$) and five ($R_5Sb$). Early work has been reviewed by *W. G. Christiansen* ("Organic Derivatives of Antimony", Chemical Catalog Co., New York, 1925).

Derivatives of tervalent antimony

(i) Mono- and di-alkylstibines

Mono- and di-alkylstibines are unstable but **methylstibine,** CH_3SbH_2, b.p. 41° extrap. and *dimethylstibine,* $(CH_3)_2SbH$, b.p. 60° extrap., have been obtained by reduction of dimethylbromostibine with sodium borohydride (*A. B. Burg* and *L. R. Grant,* J. Amer. chem. Soc., 1959, **81**, 1).

(ii) Trialkylstibines

Preparation. Antimony trialkyls are conveniently synthesized by the action of a Grignard reagent or a lithium alkyl on antimony trichloride (*W. J. C. Dyke, W. C. Davies* and *W. J. Jones,* J. chem. Soc., 1930, 463). Aluminium trialkyls have also been used (*Kali-Chemie A.G.*, B.P., 768,765/1957; *L. I. Zakharkin* and *O. Yu. Okhlobystin*, Izvest. Akad. Nauk S.S.S.R., Otdel Khim. Nauk 1959, 1942).

Trivinylstibine has been prepared from vinylmagnesium bromide and antimony trichloride using tetrahydrofuran as solvent (*L. Maier et al.*, J. Amer. chem. Soc., 1957, **79**, 5884).

Properties and reactions. (For boiling points of lower members of the series see Table 10). Trialkylstibines are readily oxidised in air to trialkylstibine oxides, R_3SbO, some of the lower members being spontaneously inflammable. Pentavalent halides, R_3SbX_2, are formed by the addition of halogens.

TABLE 10

BOILING POINTS OF TRIALKYLS R_3M, M=As, Sb, Bi

R	As	Sb	Bi
CH_3	51·9°	79°	110°
C_2H_5	140°	160°	96°/50 mm.
n-C_3H_7	81–2°/16 mm.	100°/25 mm.	127°/50 mm.
n-C_4H_9	116°/10 mm.	131°/12 mm.	124°/7 mm.

Stibines can be stored and purified as the dibromo derivatives; they are regenerated by reduction with zinc (*G. T. Morgan* and *V. E. Yarsley*, J. chem. Soc., 1925, 184). Addition reactions occur with sulphur, selenium and cyanogen halides; methyl iodide forms the quaternary salt. Hydrogen halides do not add to alkylstibines; the latter act as reducing agents and trialkylstibine dichlorides are formed.

Tristrifluoromethylstibine, $(CF_3)_3Sb$, b.p. 72°, is prepared from trifluoroiodomethane and antimony. It is hydrolysed by aqueous alkali, and unlike trimethylstibine it shows no donor properties (*H. J. Emeléus et al.*, *ibid.*, 1957, 3708).

(iii) Halides, R_2SbCl and $RSbCl_2$

Preparation. Both types of halides can be formed (1) by thermal decomposition of the appropriate pentavalent halide.

(2) By equilibration reactions between a tertiary stibine and antimony trichloride.

(3) Addition of a Grignard reagent to antimony trichloride gives a mixture of products.

(4) Direct reaction of methyl bromide with antimony metal at 350° using a copper powder catalyst gives a mixture of dimethylbromostibine and methyldibromostibine (*L. Maier, E. G. Rochow* and *W. C. Fernelius*, J. inorg. nucl. Chem., 1961, 16, 213).

Properties. Mono- and di-halides are very easily oxidised, and easily hydrolysed under alkaline conditions.

Dimethylchlorostibine, $(CH_3)_2SbCl$, b.p. 155–160°, *methyldichlorostibine*, CH_3SbCl_2, b.p. 115–120°/60 mm., and the corresponding bromo- and iodo-derivatives have been prepared by these methods (*G. T. Morgan* and *G. R. Davis*, Proc. roy. Soc., 1926, **110A**, 523). *Ethyldichlorostibine*, $C_2H_5SbCl_2$, b.p. 62–83°/1 mm., has been obtained from antimony trichloride using tetraethyl-lead as a mild alkylating reagent. (*M. S. Kharasch, E. V. Jensen* and *S. Weinhouse*, J. org. Chem., 1949, **14**, 429).

(iv) Alkylstibine oxides, $RSbO$, $(R_2Sb)_2O$

The primary oxides, RSbO, are colourless polymeric solids of low volatility obtained by alkaline hydrolysis of alkylstibine dichlorides. Secondary oxides, $(R_2Sb)_2O$, are generally monomeric and crystalline.

Bis-dimethylstibine oxide $[(CH_3)_2Sb]_2O$ is a spontaneously inflammable liquid obtained by alkaline hydrolysis of the monohalide (*Morgan* and *Davies, loc. cit.*).

(v) Compounds containing Sb—Sb bonds

Tetramethyldistibine $(CH_3)_2Sb—Sb(CH_3)_2$ m.p. 17°, and *tetraethyldistibine*, m.p. —61°, were obtained by the reaction of free methyl and ethyl radicals on an antimony mirror (*F. A. Paneth* and *H. Loleit*, J. chem. Soc., 1935, 366). The former is best prepared by reduction of dimethylbromostibine using sodium in liquid ammonia (*Burg* and *Grant, loc. cit.*).

Derivatives of quinquevalent antimony

(vi) Penta-alkyls, R_5Sb

Pentamethylantimony, $(CH_3)_5Sb$, a stable liquid, b.p. 126–7°, is obtained from the reaction of methyl-lithium with trimethylstibine dibromide or tetramethyl-stibonium bromide. It is decomposed by water and oxidised in air although it is not spontaneously inflammable. With excess methyl-lithium it forms the unstable complex $Li^{\oplus}(CH_3)_6Sb^{\ominus}$ (*G. Wittig* and *K. Torssell*, Acta. chem. Scand., 1953, 7, 1293).

(vii) Quaternary stibonium salts, $R_4Sb^{\oplus}X^{\ominus}$

These are formed when an alkyl iodide is added to the lower trialkylstibines. They are water soluble and the *hydroxides*, formed by reaction with moist silver oxide, are strong bases like quaternary ammonium hydroxides.

(viii) Trialkylstibine dihalides, R_3SbX_2

Preparation. (1) By addition of halogen to a trialkylstibine (*Dyke* and *Jones*, J. chem. Soc., 1930, 1921).

(2) By the action of a Grignard reagent on an antimony pentahalide.

(3) From a dialkylmercury and antimony trichloride:

$$2(CH_3)_2Hg + SbCl_3 \longrightarrow (CH_3)_3SbCl_2 + CH_3HgCl + Hg$$

(4) From a trialkylstibine and a hydrogen halide:

$$(C_2H_5)_3Sb + 2HCl \longrightarrow (C_2H_5)_3SbCl_2 + H_2$$

Trimethylstibine dichloride, $(CH_3)_3SbCl_2$, forms colourless plates moderately soluble in hot water.

(ix) Dialkylstibine trihalides, R_2SbX_3

These unstable compounds, which decompose very easily with loss of alkyl halide, are formed when halogen is added to a dialkylhalogenostibine:

$$(CH_3)_2SbCl + Cl_2 \longrightarrow (CH_3)_2SbCl_3 \longrightarrow CH_3SbCl_2 + CH_3Cl$$

(x) Trialkylstibine oxides, R_3SbO

These compounds can be obtained by oxidation of trialkylstibines, preferably with mercuric oxide. They are soluble in water and hydroxylic solvents. Hydrogen sulphide converts them to the corresponding *trialkylstibine sulphides*.

(xi) Dialkylstibinic acids, $R_2SbO(OH)$

Stibinic acids are very weak acids having polymeric structures.

Dimethylstibinic acid, $(CH_3)_2SbO(OH)$, has been obtained by oxidation and hydrolysis of dimethylchlorostibine (*Morgan* and *Davies*, Proc. roy. Soc., 1926, **110A**, 523).

(xii) Alkylstibonic acids, $RSbO(OH)_2$

These appear to be unknown.

(d) *Bismuth*

Bismuth is the most electropositive of the Group V elements and its organic derivatives have the most metallic character. They are comparable in reactivity with those of lead. Organobismuth compounds have been widely studied because of their possible pharmacological activity. The literature prior to 1941 has been reviewed (*H. Gilman* and *H. L. Yale*, Chem. Reviews, 1942, 30, 281). Alkyl derivatives of bismuth(V) are unknown.

(i) *Bismuth(III) hydrides*

Methylbismuth hydride, CH_3BiH_2, b.p. 72° extrap., and *dimethylbismuth hydride,* $(CH_3)_2BiH$, b.p. 103° extrap., have recently been prepared by the reduction of methylbismuth dichloride and dimethylbismuth bromide by lithium aluminium hydride at —110°. These compounds are colourless liquids, stable at —60°, but disproportionate at —45° and —15° respectively forming trimethylbismuth and bismuth hydride (*E. Amberger*, Ber., 1961, **94**, 1447).

(ii) *Bismuth(III) alkyls*

These are best prepared by interaction of Grignard reagents with bismuth trichloride (*H. Gilman* and *J. F. Nelson*, J. Amer. chem. Soc., 1937, 59, 935).

They are colourless oily liquids, the lower members being spontaneously inflammable in air (for b.ps. see Table 10, p. 237). Chlorine and bromine react vigorously but, unlike the other alkyls of Group V, substitution, not addition, occurs:

$$R_3Bi + Cl_2 \longrightarrow R_2BiCl + RCl$$

Bismuth trialkyls do not react with methyl iodide to form quaternary bismuthonium salts; neither do they show donor properties. The carbon-bismuth bond is cleaved by mineral acids.

(iii) *Bismuth(III) monohalides and dihalides,* R_2BiCl and $RBiCl_2$

These derivatives can be formed by equilibration reactions between a bismuth halide and a bismuth trialkyl.

The monohalides are usually crystalline solids, soluble in alcohol and very sensitive to air and moisture. Alkylbismuth dihalides are much more stable.

Dimethylbismuth chloride, $(CH_3)_2BiCl$, m.p. 116°, has also been obtained by direct halogenation of trimethylbismuth at low temperatures (*A. Marquardt*, Ber., 1887, **20**, 1516; 1888, **21**, 2035). Alkylation of bismuth trichloride with tetraethyl-lead gives a mixture of triethylbismuth and diethylbismuth chloride (*H. Gilman* and *L. D. Apperson*, J. org. Chem., 1939, **4**, 162). Small amounts of methylbismuth dibromide have been obtained by direct reaction of methyl bromide with bismuth metal at 250° (*L. Maier et al.*, J. inorg. nucl. Chem., 1961, **16**, 213).

7. Group VIB: Selenium, Tellurium, Polonium

The derivatives of selenium and tellurium closely resemble those of sulphur and also of arsenic and antimony in Group VB. Covalencies of 2, 4 and 6 are common. The alkyls form coordination compounds with electron acceptor groups and readily add one molecule of alkyl halide to form the "onium" salts. Organo-derivatives of selenium are toxic although it may be an essential trace element in animal metabolism (*K. Schwarz* and *C. M. Foltz*, J. Amer. chem. Soc., 1957, **79**, 3292). The methyl derivatives of selenium and tellurium can be formed by biological methylation of their oxides or acids by certain moulds, particularly *Scopulariopsis brevicaulis* (*F. Challenger*, Chem. Reviews, 1945, **36**, 315). The nomenclature of these compounds is similar to that of sulphur (see Chapter 5, p. 74 and C.A. 1945, **39**, 5942, 5946).

(a) Selenium

The organic chemistry of selenium has been reviewed (*T. W. Campbell*, *H. G. Walker* and *G. M. Coppinger*, Chem. Reviews, 1952, **50**, 279).

(i) Alkaneselenols, RSeH

Preparation. (1) By reaction of selenium with a Grignard reagent:

$$RMgX + Se \longrightarrow RSeMgX \xrightarrow{HX} RSeH + MgX_2$$

(2) By reaction of an alkyl halide with sodium hydrogen selenide (*L. Tschugaeff*, Ber., 1909, **42**, 49):

$$NaSeH + RX \longrightarrow RSeH + NaX$$

(3) By reduction of dialkyl diselenides, RSe_2R.

This is conveniently carried out in liquid ammonia by reacting the diselenide with sodium (atomic ratio 1:1) to form the selenolate, and then acidifying (*L. Brandsma* and *H. E. Wijers*, Rec. Trav. chim., 1963, **82**, 68):

$$R_2Se_2 \xrightarrow[\text{liq. NH}_3]{\text{2Na}} RSeNa \xrightarrow{\text{acid}} RSeH$$

Air must be excluded as selenols are easily oxidised to diselenides.

Properties. Alkaneselenols are dense, evil smelling liquids, which must be handled with care. They are insoluble in water but soluble in organic solvents and aqueous alkali. The reactions of ethaneselenol with ketones have been studied (*E. H. Shaw* and *E. E. Reid*, J. Amer. chem. Soc., 1926, **48**, 520).

Methaneselenol, b.p. 12°/758 mm.; *ethaneselenol,* b.p. 53·5°, d^{24} 1·3954, n_D^4 1·47715; *propane-1-selenol,* b.p. 84°; *butane-1-selenol,* b.p. 114°.

(ii) Alkylselenoalkanes, dialkylselenides, R_2Se

Preparation. (1) Symmetrical alkylselenoalkanes can be prepared by reacting powdered selenium, sodium formaldehydesulphoxylate, sodium hydroxide and an alkyl iodide at 50° (*M. L. Bird* and *F. Challenger,* J. chem. Soc., 1942, 570):

$$Se + RI + H_2O + NaOH + HCHO \cdot NaHSO_3 \longrightarrow R_2Se$$

(2) Symmetrical dialkylselenides are easily prepared in good yield by treating sodium selenide (from sodium and selenium in the correct proportions) in liquid ammonia with an alkyl halide (*Brandsma* and *Wijers, loc. cit.*):

$$2Na + Se \xrightarrow[NH_3]{liq.} Na_2Se \xrightarrow{RX} R_2Se$$

(3) Unsymmetrical derivatives are prepared by the reaction of sodium alkaneselenolate, RSeNa, with an alkyl iodide. The former can be obtained from alkaneselenols and sodium or, more conveniently, by the action of sodium in alcohol on a diselenide (*Bird* and *Challenger, loc. cit.*):

$$R_2Se_2 + 2Na \longrightarrow 2RSeNa \xrightarrow{RI'} RR'Se + NaI$$

Properties. Dialkyl selenides are unassociated, highly refractive liquids with unpleasant odours. Halogens add readily to form the dihalides, R_2SeX_2.

Methylselenomethane, *dimethyl selenide,* b.p. 55°; *ethylselenoethane,* b.p. 108°; 1-(1′-*propylseleno*)*propane, di-n-propyl selenide,* b.p. 159°.

(iii) Alkyldiselenoalkanes, dialkyl diselenides, RSeSeR

Preparation. (1) By oxidation of selenols:

$$2RSeH \xrightarrow{O_2} RSeSeR + H_2O$$

(2) By interaction of alkyl sulphates or bromides and potassium diselenide, K_2Se_2 (*H. J. Backer* and *W. van Dam,* Rec. Trav. chim., 1935, 54, 531).

(3) As described above for dialkyl selenides, using two equivalents of selenium (*Bird* and *Challenger, loc. cit.*; *Brandsma* and *Wijers, loc. cit.*).

(4) Di-(hydroxyalkyl)diselenides, $[HO(CH_2)_nSe]_2$, are obtained by Challenger's method using selenium, sodium hydroxide, sodium formaldehydesulphoxylate and a halohydrin. The red-yellow oils have characteristic unpleasant odours (*L. Agenas,* Acta Chem. Scand., 1962, 16, 1809).

Properties. Alkyldiselenoalkanes are bright yellow oily liquids with disagreeable odours.

Methyldiselenomethane, *dimethyl diselenide,* b.p. 155–157°; *diethyl diselenide,* b.p. 85°/21 mm.; *di-n-propyl diselenide,* b.p. 103–104°/15 mm.

(iv) Acids

(1) *Alkaneseleninic acids, seleninoalkanes,* RSeOOH, have been prepared by the oxidation of the corresponding diselenides with nitric acid (*d.*1·4) at 50–100° (*Backer* and *van Dam, loc. cit.*) and also by the action of moist silver oxide on dialkylselenide dihalides (*Bird* and *Challenger, loc. cit.*).

They are hygroscopic and soluble in water. They are weaker than the corresponding carboxylic acids. With strong acids they form compounds of the type $RSe(OH)_2X$ where X = univalent acid radical.

Seleninomethane, m.p. 134°; *seleninoethane,* m.p. 47°; *1-seleninopropane,* m.p. 75°; *1-seleninobutane,* m.p. 68°.

(2) *Alkaneselenonic acids, selenonoalkanes,* $RSeO_2OH$, have been obtained as their potassium salts by careful permanganate oxidation of the corresponding seleninic acids (*Bird* and *Challenger, loc. cit.*).

The salts are unstable and eliminate selenious acid, H_2SeO_3, on heating with acid or alkali.

(v) Selenocarbonyl compounds, RCSeR′

Preparation. By treatment of aldehydes and ketones with hydrogen selenide in concentrated hydrochloric acid or alcoholic hydrogen chloride.

Selenoformaldehyde, *methaneselenal,* HCHSe, m.p. 215°; *selenoacetaldehyde, selenoformylmethane,* CH_3CHSe, m.p. 139° (*L. Vanino* and *A. Schinner,* J. pr. Chem., 1915, **91**, 116). The selenoketones, R_2CSe, are dense, red oils which are dimeric in benzene solution (*R. E. Lyons* and *W. E. Bradt,* Ber., 1927, **60**, 824). *Selenoacetone,* $[(CH_3)_2CSe]_2$, b.p. 220–230°.

(vi) Organoselenium halides

(1) *Dialkylselenium dihalides,* R_2SeX_2, are formed by direct addition of halogen to dialkyl selenides.

Dimethylselenium dichloride, m.p. 59·5° and the *dibromide* m.p. 82°. The closely related alkylarylselenium dihalides are thermally unstable and decompose according to the equation

$$RR'SeX_2 \longrightarrow R'SeX + RX \qquad R{=}\text{alkyl, } R'{=}\text{aryl}$$

(*O. Behaghel* and *K. Hofmann,* Ber., 1939, **72**, 697; *O. K. Edwards et al.,* J. chem. Soc., 1928, 2293).

(2) *Alkylselenium trihalides,* $RSeX_3$.

Ethylselenium tribromide, bright red crystals, m.p. 74°, has been obtained from the bromination of benzylselenoethane in chloroform (*Behaghel* and

Hofmann, loc. cit.), and also by treatment of seleninoethane nitrate, $C_2H_5SeOOH \cdot HNO_3$, with hydrobromic acid (*Shaw* and *Reid, loc. cit.*).

(3) *Alkylselenonium halides*, R_3SeX, are prepared by direct addition of an alkyl halide to a dialkyl selenide. Moist silver oxide liberates the *hydroxide*, R_3SeOH, a strong base which forms neutral salts with acids.

Trimethylselenonium iodide, m.p. 150–1° (*W. von E. Doering* and *A. K. Hoffmann*, J. Amer. chem. Soc., 1955, **77**, 521).

(vii) Alkylseleninylalkanes, dialkylselenoxides, R_2SeO

These derivatives are unstable, decomposing to give a selenol and an aldehyde (*Edwards et al., loc. cit.*), but some halogenated derivatives *e.g.* di-(β-chloroethyl)-selenoxide, $(ClCH_2CH_2)_2SeO$, have been described (*T. Smedsland*, C.A., 1932, **26**, 5905). *Dibenzyl selenoxide*, $(C_6H_5CH_2)_2SeO$, m.p. 136°, is soluble in water (*W. Strecker* and *A. Willing*, Ber., 1915, **48**, 196). Dialkyl selenones, R_2SeO_2, have not been reported.

(b) Tellurium

The properties and reactions of organotellurium derivatives resemble those of selenium. Infra-red and nuclear magnetic resonance spectra of some tellurium compounds have been reported (*H. P. Fritz* and *H. Keller*, Ber., 1961, **94**, 1524; *G. Close*, Z. Naturforsch., 1961, **16a**, 528).

(i) Alkanetellurols, RTeH

Alkanetellurols have been prepared by the action of hydrogen telluride on alkyl bromides in an alcoholic solution of sodium ethoxide under hydrogen. They are yellowish liquids with strong disagreeable odours. (*A. Baroni*, Atti Accad. Lincei, 1938, **27**, 238).

Methanetellurol, CH_3TeH b.p. 57°; *ethanetellurol*, b.p. 90°; *1-propanetellurol*, b.p. 121°; *1-butanetellurol*, b.p. 151°.

(ii) Alkyltelluroalkanes, dialkyl tellurides, RTeR

Preparation. (1) From an alkyl halide and tellurium in the presence of sodium formaldehydesulphoxylate and sodium hydroxide (*M. P. Balfe*, *C. A. Chaplin* and *H. Phillips*, J. chem. Soc., 1938, 341).

(2) By alkylation of sodium telluride (Na_2Te) in liquid ammonia (*Brandsma* and *Wijers, loc. cit.*).

Halogens, metallic salts and alkyl halides add in the same way as to the selenides. Oxidation under alkaline conditions gives dialkyltellurium dihydroxide, $R_2Te(OH)_2$ (*Balfe et al., loc. cit.*; *Balfe* and *K. N. Nandi, ibid.*, 1941, 70).

Dimethyl telluride, $(CH_3)_2Te$, a yellow oil, b.p. 94°, is obtained in good yield by reduction of its di-iodide with sodium sulphite (*R. H. Vernon, ibid.*, 1920, **117**, 889); *diethyl telluride*, b.p. 38°/14 mm.; *dibutyl telluride*, b.p. 109–112°/12 mm., d_4^{25} 1·2923.

(iii) Organotellurium halides

(1) *Dialkyltellurium dihalides*, R_2TeX_2, are formed by addition of halogen to a dialkyl telluride or by direct reaction from alkyl halide and tellurium.

There are two forms of the methyl and ethyl derivatives which were formerly thought to be isomers (*Vernon, ibid.*, 1920, **117**, 86, 889). *H. D. K. Drew* (*ibid.*, 1929, 560) later showed that one form is a dimer which is highly ionised $[(CH_3)_3Te]^{\oplus}$ $[CH_3TeX_4]^{\ominus}$. The corresponding *dihydroxides*, $R_2Te(OH)_2$, are produced by the action of moist silver oxide on the dihalide. Dimethyltellurium dihydroxide easily loses water to give the compound $(CH_3)_3Te{\cdot}O{\cdot}TeO(CH_3)$, from which the dimeric dihalide above is formed by reaction with HX. Dimethyltellurium dichloride is quantitatively hydrolysed in aqueous solution to dimethyltellurium oxychloride $(CH_3)_2Te(OH)Cl$ (*P. Nylén*, Z. anorg. Chem., 1941, **246**, 227).

(2) *Alkyltellurium trihalides*, $RTeX_3$, are formed along with the dihalides in the reaction of dialkyl tellurides with halogen, or treatment of dialkyltellurium dihydroxide with halogen acid (*Balfe et al., loc. cit.*).

Methyltellurium triiodide, decomposes on heating with evolution of iodine.

(3) *Alkyltelluronium halides*, R_3TeX, are produced by direct addition of an alkyl halide to a dialkyl telluride (*Balfe et al., loc. cit.*).

Trimethyltelluronium fluoride is deliquescent (*H. J. Emeléus* and *H. G. Heal*, J. chem. Soc., 1946, 1126); the *iodide* melts at 240–1° (*W. von E. Doering* and *A. K. Hoffmann*, J. Amer. chem. Soc., 1955, **77**, 521).

(iv) Dialkyltelluroxides (alkyltellurinylalkanes) and alkyltellurinic acids (tellurinoalkanes)

Aliphatic telluroxides, R_2TeO, the primary products of the oxidation of the tellurides, are unstable and decompose giving aldehydes and alkanetellurols. In the presence of excess oxidising agent the alkanetellurol is further oxidized to give an alkyltellurinic acid. Under neutral conditions di-*n*-butyl telluride gives a complex compound containing di-*n*-butyltelluroxide and n-*butyltellurinic acid*, C_4H_9TeOOH, (decomp. 250°), which cannot be separated. The latter is an infusible, sparingly soluble solid and is probably polymeric (*Balfe et al., loc. cit.*).

(c) Polonium

Dimethyl and *dibenzyl polonides* have been reported (see *K. W. Bagnall*, Quart. Reviews, 1957, **11**, 45).

8. Transition metals

The transition metals, *i.e.* all the elements in Groups IIIA, IVA, VA, VIA, VIIA, VIII and IB of the Periodic Table (Table 2, p. 169), are classed together. Most of these elements resisted attempts to prepare organic derivatives by conventional means, only platinum and the coinage metals giving stable sigma (σ-) bonded derivatives. Treatment of the anhydrous halides of most of these elements with Grignard reagents gave only decomposition products from the alkyl group, although the formation and decomposition of an intermediate metal alkyl was often suspected. Under these conditions metal aryl reagents often give the dimer of the hydrocarbon radical in good yield *e.g.* biphenyl from phenylmagnesium bromide, but this is not true of alkylmetals, which give complex products (*F. A. Cotton*, Chem. Reviews, 1955, 55, 551). Since 1953, however, σ-bonded alkyl derivatives of a number of transition metals have been prepared. These successes have been due in some cases to great care over the experimental conditions, as in the preparation of methyltitanium compounds at low temperatures, but mainly it has been due to a greater understanding of carbon-metal bonding which has made it possible to find conditions (discussed below), which allow of the preparation of stable σ-bonded derivatives. Unlike the metals and metalloids discussed in earlier sections of this chapter, there is usually no relation between the electronegativity values given for transition metals and the chemical properties of their alkyl derivatives.

Metal-ligand bonding

A *ligand* is any radical, atom, or molecule which is attached to the metal and in some way bonded to it. There are two main types (1) negative or anionic ligands *e.g.* $Cl^{\ominus}$, $Br^{\ominus}$, $CN^{\ominus}$, $NO_2^{\ominus}$, $R^{\ominus}$ and (2) neutral molecules *e.g.* R_3N, R_3P, CO, NO, H_2O and unsaturated hydrocarbons such as olefins, acetylenes and benzene. A ligand having two or more atoms which can serve as donors is classed as a *polydentate ligand* and if these donor atoms attach themselves to the same metal atom it is also a *chelating ligand*. The latter often form complexes of greater stability because of the decrease in entropy.

Several different approaches have been made towards the theoretical treatment of metal-ligand bonding. The most recent, and probably the most useful for transition metal complexes, is the ligand-field theory developed around 1956 (see *R. J. Gillespie* and *R. S. Nyholm*, Quart. Reviews, 1957, **11**, 354 and *J. S. Griffith* and *L. E. Orgel*, *ibid.*, p. 38). Non-mathematical discussions of these theories are given by *L. Pauling* ("The Nature of the Chemical Bond", 3rd Edn. Cornell University Press, New York, 1960), *L. E. Orgel* ("An Introduction to Transition Metal Theory: Ligand Field Theory", Methuen, London, 1960) and by *F. A. Cotton* and *G. Wilkinson* ["Advanced Inorganic Chemistry", Interscience Publishers (Wiley) 1962, Chapter 26].

Neutral ligands such as CO, R_3P, R_3Sb, R_3As and NO are capable of stabilizing low oxidation states of the transition metals. A σ-type bond is formed by the donation of the lone pair of electrons and this is supplemented by a π-type bond formed by back-donation of electron density from the filled metal orbitals to the vacant orbitals in these ligands. Although these molecules are termed *donor ligands,* they function by also acting as electron acceptors.

A similar situation occurs in the bonding of unsaturated hydrocarbons to transition metals. A σ-type bond is formed by overlap of olefin π-electron orbitals with σ-type acceptor orbitals of the metal. This is reinforced by a "back bond" resulting from overlap of the filled metal *d*-orbitals with the anti-bonding (π^*) orbitals of the carbon atoms (see Fig. 1, p. 168). Many, but not all, transition metal complexes have the number of electrons (*i.e.* metal valency electrons plus electrons contributed by the ligands) corresponding to the next inert gas structure. For the purposes of electron counting, CO, R_3P, R_3Sb and olefins donate two electrons, NO donates three electrons, and the cyclopenta-dienyl group, five electrons.

Elucidation of the nature of metal-ligand bonding has been made possible only by the use of physical methods of structural analysis such as infra-red spectroscopy, proton magnetic resonance, dipole measurement and particularly the complete X-ray analysis of representative compounds.

In the account which follows, after the section on nomenclature, the organic chemistry of the transition metals is discussed in sections classified by the number of carbon atoms in the ligand involved in bonding to the metal: *viz.* (*a*) dealing with σ-bonded alkyls, (*b*) with olefin and acetylene compounds, (*c*) with π-allylic complexes and (*d*) with diene and some cyclobutadiene complexes. Some of the compounds in sections (*c*) and (*d*) involve metal bonding not just to olefin bonds but to the whole allyl or diene system of bonds: these are the simplest examples of the π-bonding with occurs in cyclopentadienyl and arene complexes. However, the latter cannot be strictly classed as alkyl derivatives and their discussion is not included therefore in this chapter.

Nomenclature

The nomenclature of complexes, including most of the transition metal compounds described here, follows the recommendations of the International Union of Pure and Applied Chemistry as set out briefly below. (For more details, see the Handbook for Chemical Society Authors, 1960, p. 36 *et seq.*). These rules have not been consistently followed in the literature so that the names and formulae used here for some compounds depart formally from those used elsewhere.

In the *formulae,* the central atom of a complex is written first, followed by all the anionic ligands, and then the neutral ligands, the whole being enclosed in square brackets. Formulae may be prefixed *cis*, *trans*, *sym*, *asym*, etc., as appropriate. In listing ligands, inorganic precede organic, the latter being in alphabetical order. In ionic complexes, the cation precedes the anion, either or both being bracketed separately if complex, the order of the groups inside the brackets being as for non-ionic complexes. (This order is not obligatory if the formula is primarily structural).

In *naming*, the central atom or atoms are placed after the ligands, the order being otherwise the same. The termination *-ate* is used for anions. In *di-* or *poly*-nuclear complexes, the bridging group or groups are prefixed by μ- and separated from the rest of the complex by a hyphen. The valency of the central atom may be indicated by the Stock nomenclature (a Roman numeral in parenthesis following the symbol or name of the element). Some examples may clarify the system *e.g.*

$[PtCl(C_2H_5)\{P(C_2H_5)_3\}_2]$, chloro(ethyl)bis-(triethylphosphine)platinum(II).
$[Cr(CH_3)(C_5H_5)(CO)_3]$, (cyclopentadienyl)methyltricarbonylchromium(II).
$K[PtCl_3(C_2H_4)]$, potassium trichloro(ethylene)platinate(II),
or potassium trichloromonoethyleneplatinate(II).

Much of the literature on organic transition metal compounds is reviewed by *G. E. Coates* ("Organo-Metallic Compounds", Methuen, London, 1960) and in "Organometallic Chemistry", A.C.S. Monograph No. 147, edited by *H. Zeiss*.

(a) Compounds with metal bonded to one carbon atom in an aliphatic compound or radical

The alkyl compounds of the elements discussed in earlier sections contain metal-carbon bonds of very variable polarity. However, they normally either do not dissociate or their dissociation is ionic and reversible, as with methyl-lithium. Homolysis, *e.g.* of tetraethyl-lead, normally occurs on heating only. The cyclopentadienyl derivatives of Group IIIA elements such as scandium are probably ionic and of this type, but it is uncommon among organic derivatives of the transition elements. Normally transition metal to carbon bonds break homolytically, giving a free alkyl radical which then reacts further with the solvent rather than by recoupling with the metal atom. On the other hand, the corresponding halide of the metal will give a reversible ionization. Thus even compounds containing relatively strong metal-carbon bonds appear less stable than the analogous halides:

$$2M{-}R \longrightarrow 2M + 2R\cdot \longrightarrow R_2$$

$$M{-}X \rightleftharpoons M^{\oplus} + X^{\ominus}$$

The weakness of such metal-carbon bonds is the result of the energy levels of the orbitals of transition metals. All these elements have *d*- orbitals which are either partially filled or just complete. The difference in energy (ΔE) between these *d*- orbitals and the valency *s*- and *p*- orbitals used in bonding, is comparatively small; *e.g.* the 3*d*, 4*s* and 4*p* orbitals of metals in the iron transition series have about the same energy. High energy *d*- electrons can therefore move into the antibonding σ-orbitals of the metal-carbon bond, or if *d*- orbitals are vacant, electrons can move into them from the metal-carbon bonding orbitals. In either case the metal-carbon bond is weakened. Therefore to stabilize the metal-carbon bond (1) the metal should be complexed so that it has the effective atomic number (E.A.N.) of the next inert gas, (or is two electrons short of it in the case of nickel, palladium and platinum) and (2) the ligands used to attain this should be ones of high ligand field strength; *i.e.* they should be capable of forming not merely a donating σ-bond to a metal but also of forming π-bonds by back donation of electrons from the *d*- orbitals of the metal to the ligand, so that ΔE is increased. Sulphides, arsines, stibines and, particularly, phosphines have been most valuable in preparing σ-alkyl derivatives of nickel, palladium, platinum and the coinage metals. In earlier groups, where more electrons are required to reach the E.A.N. of the next inert gas, groups such as carbonyl, nitrosyl or cyclopentadienyl have been more useful, although mixed ligands are common *e.g.* (cyclopentadienyl)methyltricarbonylmolybdenum, $[Mo(CH_3)(C_5H_5)(CO)_3]$ Provided the above factors are considered, the formal oxidation state of the metal is of subsidiary importance in determining the stability of an alkyl derivative.

(i) Group IIIA

Alkyl derivatives of these elements have been reported (*V. M. Plets*, Compt. rend. acad. Sci. U.S.S.R., 1938, **20**, 27) but the preparations could not be repeated. (*B. N. Afanas'ev* and *P. A. Tsyganova*, J. gen. Chem. U.S.S.R., 1948, **18**, 306). Since the cyclopentadienyl derivatives are apparently ionic, it is unlikely that σ-bonded alkyls will be stabilized by complexing in this way, except possibly in the case of **uranium(IV)** which is π-bonded to the three arene rings in tri(cyclopentadienyl)uranium(IV) chloride. (*J. M. Birmingham* and *G. Wilkinson*, J. Amer. chem. Soc., 1956, **78**, 42; *L. T. Reynolds* and *Wilkinson*, J. inorg. nucl. Chem., 1956, **2**, 246).

(ii) Group IVA: Titanium, (Zirconium), (Hafnium)

Titanium(IV) is now known to form a number of σ-bonded alkyl compounds. Although the only known tetra-alkyl, tetramethyltitanium is very unstable, the metal-alkyl bond can be stabilized by the presence of electronegative substituents as in the alkylmetal halides, or by an increase in the electronegativity of the alkyl group. Thus the stability increases in the order: butyl < methyl < ethynyl < aryl. Neopentyl groups are apparently stabilized by the absence of a β-hydrogen atom (*D. F. Herman* and *W. K. Nelson*, J. Amer. chem. Soc., 1953, 75, 3882; *H. de Vries*, Rec. Trav. chim., 1961, 80, 866). However, the most stable alkyl compound yet obtained appears to be bis-(cyclopentadienyl) dimethyltitanium (IV), $[Ti(CH_3)_2(C_5H_5)_2]$, where the cyclopentadienyl groups presumably stabilize the σ-bonds. It has been suggested that the contribution of a small amount of sp^3 hybridization to the lowest d^3s state is responsible for the stabilization of titanium-carbon bonds (*G. Pilcher* and *H. A. Skinner*, J. inorg. nucl. Chem., 1958, 7, 8).

Methyltitanium trichloride, m.p. 28°, is obtained as violet crystals from trimethylaluminium and titanium tetrachloride in hexane at low temperatures, followed by distillation at 25°/2 mm. (*G. L. Karapinka, J. J. Smith* and *W. L. Carrick*, J. Polymer Sci., 1961, **50,** 143). The *tribromide* was prepared from dimethylzinc and titanium tetrabromide. *Ethyltitanium trichloride* is obtained from tetraethyl-lead at —80°, using excess titanium tetrachloride to avoid decomposition (*C. E. H. Bawn* and *J. Gladstone*, Proc. chem. Soc., 1959, 227). *Dimethyltitanium dichloride* is obtained as black crystals from titanium tetrachloride and two moles of trimethylaluminium. Complete alkylation to *tetramethyltitanium* occurs when titanium tetrachloride is treated with methyl-lithium, in diethyl ether at —80°. It can be purified by distillation at low temperatures, but decomposes at room temperature. It reacts with water, alcohols, iodine and ketones, and alkylates tin tetrachloride (*C. Beerman*, Angew. Chem., 1959, **71,** 195; *Beerman* and *H. Bastian, ibid.*, 618; *K. Clauss* and *Beerman, ibid.*, 627; *H. J. Berthold* and *G. Groh*, Z. anorg. Chem., 1963, **319,** 230).

Catalysts formed from aluminium alkyls and titanium halides are used in olefin polymerisation. Since they are often formed under conditions similar to those known to form titanium alkyls, they could contain titanium-carbon bonds, and it has been suggested that they function by insertion of an initially π-bonded olefin into a Ti—C bond (*P. Cossee*, Trans. Faraday Soc., 1962, **58,** 1226; *C. E. H. Bawn* and *A. Ledwith*, Quart. Reviews, 1962, **16,** 361).

(iii) Group VA: (Vanadium, Niobium, Tantalum)

Attempts to prepare alkyl derivatives of these elements have so far been unsuccessful. (*Cotton, loc. cit.*; *H. Zeiss* and *G. N. Schrauzer*, Chem. and Ind., 1959, 540).

(iv) Group VIA: Chromium, Molybdenum, Tungsten

The organic chemistry of σ-alkyl derivatives of these elements is just beginning, largely as a result of attention to stabilization by coordination.

Chromium. Thus, although anhydrous chromium(III) halides give only decomposition products from reaction with lithium alkyls in ether, the use of tetrahydrofuran as solvent gives trialkyl(tris-tetrahydrofuran)chromium, $[CrR_3(C_4H_8O)_3]$, where $R{=}CH_3,C_2H_5$, of finite life, although it has not been possible to isolate them (*H. Zeiss* and *M. Tsutsui*, J. Amer. chem. Soc., 1959, 81, 6090). These solutions polymerize disubstituted acetylenes, probably by the formation and rearrangement of π-bonded acetylene complexes. In contrast to these compounds are two organic ions produced in aqueous solution, *viz.* penta-aquo(benzyl)chromium, $[Cr(CH_2C_6H_5)(H_2O)_5]^{2\oplus}$, and penta-aquo(dichloromethyl)chromium, $[Cr(CHCl_2)(H_2O)_5]^{2\oplus}$, produced by the reduction of benzyl chloride and chloroform, respectively, with aqueous acid chromous perchlorate. The former gives 1,2-diphenylethane on heating, and reacts with mercuric chloride to give benzylmercuric chloride (*F. A. L. Anet* and *E. Leblanc*, J. Amer. chem. Soc., 1957, 79, 2649; *Anet*, Canad. J. chem., 1959, 37, 58).

More stable alkyls are obtained when the metal is complexed with cyclopentadienyl and nitrosyl or carbonyl groups. Thus alkyl(cyclopentadienyl)-tricarbonyl-chromium, -molybdenum and -tungsten, $[M(R)(C_5H_5)(CO)_3]$, have been obtained (*T. S. Piper* and *G. Wilkinson*, J. inorg. nucl. Chem., 1956, 3, 104). Their stability increases in this order.

(Cyclopentadienyl)methyltricarbonylchromium is an unstable oil, while the *molybdenum* and *tungsten* analogues melt at 124° (dec.) and 145° (without decomposition). *Ethyl* and *isopropyl* analogues are less stable. These alkyls are prepared either by methylenation of the hydride with diazomethane or from the alkyl iodide and the sodium salt:

$$[MoH(C_5H_5)(CO)_3] + CH_2N_2 \xrightarrow{-N_2} [Mo(CH_3)(C_5H_5)(CO)_3] \xleftarrow{-NaI} CH_3I + Na[Mo(C_5H_5)(CO)_3]$$

Tetrafluoroethylene adds to the hydride directly to give *(tetrafluoroethyl)tricarbonyl(cyclopentadienyl)molybdenum*, $[Mo(C_2F_4H)(C_5H_5)(CO)_3]$, m.p. 53–54°. The tungsten analogue is less stable (*P. M. Treichel, J. H. Morris* and *F. G. A Stone*, J. chem. Soc., 1963, 720). The dinitrosyl series of chromium compounds, the alkyl(cyclopentadienyl)dinitrosylchromiums $[Cr(R)(C_5H_5)(NO)_2]$, are more stable than the carbonyls with which they are isoelectronic. *(Cyclopentadienyl)-methyldinitrosylchromium* is a green solid, melting at 83° (*Piper* and *Wilkinson, loc cit.*).

(v) Group VIIA: Manganese, (Technetium), Rhenium

Manganese. Simple alkyl derivatives of manganese and possibly rhenium, have been prepared, but stable compounds are only formed when the metal attains the effective atomic number of the next inert gas by coordination, normally with carbonyl or cyclopentadiene ligands.

Dimethylmanganese(II), prepared from manganese(II) iodide and methyl-lithium, is an unstable yellow powder, which ignites in air. It is insoluble in ether and probably polymeric, but it dissolves in methyl-lithium to give lithium trimethylmanganese, $Li[Mn(CH_3)_3]$, and reacts with more manganese(II) iodide to form *methylmanganese iodide*, CH_3MnI (*Beermann* and *Clauss*, Angew. Chem., 1959, **71**, 627). In contrast, methylpentacarbonylmanganese, $[Mn(CH_3)(CO)_5]$, obtained from methyl iodide and sodium pentacarbonylmanganese, is stable to air and water and melts only at 95° (*R. D. Closson, J. Kozikowsky* and *T. H. Coffield*, J. org. Chem., 1957, **22**, 598). The same reaction gives the decreasingly stable *ethyl-* and n-*propyl*-analogues. (*W. Hieber* and *G. Wagner*, Ann., 1958, **618**, 24). The *benzyl*-compound, which is more stable, can be prepared from benzylmagnesium chloride and the bromide, $[MnBr(CO)_5]$. Allyl bromide with sodium pentacarbonylmanganese gives initially a σ-bonded *allyl* derivative, if the reaction is kept cold, but this is converted on heating to 80° to a π-allylic compound (see p. 271), with loss of carbon monoxide:

$$CH_2{:}CHCH_2Br + Na[Mn(CO)_5] \longrightarrow [Mn(\sigma\text{-}CH_2CH{:}CH_2)(CO)_5] + NaBr$$
$$\longrightarrow [Mn(\pi\text{-}CH_2CHCH_2)(CO)_4] + CO$$

(*W. R. McClellan et al.*, J. Amer. chem. Soc., 1961, **83**, 1601). Allylpentacarbonylmanganese reacts with carbon monoxide under pressure in the cold, to give an *acyl* derivative:

$$[Mn(\sigma\text{-}CH_2CH{:}CH_2)(CO)_5] + CO \underset{\text{heat}}{\overset{\text{CO}}{\rightleftarrows}} [Mn(\sigma\text{-}COCH_2CH{:}CH_2)(CO)_5]$$

This reversible reaction also occurs with methyl- and similar alkyl-pentacarbonylmanganese compounds, the acyl products obtained being identical with those from acyl halides and sodium pentacarbonylmanganese. The kinetics of the carbonylation suggest a direct combination of carbon monoxide with methylpentacarbonylmanganese (*E. Calderazzo* and *F. A. Cotton*, Inorganic Chemistry, 1962, **1**, 30). The reaction is not reversible with σ-perfluoroacyl compounds. Heating trifluoroacetylpentacarbonylmanganese to 130° gives *trifluoromethylpentacarbonylmanganese*, which does not take up carbon monoxide again even under pressure (*McClellan*, J. Amer. chem. Soc., 1961, **83**, 1599; *R. B. King et al.*, Proc. Int. Conf. on Coordination Chem., 1961, 619; Proc. chem. Soc., 1961, 69). If the acyl-carbonyl group is labelled, as in $[(CH_3{}^{14}CO)Mn(CO)_5]$, this activity is not diluted by exchange with unlabelled carbon monoxide, although the latter exchanges readily with the metal-carbonyl groups (*Coffield, Kozikowsky* and *Closson, loc. cit.*). Treatment of alkylpentacarbonylmanganese compounds with amines such as cyclohexylamine converts them into (acylamino)tetracarbonylmanganese derivatives without loss of carbon monoxide (*K. A. Keblys* and *A. H. Filbey*, J. Amer. chem. Soc., 1960, **82**, 4204). Fluorinated alkyl derivatives of pentacarbonylmanganese can be prepared by the addition of pentacarbonylmanganese hydride to perfluoro-olefins under low pressure:

$$CF_2{:}CF_2 + [MnH(CO)_5] \longrightarrow [Mn(CF_2CF_2H)(CO)_5]$$

in contrast, ethylene is reported to give only ethane and decacarbonyldimanganese. The fluorinated alkyls are volatile, stable in air and decompose only slowly on heating. The β-chloro compound, *(2-chloro-1,1,2-trifluoroethyl)pentacarbonylmanganese*, $[Mn(CF_2CFClH)(CO)_5]$ gives a mixture of $[MnCl(CO)_5]$ and $[MnCl(CO)_4]_2$ on heating to 80° as well as trifluoroethylene and carbon monoxide. These compounds show the great relative stability of the metal—CF_2 bond as compared with the metal—CH_2 bond (*P. M. Treichel, E. Pitcher* and *F. G. A. Stone*, Inorganic Chemistry, 1962, 1, 511).

Rhenium forms stable **σ-alkylpentacarbonylrhenium** compounds similar to those of manganese (*Hieber* and *G. Braun*, Z. Naturforsch, 1959, **14B**, 132; *King et al., loc. cit.*). A reported preparation of trimethylrhenium could not be repeated (*H. Gilman et al.*, J. Amer. chem. Soc., 1941, 63, 2525).

(vi) Group VIII: Iron, Ruthenium, Osmium

Iron. Alkyl derivatives of iron have been isolated only when the metal is coordinated to cyclopentadiene or carbonyl groups. Several alkyl(cyclopentadienyl)dicarbonylirons, $[Fe(R)(C_5H_5)(CO)_2]$, have been obtained either from the alkyl iodide:

$$Na[Fe(C_5H_5)(CO)_2] + C_2H_5I \longrightarrow [Fe(C_2H_5)(C_5H_5)(CO)_2] + NaI$$

or by using a Grignard reagent:

$$[Fe(C_5H_5)(CO)_2I] + CH_3MgI \longrightarrow [Fe(CH_3)(C_5H_5)(CO)_2]$$

(Cyclopentadienyl)methyldicarbonyliron melts at 82° and is more stable than the ethyl analogue (*T. S. Piper* and *G. Wilkinson*, J. inorg. nucl. Chem., 1956, 3, 104; *idem.*, with *D. Lemal*, Naturwiss., 1956, 43, 129). Perfluoroalkyl derivatives are much more stable. *Iodo(perfluoropropyl)tetracarbonyliron*, $[Fe(I)(C_3F_7)(CO)_4]$, prepared by heating perfluoropropyl iodide with pentacarbonyliron, is a red solid melting at 69° with decomposition. It is inert to water or concentrated hydrochloric acid and soluble in organic solvents. Perfluoroethyl iodide reacts similarly:

$$C_3F_7I + Fe(CO)_5 \longrightarrow [FeI(C_3F_7)(CO)_4] + CO$$

Perfluoroethylene gives a white, air stable, heterocyclic compound,

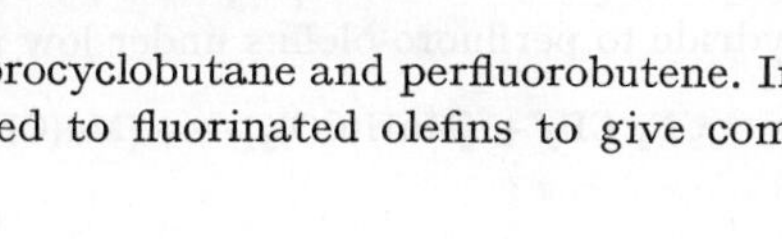

which on heating gives perfluorocyclobutane and perfluorobutene. Iron carbonyl hydrides have also been added to fluorinated olefins to give compounds like *bis-(2-chloro-1,1,2-trifluoroethyl)tetracarbonyliron,* $[Fe(CF_2CFClH)_2(CO)_4]$ (*T.A. Manuel, S. L. Stafford* and *F. G. A. Stone,* J. Amer. chem. Soc., 1961, **83**, 249; *Stone et al.*, Proc. Int. Conf. on Coordination Chemistry, 1961, 621; *P. M. Treichel,* A.C.S. Meeting Proc., Sept. 1961, 22N.; *G. Wilkinson et al.*, J. chem. Soc., 1961, 2738). These owe their stability to the greater electronegativity of perfluoroalkyl groups. However, it has been suggested that the iron–alkyl bond may involve some coupling of the β-protons or carbon atom with the metal d-orbitals. Thus *allyl(cyclopentadienyl)dicarbonyliron,* in which the allyl group is σ-bonded, can be protonated readily to give a π-bonded propene derivative; the latter can also be obtained by hydrogen abstraction from the *n*- or iso-propyl compound by triphenylmethyl perchlorate in tetrahydrofuran:

$$[Fe(\sigma\text{-}CH_2CH{:}CH_2)(C_5H_5)(CO)_2] \xrightarrow{H^{\oplus}} [Fe(\pi\text{-}CH_2{:}CHCH_3)(C_5H_5)(CO)_2]^{\oplus}$$

$$[Fe(\sigma\text{-}CH_2CH_2CH_3)(C_5H_5)(CO)_2] \xrightarrow{-H^{\ominus}} [Fe(\pi\text{-}CH_2{:}CHCH_3)(C_5H_5)(CO)_2]^{\oplus}$$

Reduction with sodium borohydride reverses the second reaction giving back the σ-bonded isopropyl alkyl (*M. L. H. Green* and *P. L. I. Nagy,* J. Amer. chem. Soc., 1962, **84**, 1310).

Ruthenium gives a diphosphine complexed methyl compound, trans-*chloro(methyl)-bis{bis(diphenylphosphine)ethyl}ruthenium, trans*-$[RuCl(CH_3)\{C_2H_4(PPh_2)_2\}_2]$, by treatment of the *trans*-dihalide with trimethylaluminium. This, and the *ethyl* and n-*propyl* analogues, can be reduced to trans-*hydrides,* $[Ru\,H(R)\{C_2H_4(PPh_2)_2\}_2]$ using lithium aluminium hydride.

Osmium appears to be less able to form alkyl compounds than ruthenium (*J. Chatt* and *R. G. Hayter,* Proc. chem. Soc., 1959, 153, J. chem. Soc., 1961, 772).

(vii) Group VIII: Cobalt, (Rhodium), (Iridium)

Cobalt. Alkyl-cobalt bonds require stabilization by coordination of the metal, usually with carbonyl groups.

However, **methyltetracarbonylcobalt,** $[Co(CH_3)(CO)_4]$, from methyl iodide and $Na[(Co(CO)_4]$, decomposes even at -44^0, being much less stable than the isoelectronic manganese compound (*W. Hieber, O. Vohler* and *G. Braun,* Z. Naturforsch., 1958, **13b**, 192). More stable compounds are obtained if one carbonyl group is replaced by triphenylphosphine, as in the *acyl compounds,* $[Co(COR)(CO)_3\{P(C_6H_5)_3\}]$, which can be distilled, while *acyltetracarbonylcobalts* decompose at -20^0. The use of phosphine complexes has allowed the isolation of products from the "OXO" or hydroformylation reaction, under low temperatures, suggesting a mechanism involving formation of tricarbonylcobalt hydride and its initial π-bonding to the olefin:

$$HCo(CO)_4 \rightleftharpoons HCo(CO)_3 + CO$$

$$RCH{:}CHR + HCo(CO)_3 \rightleftharpoons H[Co(CO)_3(\pi\text{-}RCH{:}CHR)] \rightleftharpoons [Co(CO)_3(\sigma\text{-}CHRCH_2R)]$$

$$[Co(CO)_3(CHRCH_2R)] \xrightarrow{CO} [Co(CO)_4(CHRCH_2R)] \xrightarrow[CO]{HCo(CO)_3} RCH_2CHRCHO + Co_2(CO)_7$$

(*R. F. Heck* and *D. S. Breslow*, J. Amer. chem. Soc., 1961, **83**, 4023). The conversion of methanol and other alcohols to aldehydes may involve a similar reaction:

$$CH_3OH + HCo(CO)_4 \longrightarrow [CH_3Co(CO)_4] \xrightarrow[H_2]{CO} CH_3CHO + HCo(CO)_4$$

(*W. Reppe* and *H. Vetter*, Ann., 1953, **582**, 1, etc.).

Reaction of allyl bromide with $Na[Co(CO)_4]$ gives not the expected σ-allyl compound, but a π-allylic tricarbonyl, *π-allyltricarbonylcobalt*, $[Co(\pi\text{-}C_3H_5)(CO)_3]$, one carbonyl group being displaced (see p. 272). Long chain olefinic compounds, however, are bonded by separate σ- and π-bonds. Again more stable products are obtained if a carbonyl group is replaced with triphenylphosphine. As with manganese, acyl compounds can lose carbon monoxide, to give π-allylic compounds (*Heck* and *Breslow*, J. Amer. chem. Soc., 1961, **83**, 1097; 1962, **84**, 2499). Metal-carbon bonds are again stabilized by the electron withdrawing effect of perfluoroalkyl groups. Thus stable compounds like *iodo(cyclopentadienyl)perfluoropropylcarbonylcobalt* $[Co(C_3F_7)(C_5H_5)(CO)I]$, m.p. 122°, are obtained from cyclopentadienyldicarbonylcobalt and perfluoroalkyl iodides (*R. B. King, P. M. Treichel* and *F. G. A. Stone*, Proc. chem. Soc., 1961, 69; *W. R. McClellan*, J. Amer. chem. Soc., 1961, **83**, 1598).

A σ-bond of carbon to cobalt occurs naturally in vitamin B12 coenzyme, the metal being coordinated to five nitrogen atoms (*D. M. Hodgkin* and *Lenhert*, quoted by *A. W. Johnson* and *N. Shaw*, Proc. chem. Soc., 1961, 447).

(viii) Group VIII: Nickel, Palladium, Platinum

Alkyl derivatives of these elements show a marked increase in stability in the order nickel < palladium < platinum, platinum being one of the few transition metals to form a stable σ-alkyl when it is not coordinated to other ligands.

Nickel is so far known only to form σ-bonds to alkyl (as opposed to aryl or ethynyl) groups in the unstable diethylnickel, obtained as solutions in tetrahydrofuran. This is rather similar to triethylchromium, and also polymerises disubstituted acetylenes to benzene derivatives (*M. Tsutsui* and *H. Zeiss*, J. Amer. chem. Soc., 1959, **81**, 6090). Another possible case is allyl-

(triphenylphosphine)nickel chloride, which is normally a π-allylic system, but apparently reverts to σ-bonding when dissolved in dimethyl sulphoxide. (*J. C. W. Chien* and *H. C. Dehm*, Chem. and Ind., 1961, 745).

However, perfluoroalkyl derivatives have been obtained by the reaction of perfluoroalkyl iodides with a diphosphinenickeldicarbonyl, giving air-stable, reddish brown *complexes* of formula $(Ph_2PC_2H_4PPh_2)_2NiR_fI$, where $R_f = C_2F_5$ (decomp. 260°), n-C_3F_7 (m.p. 220° decomp.). These apparently have a square-planar arrangement of ligands round the nickel atom (*D. W. McBride, F. L. Stafford* and *Stone*, J. chem. Soc., 1963, 723).

Palladium. *Alkylpalladium complexes* of formula $[PdXRD_2]$ and $[PdR_2D_2]$ have been obtained where the donor groups (D) are phosphines, arsines, dipyridyl, ethylthio groups, and cyclo-octa-1,4-diene and X is an anion *e.g.* halogen. In the complexes $[PdX_2D_2]$, Grignard reagents replaced one halogen, and methyl-lithium replaced two. Only *methyl-* and *ethynyl-* compounds were isolable, although attempts were made to prepare ethyl and propyl analogues. A binuclear compound,

$$[(n\text{-}C_4H_9)_3PPd(CH_3)SC_2H_5]_2$$

containing Pd—S—Pd bridges, was also prepared. Most of these compounds decomposed readily in solution, but were more stable when crystalline (*G. Calvin* and *G. E. Coates*, J. chem. Soc., 1960, 2008).

Platinum. The analogous *platinum(II) alkyls* are more stable, but strongly π-bonding ligands must again be used. If the *cis-*, or *trans-* bis(tertiary-phosphine)platinum dihalide is treated with a methylmagnesium halide, a mixture of the *trans-* mono- and *cis-*di-methyl derivatives is obtained:

$$[PtX_2(PR_3)_2] \xrightarrow{CH_3MgX} trans\text{-}[PtX(CH_3)(PR_3)_2] + cis\text{-}[Pt(CH_3)_2(PR_3)_2]$$

Methyl-lithium gives complete alkylation to the *cis-* dimethyl compound, whence the *cis-* chloro(methyl) compound can be obtained on treatment with hydrogen chloride. The order of stabilities is ethynyl $<$ alkyl $<$ methyl $<$ aryl $<$ arylethynyl, but in general, they are unusually stable. The halogen is readily exchanged with other anions, without change in configuration. Alkyl groups can be removed by treatment with iodine, acid, or magnesium halides, but are unaffected by air or water.

cis-*Dimethylbis(triethylphosphine)platinum*, $[Pt(CH_3)_2\{P(C_2H_5)_3\}_2]$, can be distilled at 130°/12 mm., with only slight decomposition. Some platinum(II) compounds of this type can be oxidized to platinum(IV) compounds by treatment with methyl iodide at 100°:

$$trans\text{-}[PtI(CH_3)\{P(C_2H_5)_3\}_2] \xrightarrow{CH_3I} [PtI_2(CH_3)_2\{P(C_2H_5)_3\}_2]$$

Carbon monoxide has recently been shown to insert into the Pt—C bond under pressure to give an acyl derivative (*J. Chatt* and *B. L. Shaw, ibid.*, 1959, 705, 4020; *Chatt* and *G. Booth*, Proc. chem. Soc., 1961, 67).

Platinum(IV) methyl compounds have been known for some time, since they can be prepared directly from methylmagnesium iodide and dry platinum(IV) halides.

The main product of such a reaction is **trimethylplatinum iodide,** $[(CH_3)_3PtI]_4$, m.p. 215° dec., together with smaller amounts of methylplatinum tri-iodide, CH_3PtI_3, dimethylplatinum di-iodide, $(CH_3)_2PtI_2$, tetramethylplatinum, $(CH_3)_4Pt$, and methylplatinum pentaiodide, CH_3PtI_5. Trimethylplatinum iodide forms complexes with amines, *e.g.* bis{iodotrimethyl(pyridine)platinum}, $[Pt(I)(CH_3)_3(C_5H_5N)]_2$, with pyridine. This complex is also the product of the preferred preparation of trimethylplatinum iodide by alkylation of *cis*-dipyridinoplatinum tetrachloride (*M. E. Foss* and *C. S. Gibson*, J. chem. Soc., 1951, 299). Trimethylplatinum iodide can be alkylated with methylsodium, to give **tetramethylplatinum,** while powdered potassium converts it to the explosive **hexamethyldiplatinum,** a compound which is apparently monomeric in benzene solution, but polymerized in the solid state (*H. Gilman, M. Lichtenwalter* and *R. A. Benkeser*, J. Amer. chem. Soc., 1953, **75**, 2063). X-ray structural analysis shows that trimethylplatinum chloride, tetramethylplatinum and probably trimethylplatinum iodide have similar structures, based on a tetramer unit with alternate platinum and halogen (or methyl) groups at the corners of a cube (*R. E. Rundle* and *J. H. Sturdivant, ibid.*, 1947, **69**, 1561).

Trimethylplatinum iodide reacts with cyclopentadienylsodium to give the stable and volatile *cyclopentadienyltrimethylplatinum*, m.p. 65°, which is monomeric (*G. D. Robinson* and *B. L. Shaw*, Z. Naturforsch., 1963, **18b**, 507).

Platinum(IV) also forms some heterocyclic compounds. Cyclopropane reacts with hexachloroplatinic acid to give a brown solid, $[PtCl_2(C_3H_6)]$, whose insolubility suggests that it is polymeric. This has been shown to be built up of dichloro(trimethylene)platinum(IV) units, pyridine giving a monomeric dipyridine complex, $[PtCl_2(C_3H_6)(C_5H_5N)_2]$ (*Chatt et al.*, J. chem. Soc., 1961, 738). A cyclic structure has also been demonstrated for the dimeric product, $[Pt(CH_3)_3(C_5H_7O_2)_2]_2$ obtained from trimethylplatinum iodide and thallous acetylacetonate: analogous products, also having Pt—C bonds linking the rings, are formed by *β*-diketones, including nona-4,6-dione (*A. K. Chatterjee et al.*, J. chem. Soc., 1958, 1706; *A. C. Hazell, A. G. Swallow* and *M. R. Truter*, Chem. and Ind., 1959, 564).

(ix) Group IB: Copper, Silver, Gold

All the simple alkyl derivatives of Group IB metals are rather unstable, the stability decreasing in the order copper > silver > gold. No copper(II) alkyls are known: reduction to copper(I) always takes place when copper(II) halides are treated with metal alkyls. Alkyl derivatives of gold(I) and gold(III) exist, however, when the bonding is stabilized by complexing with ligands such as phosphines, amines or halide bridges.

Copper. Methylcopper, CH_3Cu, is best prepared from methyl-lithium and copper(I) iodide at —15° (*H. Gilman, R. G. Jones* and *L. A. Woods*, J. org. Chem., 1952, **17**, 1630). It can also be formed from copper(II) nitrate and tetra-methyl-lead at —45° (*C. E. H. Bawn* and *F. J. Whitby*, J. chem. Soc., 1960, 3926):

$$Cu^{2\oplus} + Pb(CH_3)_4 \longrightarrow Cu^{\oplus} + Pb(CH_3)_3{}^{\oplus} + \cdot CH_3$$
$$Cu^{\oplus} + Pb(CH_3)_4 \longrightarrow CuCH_3 + Pb(CH_3)_3{}^{\oplus}$$

It is a yellow solid, insoluble in ether and although it is stable in nitrogen below 0°, it explodes in air at room temperature, giving metallic copper, methane and ethane. It gives a colourless solution with excess methyl-lithium, perhaps due to the formation of a salt $Li[Cu(CH_3)_2]$. Methylcopper is also formed as an unstable intermediate in the preparation of methylsilicon chlorides from methyl chloride and a copper-silicon alloy at 300° (*D. T. Hurd* and *E. G. Rochow*, J. Amer. chem. Soc., 1945, **67**, 1057).

Ethylcopper is apparently similar but appreciably less stable (*C. E. H. Bawn* and *R. Johnson*, J. chem. Soc., 1960, 3923).

Silver. Methylsilver, *ethyl-* and n-*propyl-silver* have been prepared from silver nitrate and tetra-alkyl-leads at low temperatures (*G. Semerano* and *L. Riccoboni*, Ber., 1941, **74**, 1089; *Bawn* and *Johnson, loc. cit.*):

$$R_4Pb + AgNO_3 \longrightarrow RAg + R_3PbNO_3$$

Simple silver alkyls decompose rapidly even at —50°, methylsilver giving silver and ethane. Alk-1-enyl derivatives may be more stable, since *isobut-1-enylsilver* decomposes only at —30°. The stability of the alkylsilver is determined partly by the stability of the radical formed on homolysis: thus *benzylsilver* is unstable above —100° (*F. Glockling* and *D. Kingston*, J. chem. Soc., 1959, 3001). A new preparative method, using trialkylboranes in the presence of sodium hydroxide, allows the preparation of silver (and gold) alkyls in a medium free from other organometallic by-products (*H. C. Brown, N. C. Hebert* and *C. H. Snyder*, J. Amer. chem. Soc., 1961, **83**, 1001):

$$Et_3B + 3AgOH + NaOH \longrightarrow 3EtAg + NaB(OH)_4$$

Gold. Phosphinegold(I) halides react with lithium or Grignard reagents to give gold(I) complexes of the type $[Au(R)P(C_2H_5)_3]$, where $R{=}CH_3$, C_2H_5 or $C{:}CC_6H_5$. These are stable, colourless compounds (*G. Calvin, G. E. Coates* and *P. S. Dixon*, Chem. and Ind., 1959, 1628). In these, gold has a coordination number of two.

Most alkyl-gold compounds, however, are derived from gold(III), with a coordination number of four. **Trimethylgold,** $(CH_3)_3Au$, has been prepared from methyl-lithium and gold(III) bromide in ether at —65°. It decomposes at —40°

and has not been obtained free from ether, suggesting a formula $[Au(CH_3)_3(OEt_2)]$. The corresponding amine complexes are more stable: *trimethylbenzylaminegold*(III) melts at 51·5–53°. Ethylenediamine gives a complex, *bis(trimethylgold)ethylenediamine,* $[Au_2(CH_3)_6(NH_2CH_2)_2]$, which decomposes at about 95°, and reacts with hydrogen chloride in ether to give dimethylgold chloride (*H. Gilman* and *L. H. Woods*, J. Amer. chem. Soc., 1948, **70**, 550).

Dialkylgold halides are normally prepared from aliphatic Grignard reagents and gold(III) halides:

$$AuBr_3 + C_2H_5MgBr \longrightarrow [Au_2(Br_2)(C_2H_5)_4]$$

Diethylgold bromide melts at 58° and sublimes *in vacuo*: it explodes at 70°. It is dimeric, with Au–Br–Au bridges (*W. J. Pope* and *C. S. Gibson*, J. chem. Soc., 1907, 2061; *Gibson et al., ibid.*, 1937, 1690). *Dimethylgold iodide,* $[Au_2I_2(CH_3)_4]$, obtained as colourless needles, m.p. 78·5°, from dichlorodipyridinogold chloride and methylmagnesium iodide, is also dimeric (*F. H. Brain* and *Gibson, ibid.*, 1939, 762). The halide bridges are broken by donor groups such as amines, and thallous acetylacetonate gives a monomeric dimethylacetylacetonatogold:

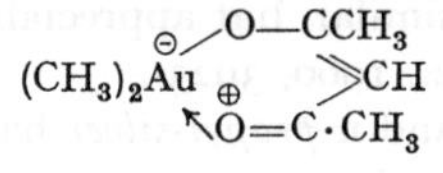

Treatment of dialkylgold(III) halide with silver cyanide gives the corresponding dialkylgold cyanide. The suggested tetrameric structure

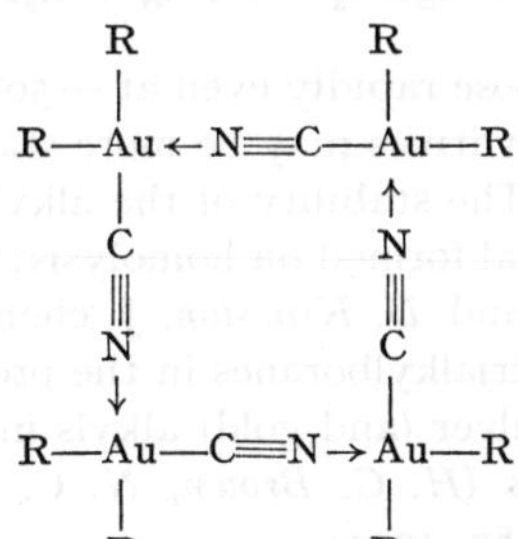

has been confirmed for di(*n*-propyl)gold cyanide (*A. Burawoy*, *Gibson* and *S. Holt*, J. chem. Soc., 1935, 1024; *R. F. Phillips* and *H. M. Powell*, Proc. Roy. Soc., 1939, **173**, 147).

Ethylgold dibromide can be made by careful bromination of diethylgold bromide. It is much less stable than the latter, decomposing at 80°.

Propylgold dibromide is dimeric, with a high dipole moment, suggesting an asymmetrical structure:

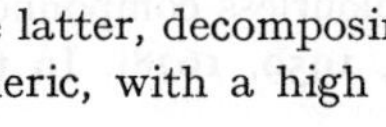

Ethylenediamine gives a mixture of $[Au(en)_2]Br_3$ and $[Au(C_2H_5)_2(en)]$ Br in agreement with this (*Burawoy* and *Gibson*, J. chem. Soc., 1934, 860).

(b) *Compounds with metal bonded to two carbon atoms in an aliphatic compound or radical*

(i) *Transition metal olefin complexes*

Although olefin complexes of platinum, such as Zeise's salt, $K[PtCl_3(C_2H_4)]H_2O$, have been long known, the nature of their bonding as coordination compounds was not clarified until 1953. As ligands, olefins behave similarly to carbonyl groups, since (1) both donate two electrons, and also undergo donation from the metal, (2) both tend to stabilise the metal in a low valency state and (3) in these complexes, the metal usually attains the effective atomic number of the next higher inert gas. This similarity is underlined by the common preparation of olefin compounds *e.g.* of molybdenum, from the di-olefin and a metal carbonyl. However, the analogy is limited, since some metals, especially platinum, form olefin complexes much more readily than they do carbonyls. Also, mono-olefins rarely replace carbonyl groups.

The formulation of olefin compounds as coordination compounds is in agreement with their physical and chemical properties. The older formulations for example, of Zeise's salt, would have given dimers, or involved σ-metal-to-carbon bonds. In fact, ethylene is liberated when the salt is treated with more powerful donor groups, such as cyanide. The infra-red spectra suggest that the ethylenic C—H and C=C frequencies are little affected (*J. Chatt* and *L. A. Duncanson*, J. chem. Soc., 1953, 2939) and the X-ray structures of analogous compounds, such as $[PdCl_2(C_6H_5CH{:}CH_2)]_2$, show that the olefin group is bonded at right angles to the plane of the metal bonds (*J. N. Dempsey* and *N. C. Baenziger*, J. Amer. chem. Soc., 1955, 77, 4984, 4987; *J. A. Wunderlich* and *D. P. Mellor*, Acta Cryst., 1954, 7, 130; 1955, 8, 57). Dipole measurements suggest that the bond is not very polar, (probably about 4D for C_2H_4-Pt) confirming that back-donation from the metal to the olefin can occur.

Most of the known compounds of this type are formed by metals with filled or almost filled *d*-orbitals, such as copper(I), silver(I), nickel(II), palladium(II), and platinum(II) but the same type of bond occurs in earlier groups.

Reviews of olefin complexes have been made by *M. A. Bennett* (Chem. Reviews, 1962, **62,** 611), *E. O. Fischer* and *H. Werner* (Angew. Chem. internat. Edit. 1963, **2,** 80) and *R. G. Guy* and *B. L. Shaw* (Adv. in inorg. Chem. and Radiochem., 1962, **4,** 78).

In *Groups IVa and Va* the metals do not form stable complexes with mono-olefins, but the formation of high olefin polymers using titanium catalysts probably involves π-bonding of the olefins as the first stage (*C. E. H. Bawn* and *A. Ledwith*, Quart. Reviews, 1962, 16, 361).

In *Group VIa*, **molybdenum** and **tungsten** form complexes with ethylene in the cation $[Mo(C_5H_5)(CO)_3(C_2H_4)]^{\oplus}$, formed from the halide with ethylene and aluminium halide (*E. O. Fischer* and *K. Fichtel*, Ber., 1961, **94**, 1200). These cations form stable insoluble salts with hexafluorophosphoric acid.

In *Group VIIa* **manganese** similarly forms a complex ion (ethylene)pentacarbonylmanganese, $[Mn(CO)_5(C_2H_4)]^{\oplus}$, (*H. P. Kögler* and *E. O. Fischer*, Z. Naturforsch., 1960, **15b**, 676) but a neutral complex, ethylene(cyclopentadienyl)dicarbonylmanganese, $[Mn(C_5H_5)(CO)_2(C_2H_4)]$, has also been prepared. Intramolecular reaction of an unsaturated acyl group attached to a pentacarbonylmanganese group leads to displacement of carbon monoxide:

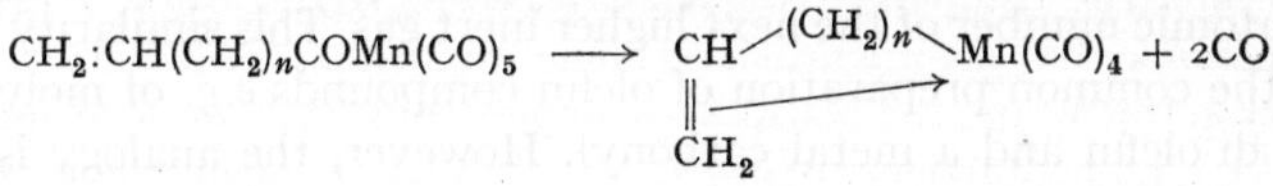

The product apparently contains independent σ- and π-bonds, unlike the π-allylic compouds discussed on page 271 (*R. F. Heck* and *D. S. Breslow*, J. Amer. chem. Soc., 1961. 83, 1097).

In *Group VIII*, only **platinum** and **palladium** readily form π-bonded complexes with mono-olefins, although complexes of **iron** and **cobalt** are known. Such an **iron** complex is the cation, ethylene(cyclopentadienyl)dicarbonyliron, $[Fe(C_5H_5)(CO)_2(C_2H_4)]^{\oplus}$, formed similarly to the isoelectronic molybdenum cation (*Fischer* and *Fichtel*, *loc. cit.*). A π-olefin complex is also formed by the protonation of σ-allyl(cyclopentadienyl)dicarbonyliron, $[Fe(CH_2CH{:}CH_2)(C_5H_5)(CO)_2]$, with hydrogen chloride to give the cation, (cyclopentadienyl)dicarbonyl(π-propylene)iron, $[Fe(\pi\text{-}CH_3CH{:}CH_2)(C_5H_5)(CO)_2]^{\oplus}$ (*M. L. H. Green* and *P. L. I. Nagy*, J. chem. Soc., 1963, 189). However, pentacarbonyliron reacts directly with acrylonitrile in sunlight to give a complex, $[Fe(CO)_4(CH_2{:}CHCN)]$ in which the iron-olefin bond is apparently stabilized by the electronegative nitrile group (*S. F. A. Kettle* and *L. E. Orgel*, Chem. and Ind., 1960, 49).

Cobalt is apparently joined by independent σ- and π-bonds to the allyl group in (allyl)tricarbonylcobalt, $[Co(CO)_3(CH_2CH{:}CH_2)]$, which is obtained as shown below, since the proton magnetic resonance spectrum shows that the methylene protons are not equivalent, in contrast to the π-allylic compounds discussed on page 272 (*R. F. Heck* and *D. S. Breslow*, J. Amer. chem. Soc., 1960, **82**, 750):

$$CH_2{:}CHCH_2COCl + Na[Co(CO)_4] \longrightarrow [Co(CO)_4(COCH_2CH{:}CH_2)]$$
$$\longrightarrow [Co(CO)_3(CH_2CH{:}CH_2)] + 2CO$$

The mode of decomposition of compounds of the type $Co(CO)_4[CO \cdot (CH_2)_n \cdot CH{=}CHR]$ has been shown to depend on the value of n and the nature of the substituent R (*Heck, ibid.*, 1963, 85, 3116).

In the series nickel, palladium and platinum there is a steady increase in the stability and ease of formation of π-olefin compounds similar to that found with σ-bonded alkyls.

Nickel forms only one type of olefin complex, of the type $[Ni(olefin)_2]$ where the olefins include acrylonitrile, acrolein and analogues with electronegative α-substituents.

They are obtained by treating nickel carbonyl with the olefin. They are probably polymeric but the infrared spectra indicate that the olefin group is π-bonded to the metal. Triphenylphosphine gives a more stable derivative *e.g.* of the acrylonitrile complex, $[Ni(CH_2{:}CHCN)P(C_6H_5)_3]_n$, (n = 1 or 2). It is therefore possible that the nitrile is also bound to metal. It polymerises acetylene to cyclo-octatetraene and the phosphine complex converts acetylene-acrylonitrile mixtures to 2,4,6-heptatrienenitrile (*G. N. Schrauzer*, J. Amer. chem. Soc., 1959, **81**, 5310; Ber., 1961, **94**, 642; *idem.* with *H. P. Fritz, ibid.*, 650).

Palladium gives complexes with a number of olefins, including cyclohexene, styrene, ethylene, isobutene, pinene and camphene (in decreasing order of stability). Since they are sensitive to water, they are best prepared, not from palladium(II) chloride, but by displacing benzonitrile from the complex, $[PdCl_2(C_6H_5CN)_2]$:

$$2C_2H_4 + 2[PdCl_2(C_6H_5CN)_2] \longrightarrow [PdCl_2(C_2H_4)]_2 + 4C_6H_5CN$$

In water, *bis[dichloro(ethylene)palladium]*, $[PdCl_2(C_2H_4)]_2$, hydrolyses to acetaldehyde. This is the simplest example of a process for converting a wide range of olefins to the corresponding aldehydes and ketones:

$$[PdCl_2(C_2H_4)]_2 + H_2O \longrightarrow 2CH_3CHO + 2Pd + 2HCl$$

The palladium can be recycled by carrying out the process in the presence of air and copper(II) chloride (*J. Smidt et al.*, Angew. Chem., 1959, **71**, 176):

$$2CuCl_2 + Pd \longrightarrow 2CuCl + PdCl_2$$

$$4CuCl + 4HCl + O_2 \longrightarrow 4CuCl_2 + 2H_2O$$

In contrast, ethylene reacts directly with an acidified aqueous solution of potassium tetrachloro**platinate**(II) to give initially, *potassium trichloroethyleneplatinate*(II), (Zeise's salt). Extraction of this solution with ether, or concentration, gives the neutral compound, *di-μ-chloro-tetrachloro-bis(ethylene)diplatinum*(II), $[PtCl_3(C_2H_4)]_2$, as a rose-coloured powder, melting at 165°. This can react in acetone solution at —70° with two more molecules of ethylene to give trans-*dichloro(diethylene)platinum*(II), $[PtCl_2(C_2H_4)_2]$, which is unstable and loses ethylene at —6°. A series of complexes of general formula $K[PtCl_3L]$ where L is a mono-olefin are known. The strength of the bond decreases in the order: ethylene > styrene > pentene > indene > cyclohexene > 1,1-diphenylethylene > α-methylstyrene (*J. S. Anderson,* J. chem. Soc., 1934, 971). Complexes are also formed with substituted olefins, such as unsaturated acids, alcohols and esters, the olefin group apparently always being the group bonded to the metal. (*P. Pfeiffer* and *H. Hoyer,* Z. anorg. Chem., 1933, **211,** 241). The configuration of the olefin is not destroyed in complexing; thus *cis*- and *trans*-2-butene give complexes with different melting points and infra-red spectra and the olefins are recovered unchanged on treatment with aqueous cyanide (*H. B. Jonasson* and *W. B. Kirsch,* J. Amer. chem., Soc. 1957, **79,** 1279). Pyrolysis of Zeise's salt gives some ethylene, but less than is obtained by reaction with cyanide. Hydrolysis gives acetaldehyde, and reduction to ethane and platinum is easy. The chlorine bridge in $[PtCl_2(C_2H_4)]_2$ can be broken by amines such as pyridine to give amine complexes, *trans*-$[PtCl_2(C_2H_4)(amine)]$, without liberation of ethylene.

(ii) Alkyne derivatives

Acetylenes can bond with transition metals in four ways: (1) by the replacement of hydrogen to give an ethynyl compound; (2) by π-bonding of the acetylene group in a similar way to olefins; (3) by the formation of not one, but two such π-bonds, so that the acetylene group acts as a bridging group and (4) by undergoing reaction, with the formation of an olefin derivative or an aromatic compound, often involving the carbonyl groups of a metal carbonyl. Since these products are of very different types, they are discussed separately.

(1) *σ-Ethynyl compounds.* In these, there is a σ-metal-to-carbon bond, similar to that in the alkyl compounds discussed above, but the increased stability and ease of preparation of ethynyl derivatives suggests that appreciable back coordination from the *d*-orbitals of the metal to the antibonding orbitals of the acetylene carbon atoms must occur. Thus the ethynyl group has considerable similarity to cyanide, carbonyl or isonitrile groups. Such charge spreading must be necessary to avoid a high charge on the metal in anions such as the *tetraethynylnickel ion* $[Ni(C{\vdots}CH)_4]^{4\ominus}$. The preparation and properties of acetylides has been reviewed (*R. Nast,* Angew. Chem., 1960, **742,** 26).

Iron, cobalt and nickel have given anionic acetylide complexes by the reaction of sodium or potassium acetylides with the thiocyanates, in liquid ammonia.:

$$Fe(CNS)_2 + 6NaC{\vdots}CH \longrightarrow Na_4[Fe(C{\vdots}CH)_6] + 2NaSCN$$

The **iron** compounds are very sensitive to air and hydrolysis. Oxidation in liquid ammonia gives an unstable iron(III) complex, *sodium hexaethynylferrate(III)*, which explodes at —30°:

$$2Na_4[Fe(C{:}CH)_6] + 2NH_3 + \tfrac{1}{2}O_2 \longrightarrow 2Na_3[Fe(C{:}CH)_6] + 2NaNH_2 + H_2O$$

The similarity of these compounds to complex cyanides is underlined by the preparation of mixed complexes such as $K_4[Fe(CN)_5(C{:}CH)]$, *potassium pentacyanoethynylferrate*, by the reaction of alkali acetylides with $K_2[Fe(CN)_5NO]$ (*Nast, loc. cit.*, *Nast* and *F. Urban*, Z. anorg. Chem., 1956, **287**, 17; 1957, **289**, 244).

Cobalt similarly forms *hexaethynylcobalt*(II) complex ions, *e.g.*, $Na_4[Co(C{:}CH)_6]$ which readily oxidise to give the more stable cobalt(III) complex ions, *e.g.*, $Na_3[Co(C{:}CH)_6]$ being isoelectronic with sodium hexacyanocobaltate(III) (*Nast* and *H. Lewinsky*, *ibid.*, 1955, **282**, 210).

Nickel initially gives an unstable salt, *potassium tetraethynylnickel*(II) $K_2[Ni(C{:}CH)_4]$, which can be reduced with potassium in liquid ammonia to a nickel(0) salt, which is very sensitive to air:

$$K_2[Ni(C{:}CH)_4] + 2K \longrightarrow K_4[Ni(C{:}CH)_4]$$

Ammine complexes, $[Ni(C{:}C{\cdot}C_6H_5)_2{\cdot}4NH_3]$, can be obtained by treating tetra(phenylethynyl)nickel(II) anions with hexamminenickel(II) thiocyanate; these lose ammonia *in vacuo*, to give polymeric black acetylides:

$$Na_2[Ni(C{:}C{\cdot}C_6H_5)_4] + [Ni(NH_3)_6](SCN)_2 \xrightarrow{NH_3} 2[Ni(C{:}C{\cdot}C_6H_5)_2{\cdot}4NH_3] + 2NaSCN$$

$$[Ni(C{:}C{\cdot}C_6H_5)_2{\cdot}4NH_3] \longrightarrow [Ni(C{:}C{\cdot}C_6H_5)_2]_n + 4NH_3$$

These polymers may be built up of $Ni[Ni(C{:}C{\cdot}C_6H_5)_4]$ units. The simple nickel acetylide, presumably $Ni(C{:}CH)_2$, is explosive and too unstable to isolate (*Nast, loc. cit.*, *Nast* and *K. Vester*, Z. anorg. chem., 1955, **279**, 146). More stable complexes are formed when phosphine ligands are present, as in $[Ni(C{:}CR)_2\text{-}\{P(C_2H_5)_3\}_2]$ (*J. Chatt* and *B. L. Shaw*, Chem. and Ind., 1959, 675).

Dimeric acetylides, apparently of nickel(I), *e.g. hexapotassium bis(tetraethynyl nickel)*, which are isoelectronic with octacarbonyldicobalt, are obtained using alkali metal acetylides in liquid ammonia in the reaction:

$$K_4[Ni_2(CN)_6(CO)_2] + 8KC{:}CH \longrightarrow K_6[Ni_2(C{:}CH)_8] + 6KCN + 2CO$$

Aqueous potassium cyanide converts these compounds to tetracyanonickel(II) salts with liberation of the alkyne:

$$[Ni_2(C{:}CH)_8]^{6\ominus} + 8CN^{\ominus} + 10H_2O \longrightarrow 2[Ni(CN)_4]^{2\ominus} + H_2 + 8HC{:}CR + 10OH^{\ominus}$$

(*Nast* and *H. Kasperl*, Ber., 1959, **92**, 2135).

Manganese and **chromium** form similar anions; the reaction of alkali metal acetylides on hexamminechromium(III) nitrate gives *tripotassium hexaethynylchromium*(III):

$$[Cr(NH_3)_6](NO_3)_3 + 6HC{:}CK \longrightarrow K_3[Cr(C{:}CH)_6] + 3KNO_3 + 6NH_3$$

while manganese(II) thiocyanate gives initially a manganese(II) complex, *dipotassium tetraethynylmanganese* $K_2[Mn(C{:}CH)_4]$ which can be oxidized to a *tripotassium hexethynylmanganese*(III) complex, $K_3[Mn(C{:}CH)_6]$, by air and acetylene in liquid ammonia.

Copper and **silver** form neutral ethynyl compounds. Ethynylpotassium gives *ethynylcopper* at low temperatures:

$$CuI + KC_2H \xrightarrow[-80^\circ]{} CuC_2H \xrightarrow[-45^\circ]{} Cu_2C_2 + C_2H_2$$

but this disproportionates on warming (*Nast* and *W. Pfab*, Z. anorg. Chem., 1957, **292**, 287). The copper and silver acetylides, Cu_2C_2 and Ag_2C_2, are normally prepared from acetylene and ammoniacal solutions of copper and silver salts. They are explosive, and probably polymeric through π-bonding of the metal to the acetylide group. Thus triethylphosphine degrades phenylethynylcopper to a dimer, $[Cu(C{:}C{\cdot}C_6H_5)P{\cdot}(CH_3)_3]_2$ (*D. Blake, G. Calvin* and *G. E. Coates*, Proc. chem. Soc., 1959, 396). Alk-1-ynes dimerize in the presence of copper(II) salts by oxidation of the acetylide (*G. Eglinton* and *A. R. Galbraith*, J. chem. Soc., 1959, 889):

$$RC{:}CH \longrightarrow RC{:}C^{\ominus} + Cu^{2\oplus} \longrightarrow RC{:}C{\cdot}C{:}CR + Cu^{\oplus}$$

(2) *π-Complexes.* Few π-bonded acetylene compounds have yet been obtained, since the acetylene often reacts further under conditions when a complex might be formed.

Manganese and **tungsten** form simple acetylenic complexes by replacement of another ligand by an acetylene. One molecule of carbon monoxide is displaced from π-cyclopentadienyl(tricarbonyl)manganese(I) by diphenylacetylene or hexafluorobut-2-yne to give complexes of the type $[Mn(C_5H_5)(CO)_2alkyne]$ (*W. Strohmeier, H. Laporte* and *D. von Hobe*, Ber., 1962, **95**, 455; *J. L. Boston, S. O. Grim* and *G. Wilkinson*, J. chem. Soc., 1963, 3468). Similarly, hex-3-yne replaces nitrile ligands in tricarbonyltriacetonitriletungsten, $[W(CO)_3(CH_3CN)_3]$, to give *carbonyltri(hex-3-yne)tungsten(O)*, $[W(CO)(CH_3{\cdot}CH_2{\cdot}C{:}C{\cdot}CH_2{\cdot}CH_3)_3]$, m.p. 55–56°, (*D. P. Tate* and *J. M. Augl*, J. Amer. chem. Soc., 1963, **85**, 2174).

Platinum. Although acetylene reacts with platinum salts, including potassium chloroplatinite, no pure compounds are obtained, and an attempt to replace ethylene in ethyleneplatinous chloride gives only reduction to platinum. But-2-yne has been claimed to form a complex, but this could not be repeated, and it now appears that platinum(II) complexes of the type $[Pt_2Cl_4\ alkyne_2]$ and $K[PtCl_3\ alkyne]$, are formed only where the alkyne has at least one tertiary carbon atom adjacent to the triple bond. They can be prepared either from the alkyne and sodium tetrachloroplatinate(II) in ethanol, or, better, from ethyleneplatinous chloride in acetone. The neutral complexes are dimeric

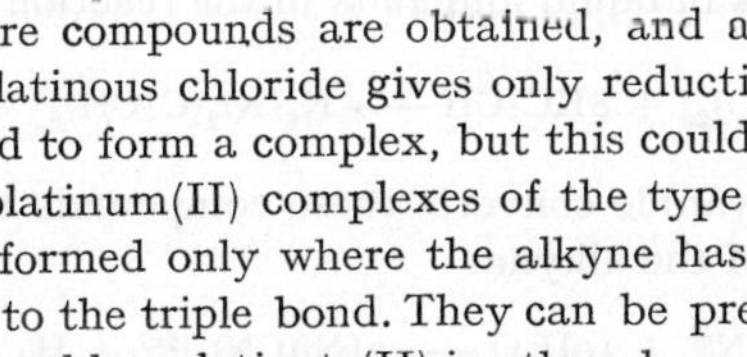

and the infra-red spectra show a lowering in the C:C bond frequency similar to that in olefin-platinum complexes. Other disubstituted acetylenes complex less strongly, since they are hydrated to ketones by ethanolic platinite solutions, diphenylacetylene giving benzyl phenyl ketone, for example. Salts of the ion $[PtCl_3(C_6H_5C{:}C{\cdot}C_6H_5)]^{\ominus}$ with large cations such as triethylammonium have been prepared (*Chatt* and *Duncanson*, J. chem. Soc., 1953, 2939; *Chatt, Guy* and *Duncanson, ibid.*, 1961, 827; *S. V. Bukhovets* and *N. K. Pukhov*, Zhur. neorg Khim., 1958, **3**, 1714). The acetylenic glycol, $(CH_3)_2C(OH)C{:}C{\cdot}C(OH)(CH_3)_2$, (*A*) also complexes with platinum(II). It is not displaced by ethylene from the complex K[$PtCl_3$(*A*)]; it is π-bonded to the metal as the other acetylene complexes are, but, in addition, with subsidiary bonding from the chlorine atoms to the hydroxyl groups (*A. Hel'man, S. V. Bukhovets* and *E. Meilakh*, C.r.Acad. Sci. U.R.S.S., 1945, **46**, 105; *Bukhovets* and *K. A. Molodova*, Zhur. neorg. Khim., 1958, **3**, 1540; *Chatt et al.*, J. chem. Soc., 1963, 5170).

Alkylacetylenes also complex with **copper** and **silver**: the X-ray structure of but-2-ynecopper(I) chloride shows the expected geometrical arrangement of the copper atom lying beside the acetylenic bond (*F. L. Carter* and *E. W. Hughes*, Acta. Cryst., 1957, **10**, 801). Silver complexes of formula [Ag(alkyne)]NO_3, *e.g. acetylene silver nitrate*, are formed by several alkynes and silver nitrate (*A. E. Comyns* and *H. J. Lucas*, J. Amer. chem. Soc., 1957, **79**, 4341).

(3) *π-Acetylenic bridging groups.* An acetylene molecule can displace carbon monoxide from a metal carbonyl to form a π-bonded complex in which the acetylene group has become bonded to two metal atoms. In these cases, the metal atoms are also joined together, so that the acetylene group is not coplanar with the metal atoms, but joined to them by two π-bonds roughly at right angles:

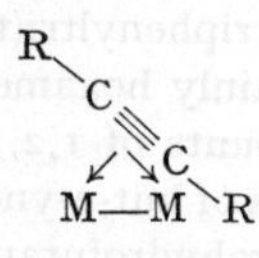

The **cobalt** compounds, alkynehexacarbonyldicobalt, $[Co_2(CO)_6(RC{:}CR)]$, are prepared from octacarbonyldicobalt in an organic solvent, treatment with an alkyne displacing the two bridging carbonyl groups. Acetylene, mono- or di-substituted acetylenes can be used. The X-ray structure of hexacarbonyldiphenylacetylenedicobalt, $[Co_2(CO)_6C_2(C_6H_5)_2]$, confirms the arrangement of the atoms (*W. G. Sly*, J. Amer. chem. Soc., 1959, **81**, 18). The infra-red spectrum shows that the bond order of the acetylenic bond has fallen, in agreement with the observed length of 1·37 Å. Reaction of the pent-1-yne complex with methanolic hydrogen chloride gives some *n*-hexanoic acid, together with a tricobalt derivative. Carbon monoxide reacts further with these compounds to give heterocyclic dicobalt derivatives, $[Co_2(CO)_9(RC{:}CR)]$ (*H. W. Sternberg et al.*, J. Amer. chem. Soc., 1956, **78**, 120; 1959, **81**, 2339).

Nickel. Similarly, mono- and di-phenylacetylenes replace the carbonyl groups in di(cyclopentadienyl)dicarbonyldinickel, $[Ni_2(CO)_2(C_5H_5)_2]$ on refluxing in toluene, to give alkyne dicyclopentadiene complexes $[Ni_2(RC{:}CR)(C_5H_5)_2]$. The acetylene complex $[Ni_2(HC{:}CH)(C_5H_5)_2]$ has been prepared from nickelocene and acetylene in tetrahydrofuran under pressure. It is a stable solid, soluble in organic solvents. Diphenylbutadiyne reacts successively with one or two nickel groupings; the monosubstituted compound must have an unattached acetylene group since it can react further with octacarbonyldicobalt to give a molecule with both nickel and cobalt groups:

$$[Ni_2(C_5H_5)_2 \cdot C_6H_5C{:}C{-}C{:}C \cdot C_6H_5 \cdot Co_2(CO)_6]$$

(*J. F. Tilney-Bassett* and *O. S. Mills*, J. Amer. chem. Soc., 1959, **81**, 4757; *Tilney-Bassett*, J. chem. Soc., 1961, 577; *M. Dubeck*, J. Amer. chem. Soc., 1960, **82**, 502).

(4) *Products derived from alkynes. Transition metal complexes in catalysis.* In contrast to compounds of type (3), many organic derivatives prepared from acetylenes and transition metal compounds contain the acetylene either as a polymerized or ethylenic residue. Metal carbonyls, in particular, often give cyclic products containing ether or keto groups derived from the carbon monoxide molecules. Transition metals have long been known to be versatile catalysts in acetylene chemistry and the isolation of these products has thrown a new light on their function.

The alkyl compounds of **chromium, nickel** and **cobalt** polymerize acetylene and monosubstituted acetylenes and are decomposed in the process but they react readily with disubstituted acetylenes to give benzene derivatives and arene complexes of the transition metal.

The relatively stable complex, triphenyltri(tetrahydrofuran)chromium, reacts with excess but-2-yne to give mainly hexamethylbenzene. Smaller mole ratios of but-2-yne give increasing amounts of 1,2,3,4-tetramethylnaphthalene, from one phenyl ring and two molecules of but-2-yne. This suggests that the acetylene initially replaces two or three tetrahydrofuran molecules from the complex and then either trimerizes or condenses with the phenyl ring. Similar reactions with trialkylchromium complexes also involve dehydrogenation of the chromium substituent. Triethylchromium solutions in tetrahydrofuran give, with diphenylacetylene, a mixture of hexaphenylbenzene and 1,2,3,4-tetraphenylbenzene. Trimethylchromium solutions give 1,2,3,4-tetraphenylcyclopentadiene. Diethylnickel gives hexaphenylbenzene and 1,2,3,4-tetraphenyl-1,3-cyclohexadiene. Both the above metals require stoichiometric amounts of metal alkyl and acetylene. Dimesitylcobalt, however, will convert up to 30 moles of but-2-yne to hexamethylbenzene, per atom of cobalt (*H. Zeiss, W. Herwig* and *W. Metlesics*, J. Amer. chem. Soc., 1959, **81**, 6203; *Zeiss* and *M. Tsutsui, ibid.*, 6090; 1961, **83**, 825). Earlier work on acetylene trimerization showed that carbonyl compounds including $[Ni(CO)_3\{P(C_6H_5)_3\}_2]$ were effective at high temperatures (*W. Reppe* and *W. J. Schweckendiek*, Ann., 1948, **590**, 104), but many unsubstituted metal carbonyls can also be used (*W. Hübel* and *C. Hoogzand*, Ber., 1960, **93**, 103).

However, side products are often formed. Thus the hexacarbonylacetylenedicobalt complex (I), (R=H) itself formed from cobalt carbonyl and acetylene, takes up three molecules of carbon monoxide under pressure to give a lactone derivative (II): such an intermediate may be involved in the formation of dimethyl succinate from acetylene, methanol, carbon monoxide and cobalt carbonyls (*H. W. Sternberg* and *I. Wender*, Proc. Int. Conf. on Coordination Chemistry, London, April 1959). Acids convert the complexes of acetylene or monosubstituted acetylenes to the σ-bonded pyramidal complexes, (III) (*L. A. Duncanson et al.*, Proc. chem. Soc., 1961, 168):

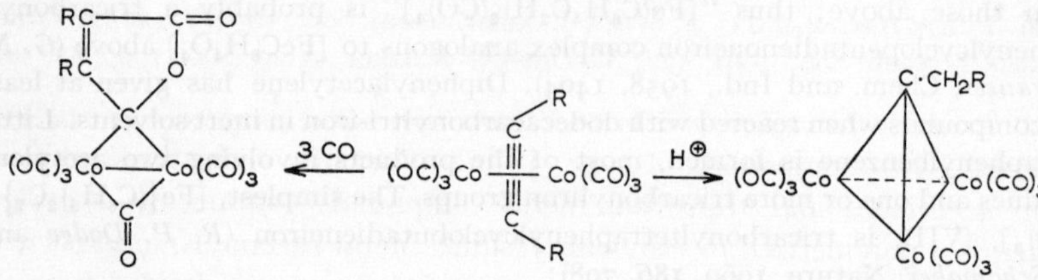

(II) (I) (III)

Iron forms a much more extensive series of compounds. Of the compounds obtained by *H. Reppe* and *H. Vetter* from acetylene and pentacarbonyliron in aqueous alcohol (Ann., 1953, **582**, 133), "$FeC_8H_4O_4$, m.p. 115°" has been shown to be *tricarbonylcyclopentadienoneiron*, $[Fe(CO)_3(C_5H_4O)]$, which can take up hydroquinone to give the dimeric *complex* "$FeC_{22}H_{14}O_{10}$, m.p. 127°". Other products include the tropone derivative (IV) and some of as yet unknown structure (*W. Hübel* and *E. Weiss*, Chem. and Ind., 1959, 703; *idem* with *R. G. Merényi*, *ibid.*, 1960, 407). In alkaline solution, the active carbonyl species is the hydride $[H\,Fe(CO)_4]^{\ominus}$ or its dimeric oxidation product, $[Fe_2(CO)_8]^{2\ominus}$ and these solutions readily react with acetylene or disubstituted acetylenes, to give complexes $[Fe_2\{C_4(OH)_2R_2\}(CO)_6]$ which have been shown by X-ray analysis of the 2-butyne product to have formulae of type (V):

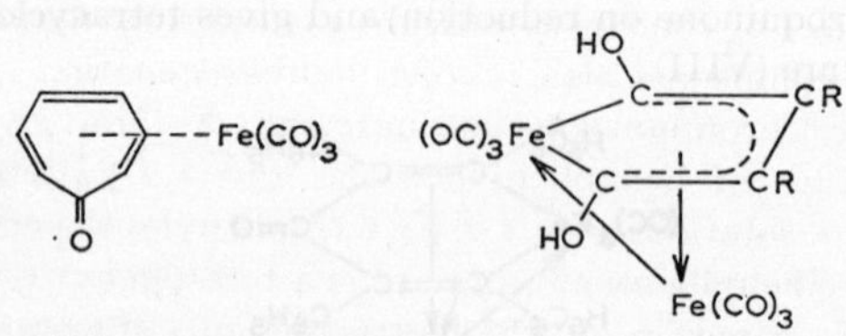

(IV) (V)

(*A. A. Hock* and *O. S. Mills*, Proc. chem. Soc., 1958, 233). The hydroxyl groups are comparable in acidity to those of phenols and can be acetylated without damage to the complex, which is quite stable. Preparation and racemization of optically active esters shows that there is a real distinction between the binding of the iron atoms in solution as well as in the solid state, but that the organic group can turn through an intermediate stage in which it is identically bonded to both atoms (*E. R. H. Jones et al.*, Proc. chem. Soc., 1962, 256). Treatment with acid removes one tricarbonyliron group to give σ-bonded compounds of type (VI) (*idem, ibid.*, 1959, 150):

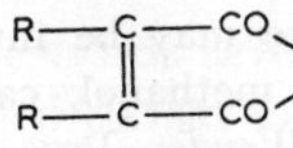

(VI)

Aryl- and diaryl-acetylenes also give complex products from both pentacarbonyliron and dodecacarbonyltri-iron. The structures of some can be inferred from those above: thus "[$Fe(C_6H_5C_2H)_2(CO)_4$]" is probably a tricarbonyldiphenylcyclopentadienoneiron complex analogous to [$FeC_8H_4O_4$] above (*G. N. Schrauzer*, Chem. and Ind., 1958, 1404). Diphenylacetylene has given at least six compounds when reacted with dodecacarbonyltri-iron in inert solvents. Little hexaphenylbenzene is formed, most of the products involving two acetylene residues and one or more tricarbonyliron groups. The simplest, [$Fe\{(C_6H_5)_2C_2\}_2(CO)_3$], (VII), is tricarbonyltetraphenylcyclobutadieneiron (*R. P. Dodge* and *V. Schomaker*, Nature, 1960, **186**, 798):

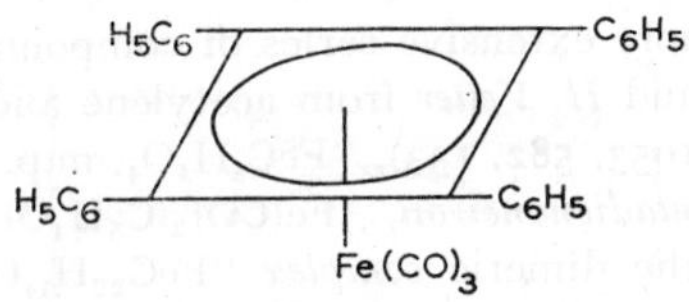

(VII)

The next, [$Fe\{(C_6H_5)_2C_2\}_2(CO)_4$], contains a tetraphenylcyclopentadienone ("tetracyclone") group, and can be synthesised independently from the ketone (*E. Weiss* and *W. Hübel*, J. inorg. nucl. Chem., 1959, **11**, 42). The major products are two di-iron complexes. The simpler, [$Fe_2\{(C_6H_5)_2C_2\}_2(CO)_6$] is the tetraphenyl analogue of (V) above: the other, [$Fe_2\{(C_6H_5)_2C_2\}_2(CO)_7$], gives some tetraphenyl-*p*-benzoquinone on reduction, and gives tetracyclone on pyrolysis, suggesting a structure (VIII):

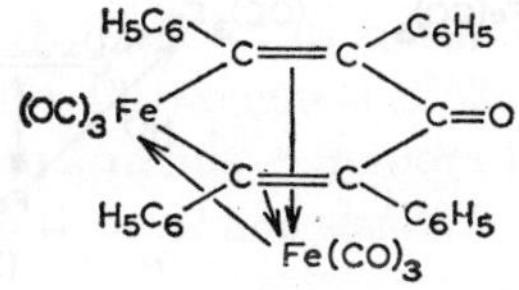

VIII

Two other products may involve true acetylene complexes, but their structures are still unproven (*W. Hubel et al., ibid.*, 1959, **9**, 204; 1959, **10**, 250).

Dimethylacetylene reacts with pentacarbonyliron in sunlight to give exclusively tricarbonylduroquinoneiron [$Fe(CO)_3(C_{10}H_{14}O_2)$]. This hydrolyses readily to tetramethylhydroquinone in acid, and gives duroquinone quantitatively in air (*H. W. Sternberg, R. Markby* and *I. Wender*, J. Amer. chem. Soc., 1958, **80**, 1009).

Since many metal carbonyls catalyse acetylene reactions, similar derivatives may be obtainable from other metals besides iron. Decacarbonyldimanganese with acetylene in tetrahydrofuran gives π-dihydropentalenyltricarbonylmanganese:

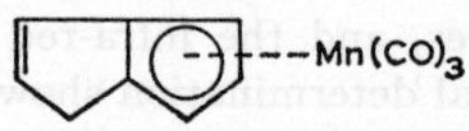

(*T. H. Coffield, K. G. Ihrman* and *W. Burns, ibid.*, 1960, **82**, 4209). Dimethyl acetylenedicarboxylate with dicyclopentadienylnickel gives a Diels-Alder addition compound of the cyclopentadiene, under conditions where acetylene gives a bridged complex (*M. Dubeck, ibid.*, 1960, **82**, 6193).

In some cases a single acetylene molecule is involved. Thus acetylene reacts with cold potassium pentacyanocobaltate(II) to give an oil, whose infra-red spectrum suggests that it is the olefin derivative (IX):

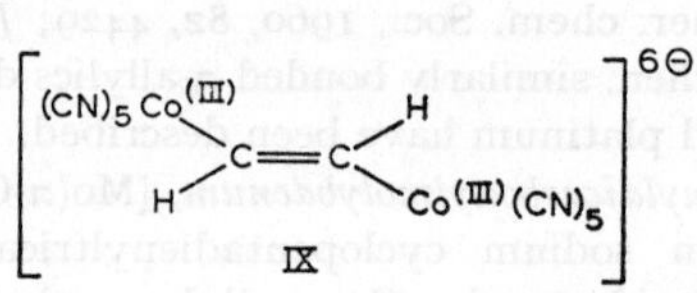

IX

(*W. P. Griffith* and *G. Wilkinson*, J. chem. Soc., 1959, 1629). Platinum(II) also forms a series of derivatives of this type. They are obtained by reducing *cis*-diphosphineplatinum(II) chlorides with hydrazine in the presence of the acetylene (or some olefins, such as stilbene). These complexes are formed by a wide range of acetylenes, the stability increasing in the order:

$$C_2H_2 < RC{:}CH < RC{:}CR < ArC{:}CH < ArC{:}CAr < (p\text{-}NO_2\cdot ArC)_2$$

Acetylene displaces all the olefins so far studied. The formulation of these complexes as platinum(II) alkyls is suggested by the infrared spectra, which show olefinic bonds at about 1700 cm^{-1}, and by a preliminary X-ray structural examination (*J. Chatt, B. L. Shaw* and *A. A. Williams, ibid.*, 1962, 3269).

(c) Compounds with metal bonded to three carbon atoms in an aliphatic compound or radical

Many allyl compounds of transition metals are much more stable than the corresponding σ-bonded alkyls. This is now known to be due to their being bonded to the metal by a delocalized π-bond involving all three carbon atoms so that they are best described as "π-allylic" compounds. A compound of this type, *bis(chloroallylpalladium)*, $[PdCl(C_3H_5)]_2$, (X) is obtained from palladium(II) chloride and allyl alcohol in acetic acid (*J. Smidt* and *W. Hafner*, Angew. Chem., 1959, **71**, 284).

It has since been shown that its nuclear magnetic resonance spectrum contains three types of protons, in ratio 1:2:2 instead of the four or five types expected of σ- or localized π-structures, and the infra-red spectra support a π-allylic structure. An X-ray structural determination shows that the plane of the allyl group is approximately perpendicular to the plane of the $(PdCl)_2$ bridge system and that the molecule has a centre of symmetry:

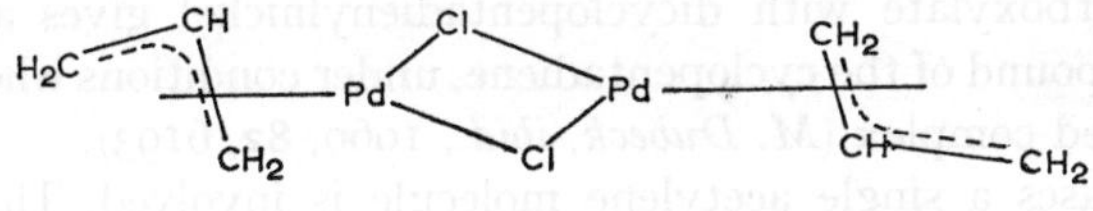

(X)

Both the C—C and Pd—C bond lengths are constant. (*H. C. Dehm* and *J. C. W. Chien*, J. Amer. chem. Soc., 1960, **82**, 4429; *J. M. Rowe*, Proc. chem. Soc., 1962, 66). Since then, similarly bonded π-allylics derived from manganese, iron, cobalt, nickel and platinum have been described.

π-Allylcyclopentadienyldicarbonylmolybdenum, $[Mo(\pi\text{-}C_3H_5)(C_5H_5)(CO)_2]$, m.p. 134°, is formed from sodium cyclopentadienyltricarbonylmolybdenum by reaction with allyl chloride to give the σ-allyl complex, which is converted to the π-allyl complex by ultra-violet radiation (*M. Cousins* and *M. L. H. Green*, J. chem. Soc., 1963, 889).

π-Allyltetracarbonylmanganese, formed by heating σ-allylpentacarbonylmanganese to 86°, is a yellow solid, boiling at 66°/16 mm. and melting at 56°. The loss of carbon monoxide and the formation of the π-allylic system is clearly shown in the simplified nuclear magnetic resonance spectrum and the disappearance of the olefinic absorption at 1620 cm^{-1}. π-Allylic compounds are also formed by the addition of pentacarbonylmanganese hydride to 1,3-dienes, butadiene giving *π-(1-methylallyl)pentacarbonylmanganese* (*F. G. A. Stone, H. D. Kaesz* and *R. B. King*, Z. Naturforsch., 1960, **15b**, 682; *W. R. McClellan et al.*, J. Amer. chem. Soc., 1961, **83**, 1601):

$$H[Mn(CO)_5] + CH_2{:}CH\cdot CH{:}CH_2 \longrightarrow [Mn(CHMe\cdot CH{:}CH_2)(CO)_5] \xrightarrow{-CO} [Mn(\pi\text{-}C_4H_7)(CO)_4]$$

π-(1-Methylallyl)tricarbonyliron, $[Fe(\pi\text{-}C_4H_7)(CO)_3]$, is obtained from (butadiene)tricarbonyliron on treatment with cold hydrogen chloride. It is a yellow brown powder, stable in the cold, but decomposing above its melting point at 60°. Its spectra are similar to those of the manganese analogues. On decomposition it gives a mixture of butenes, carbon monoxide and (butadiene)tricarbonyliron

(*F. J. Impastato* and *K. G. Ihrman*, *ibid.*, 1961, **83**, 3726). Irradiation of σ-allyl (π-cyclopentadienyl)dicarbonyliron with ultraviolet light gives *π-allyl-(π-cyclopentadienyl)carbonyliron*, $[Fe(C_3H_5)(C_5H_5)(CO)]$ (*M. L. H. Green* and *P. L. I. Nagy*, J. chem. Soc., 1963, 189).

π-Allyltricarbonylcobalt, prepared from allyl bromide and sodium tetracarbonylcobalt, is formed in place of the expected σ-allyl compound which is too unstable to isolate. The π-allyl compound is stable at room temperature: it reacts with excess allyl bromide to give 1,5-hexadiene and cobalt bromide. The corresponding acrylyl compound $[Co(COCH{:}CH_2)(CO)_3]$ is less stable, perhaps owing to its being bonded by a σ- and a localised π-bond. *π-(1-Methylallyl)tricarbonylcobalt* is also prepared from butadiene and tetracarbonylcobalt hydride (*H. B. Jonassen et al.*, J. Amer. chem. Soc., 1958, **80**, 2586; *W. R. McClellan et al.*, *ibid.*, *loc. cit.*; *R. F. Heck* and *D. S. Breslow*, *ibid.*, 1961, **83**, 1097).

1-Acylmethyl-π-allyltricarbonylcobalts have been obtained from conjugated dienes and alkyl- or acyl-tetracarbonylcobalts:

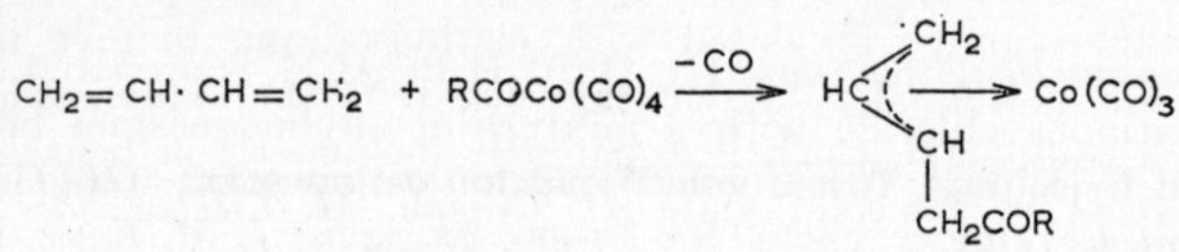

These react with triphenylphosphine to give stable derivatives, and with bases in the presence of carbon monoxide, give *1-acyl-1,3-butadiene derivatives* (*Heck*, *ibid.*, 1963, **85**, 3381, 3383, 3387):

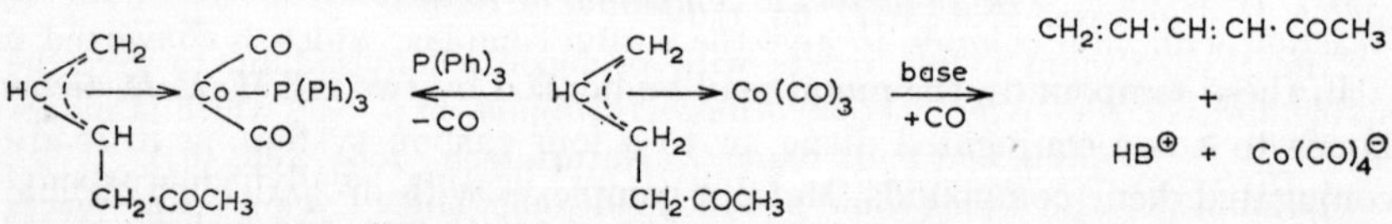

Tetracarbonylcobalt hydride reacts with epoxides and carbon monoxide at 0° to give *β-hydroxyacyltetracarbonylcobalts* (*ibid.*, p. 1460):

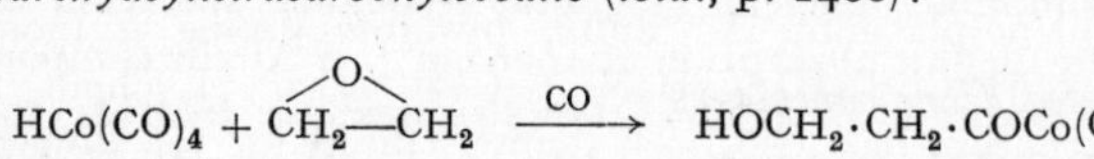

Nickel may complex either with one or two π-allylic groups. Thus *bis(allyl)-nickel*, from anhydrous nickel bromide and allylmagnesium bromide, is a yellow solid, melting about 1°, and igniting in air. Triethylphosphine gives 1,5-hexadiene and $[Ni\{P(C_2H_5)_3\}_4]$ (*G. Wilke* and *B. Bogdanovic*, Angew. Chem., 1961, **73**, 756). Single π-allyl groups have been obtained where the nickel is also bonded to cyclopentadiene or phosphine. *π-Allyl-π-cyclopentadienylnickel*, $[Ni(C_3H_5)(C_5H_5)]$, a deep red liquid boiling at 40°/0·02 mm., is prepared from nickel(II) chloride, cyclopentadienyl-lithium and allylmagnesium chloride in tetrahydrofuran. The compound, $NiC_{10}H_{12}$, which may be an intermediate in this reaction although first regarded as dicyclopentadienylnickel(0), has now been shown to have the structure *π-cyclopentenylcyclopentadienylnickel*(II), $[Ni(C_5H_7)(C_5H_5)]$

(*McClellan et al., loc. cit.*; *B. L. Shaw* and *N. Sheppard*, Chem. and Ind., 1961 517; *E. O. Fischer* and *H. Werner*, Tetrahedron Letters, 1961, **1**, 17).

Apart from π-allylpalladium chloride (above), palladium forms π-allyl compounds in *allyl(cyclopentadienyl)palladium*(II), $[Pd(C_3H_5)(C_5H_5)]$, which is stable at room temperature, and in a number of cyclohexene derivatives (*B. L. Shaw* and *N. Sheppard, loc. cit.*). A π-methylallyl compound, $[PdCl(C_4H_7)]_2$ is obtained either by boiling the isobutene complex $[PdCl_2(C_4H_8)]$ with water, or from methallyl chloride and palladium chloride in hot acetic acid. This dehydrochlorination is reversible, saturated hydrogen chloride in cold ether regenerating the isobutene complex (*R. Hüttel* and *J. Kratzer*, Angew. Chem., 1959, **71**, 456).

Butadiene (and its methyl derivatives) react with sodium chloropalladite in methanol to give stable chloro-bridged π-allylic complexes (similar to structure X, p. 271) and not, as previously suggested, a true olefin complex. *Di-μ-chlorodi-(4-methoxybut-2-enyl)dipalladium(II)*, $[Pd_2Cl_2(C_4H_6 \cdot OMe)_2]$, is obtained as yellow prims, m.p. 97·5–99·5^0 (*S. D. Robinson* and *Shaw*, J. chem. Soc., 1963, 4806).

Allyl(cyclopentadienyl)platinum(II), $[Pt(C_3H_5)(C_5H_5)]$, is prepared by treating propyleneplatinous chloride with a mixture of allylmagnesium bromide and cyclopentadienylsodium. It is a volatile, lemon yellow solid, stable in air (*Shaw* and *Sheppard, loc. cit.*).

(d) Compounds with metal bonded to four carbon atoms in an aliphatic compound or radical

In these compounds, the metal may be bonded by two independent olefin bonds to a non-conjugated diene, or to a four carbon system, as in many conjugated diene compounds. Metal π-complexes with di- and oligo-olefinic ligands have been reviewed by *E. O. Fischer* and *H. Werner* (Angew. Chem., Internat. Edn., 1963, **2**, 80).

(i) Non-conjugated diene complexes

Many examples of these are derived from alicyclic and polycyclic dienes, and are therefore left for full discussion in later volumes. The chelation effect which stabilizes complexes such as tricarbonyl(bicyclo[2,2,1]-2,5-heptadiene)-iron, (XI):

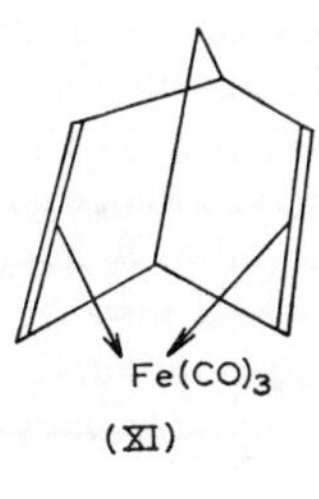

(XI)

is much less pronounced in straight chain dienes, and relatively few such aliphatic compounds occur. The isolation of biallyl and dicyclopentadiene complexes of silver does not appear to be due to their greater stability in comparison with the mono-olefin analogues, but due to their lower solubility (*S. Winstein* and *H. J. Lucas*, J. Amer. chem. Soc., 1938, **60**, 836). Platinum forms a very stable complex with the chelating diene, 1,5-hexadiene:

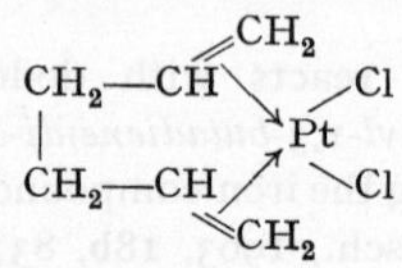

(*K. A. Jensen*, Acta Chem. Scand., 1953, **7**, 866).

(ii) Conjugated diene complexes

By treating pentacarbonyliron with butadiene under pressure, a pale yellow complex, *butadienetricarbonyliron*, $[Fe(C_4H_6)(CO)_3]$, is obtained (*H. Rheilen et al.*, Ann., 1930, **482**, 161). The thermal and oxidative stability of this complex suggests that it does not contain any σ-iron-carbon bonds; the infra-red spectrum shows no olefinic bands suggesting that all the four carbon atoms are bonded to the metal by delocalized π-bonding. X-ray analysis proves such a structure: all the butadiene carbon to carbon distances are equal (1·45 Å) with interbond angles of 118° (XII):

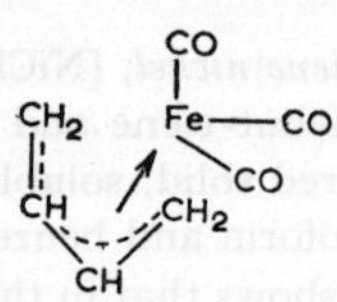

(XII)

(*O. S. Mills* and *G. Robinson*, Proc. chem. Soc., 1960, 421).

Isoprene gives an analogous complex $[Fe(C_5H_8)(CO)_3]$ on refluxing with pentacarbonyliron in benzene, but under more vigorous conditions, a compound $[Fe(C_{10}H_{16})(CO)_3]$ is formed by dimerization of the isoprene to give an α-terpinene derivative. In some cases a non-conjugated diene is isomerized to give a conjugated diene complex: 1,4-pentadiene gives *(1,3-pentadiene)tricarbonyliron*, $[Fe(C_5H_8)(CO)_3]$ derived from the conjugated 1,3-pentadiene. This is a stable, volatile liquid, which gives *trans*-1,3-pentadiene on treatment with triphenylphosphine. Similarly, 1,4-dihydromesitylene gives a 1,3-dihydromesitylene complex (*R. B. King, T. A. Manuel* and *F. G. A. Stone*, J. inorg. nucl. Chem., 1961, **16**, 233). Cyclic trienes such as cycloheptatriene readily replace only two carbonyl groups in pentacarbonyliron to give products of this type, one of the olefin groups being left unbound, but apparently with reduced reactivity (*R. Burton, L. Pratt* and *G. Wilkinson*, J. chem. Soc., 1961, 594).

A series of *butadiene complexes*, obtained from cyclopentadienylcarbonyls of *vanadium, manganese, chromium* and *molybdenum* in sunlight, have formulae such as $[V(C_4H_6)(C_5H_5)(CO)_2]$. They are quite stable, suggesting a resemblance to butadienetricarbonyliron (*E. O. Fischer, H. P. Kögler* and *P. Kuzel*, Ber., 1960, **93**, 3006).

2,3-Dimethylbutadiene-1,3 reacts with dodecacarbonyltriosmium at 200° giving *hexacarbonyl(2,3-dimethyl-1,3-butadiene)di-osmium*, m.p. 120°, in low yield. This has a structure resembling the iron compound, (V) (p. 268) (*Fischer, K. Bittler* and *H. P. Fritz*, Z. Naturforsch., 1963, **18b**, 83; *R. P. Dodge, O. S. Mills* and *V. Schomaker*, Proc. chem. Soc., 1963, 380).

Although it has not proved possible to prepare cyclobutadiene by classical methods, it has been suggested that stable complexes of it might be formed with metals such as iron, nickel or palladium, and that the formation *e.g.* of cyclo-octatetraene from acetylene in the presence of nickel might be due to the dimerization of a cyclobutadiene intermediate (*H. C. Longuet-Higgins* and *L. A. Orgel*, J. chem. Soc., 1956, 1969). Although a suggested example of a cyclobutadiene complex, $[Fe(C_4H_4)(CO)_3]$ above, proved to be a cyclopentadieneone derivative, the existence of tetrasubstituted cyclobutadiene derivatives of nickel, iron and palladium has been proved. Cyclobutadiene itself may be too reactive to isolate, even as a complex, although there is considerable evidence for a complex with silver nitrate, formed from tetrabromocyclobutane and silver nitrate in the presence of lithium amalgam (*M. Avram et al.*, Tetrahedron Letters, 1961, **1**, 21).

Dichloro(tetramethylcyclobutadiene)nickel, $[NiCl_2(C_8H_{12})]$, is obtained from 3,4-dichloro-1,2,3,4-tetramethylcyclobut-1-ene and nickel carbonyl on refluxing in benzene. It is an inert, stable, red solid, soluble in polar solvents and forming addition compounds with chloroform and benzene. It is polymerised in bromoform solution, but X-ray study shows that in the solid state the benzene adduct is dimeric (XIII) (*J. D. Dunitz et al.*, Helv., 1962, **45**, 647):

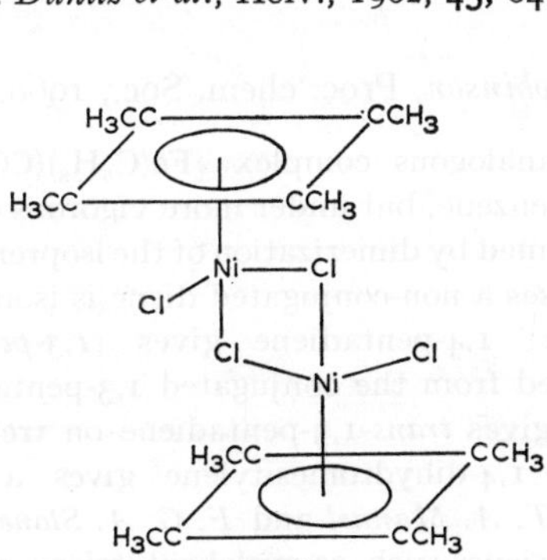

(XIII)

Pyrolysis in high vacuum gives octamethylcyclo-octatetraene (XIV), m.p. 113°, with some 3,4-dimethylhexa-2,4-diene. At higher pressures the bicyclo-octadiene (XV) is formed:

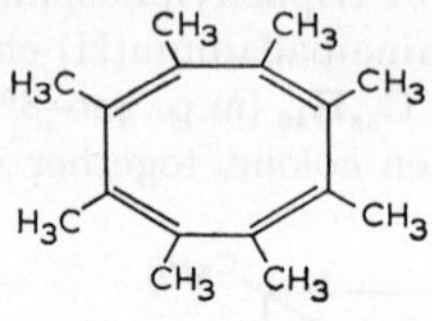

(XIV)

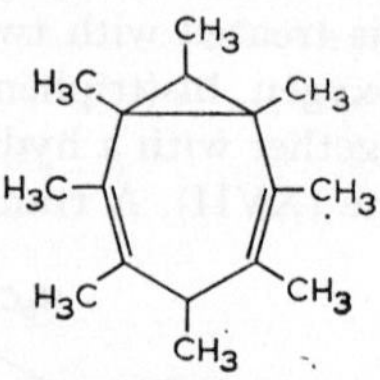

(XV)

A tetraphenylcyclobutadiene analogue has been prepared as the dibromide, $[NiBr_2\{(C_6H_5)_4C_4\}]$, from 1,1-dimethyl-2,3,4,5-tetraphenylstannole by successive treatment with bromine and nickel bromide: it is probable that free tetraphenyl-

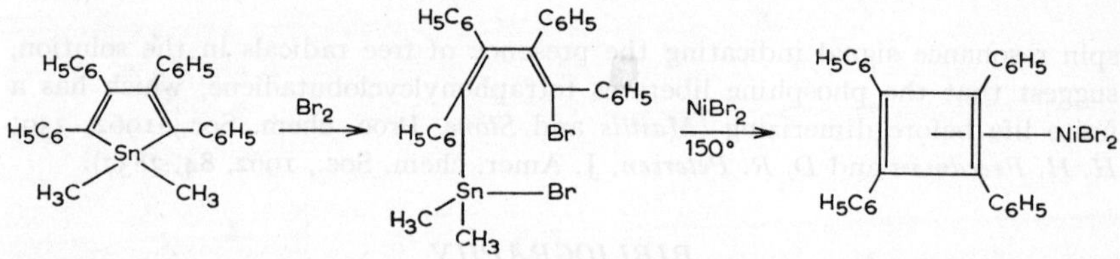

butadiene is formed as an intermediate (*H. H. Freedman*, J. Amer. chem. Soc., 1961, **83**, 2194, 2195). The iron complex $[Fe\{(C_6H_5)C\}_4(CO)_3]$ (see VII page 269) also has certainly been shown to be a cyclobutadiene derivative.

Diphenylacetylene is polymerised by bis-(benzonitrile)palladium chloride: in non-hydroxylic solvents the main product is hexaphenylbenzene, but in ethanol-chloroform solution an olefin complex is formed, which is converted to tetraphenylcyclobutadienepalladium(II) dichloride (XVI) on treatment with dry hydrogen chloride:

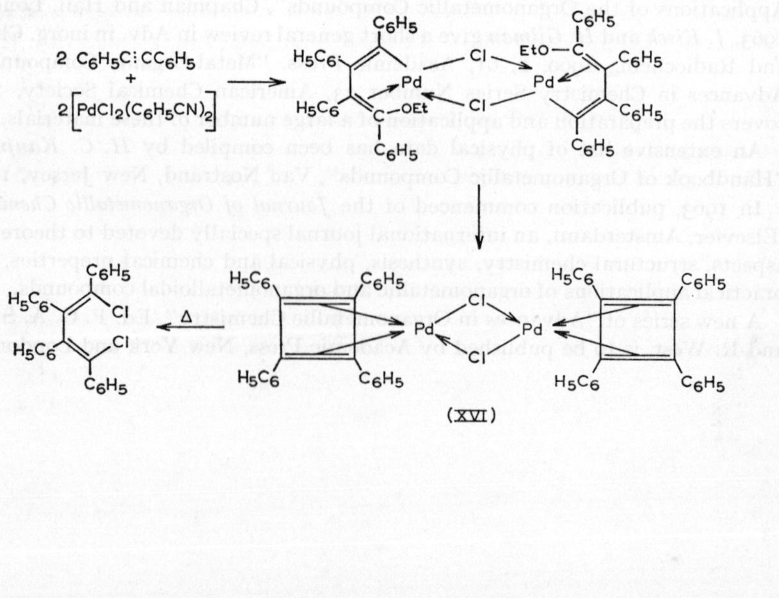

(XVI)

This complex does not undergo Diels-Alder condensations with dienophiles: pyrolysis *in vacuo* gives 1,4-dichloro-1,2,3,4-tetraphenyl-1,3-butadiene (*A. T. Blomquist* and *P. M. Maitlis*, J. Amer. chem. Soc., 1962, **84**, 2329). When the complex (XVI) is treated with two moles of triphenylphosphine in benzene in the absence of oxygen, bis(triphenylphosphine)palladium(II) chloride is formed in high yield, together with a hydrocarbon $C_{56}H_{40}$ (m.p. 426–8°), identical with octaphenylcubane (XVII). A transient green colour, together with an electron

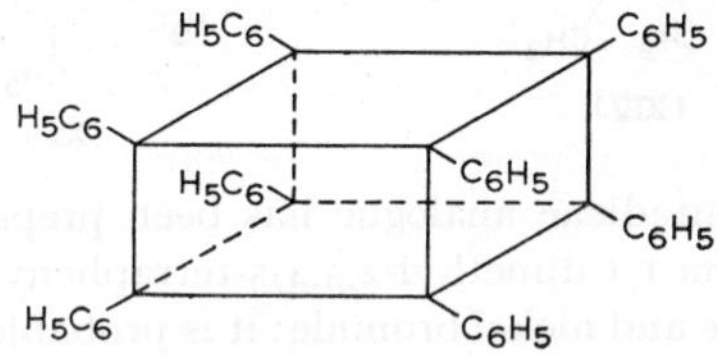

(XVII)

spin resonance signal indicating the presence of free radicals in the solution, suggest that the phosphine liberates tetraphenylcyclobutadiene, which has a finite life before dimerizing (*Maitlis* and *Stone*, Proc. chem. Soc., 1962, 330; *H. H. Freedman* and *D. R. Petersen*, J. Amer. chem. Soc., 1962, **84**, 2837).

BIBLIOGRAPHY

E. Krause and *A. Von Grosse* "Die Chemie der metall-organischen Verbindungen", Borntraeger, Berlin, 1937, give a comprehensive account of the earlier work. For later work the following may be consulted: *G. E. Coates* "Organo-Metallic Compounds", 2nd Edn., Methuen, London, 1960; "Organo-metallic Chemistry", A.C.S. Monograph No. 147, edited by *H. Zeiss*, Reinhold, New York, 1960; *E. G. Rochow, D. T. Hurd* and *R. N. Lewis*, "The Chemistry of Organometallic Compounds", Wiley, New York, 1957, and *J. H. Harwood*, "Industrial Applications of the Organometallic Compounds", Chapman and Hall, London, 1963. *J. Eisch* and *H. Gilman* give a short general review in Adv. in inorg. Chem. and Radiochem., 1960, **2**, 61, Academic Press. "Metal-Organic Compounds", Advances in Chemistry Series Number 23, American Chemical Society, 1959 covers the preparation and application of a large number of these materials.

An extensive list of physical data has been compiled by *H. C. Kaufman*, "Handbook of Organometallic Compounds", Van Nostrand, New Jersey, 1961.

In 1963, publication commenced of the *Journal of Organometallic Chemistry* (Elsevier, Amsterdam), an international journal specially devoted to theoretical aspects, structural chemistry, synthesis, physical and chemical properties, and practical applications of organometallic and organometalloidal compounds.

A new series of "Advances in Organometallic Chemistry", Ed. F. G. A. Stone and R. West, is to be published by Academic Press, New York and London.

GUIDE TO THE INDEX

This index is constructed in a similar manner to the volume indexes of the first edition of the Chemistry of the Carbon Compounds. However, to make the index easier to use more descriptive entries have been made for the commonly-occurring individual, and groups of, chemicals.

The indexes cover primarily the chemical compounds mentioned in the text, and also include reactions and techniques, where named, and some sources of chemical compounds such as plant and animal species, oils, etc.

Chemical compounds have been indexed alphabetically under the names used by authors, editing being restricted to ensuring uniformity of entries under the same heading. In view of the alternative nomenclature that can often be used, a limited amount of cross-referencing has been done where it is considered to be helpful, but attention is particularly drawn to Convention 2 below.

For this and the succeeding volumes, the indexing conventions listed below have been adopted.

1. *Alphabetisation*

(a) The following prefixes have not been counted for alphabetising:

n-	*o-*	*as-*	*meso-*	D	*C-*
iso-	*m-*	*sym-*	*cis-*	DL	*O-*
sec-	*p-*	*gem-*	*trans-*	L	*N-*
tert-		*vic-*			*S-*
		lin-			*Bz-*
					Py-

Some prefixes and numberings have been omitted in the index, when they do not usefully contribute to the reference.

(b) The following prefixes have been alphabetised:

Allo	Epi	Hetero	Pseudo
Anti	Galacto	Homo	Xylo
Arabo	Gluco	Neo	
Cyclo	Glycero	Nor	

(c) A letter by letter alphabetical sequence is followed for entries, firstly for the main entry, followed by the descriptive entry. The only exception to this sequence is the placing of plural entries in front of the corresponding individual entries to prevent these being overlooked by a strict alphabetical sequence which could lead to a considerable separation of plural from individual entries. Thus "butanes" will come before *n*-butane, *iso*-butane; "butenes" before 1-butene and 2-butene, etc.

2. *Cross references*

In view of the many alternative trivial and systematic names for chemical compounds, the indexes should be searched under any alternative names which may be indicated in the main body of the text. Only a limited amount of cross-referencing has been carried out, where it is considered that it would be helpful to the user.

3. *Esters*

In the case of lower alcohols esters are indexed only under the acid, *e.g.* propionic methyl ester, not methyl propionate. Ethyl is normally omitted, *e.g.* acetic ester.

4. *Derivatives*

Simple derivatives are not normally indexed if they follow in the same short section of text.

5. *Collective and plural entries*

In place of "— derivatives" or "— compounds" the plural entry has normally been used. Plural entries have occasionally been used where compounds of the same name but differing numbering appear in the same section of the text.

6. *Main entries*

The main entry of the more common individual compounds is indicated by heavy type. Where entries relate to sections of three pages or more, the page number is followed by "ff".

INDEX